Logical Labyrinths

Logical Labyrinths

Raymond M. Smullyan

CRC Press is an imprint of the
Taylor & Francis Group, an **informa** business
AN A K PETERS BOOK

CRC Press
Taylor & Francis Group
6000 Broken Sound Parkway NW, Suite 300
Boca Raton, FL 33487-2742

Printed in the United States of America on acid-free paper
International Standard Book Number-13: 978-1-56881-443-8 (Hardcover)

Library of Congress Cataloging-in-Publication Data

Smullyan, Raymond M.
Logical labyrinths / Raymond M. Smullyan.
p. cm.
Includes bibliographical references and index.
ISBN: 978-1-56881-443-8 (alk.paper)
1. Logic, Symbolic and mathematical. 2. Mathematical recreations. 3. Puzzles. I. Title.
QA9.S575 2009
511.3—dc22 2008039963

Visit the Taylor & Francis Web site at
http://www.taylorandfrancis.com
and the CRC Press Web site at
http://www.crcpress.com

Contents

Preface

This book serves as a bridge from all my previous recreational puzzle books to all my technical writings in the fascinating field of symbolic logic. It starts out as a typical puzzle book and ends up with highly significant results—significant for philosophy, mathematics and computer science. After having read this book, you will have the knowledge of a typical one-semester graduate course in symbolic logic, and you will then have the preparation to read, not only all of my technical writings, but much of the general literature in the field. I have written this book for the many people who have asked me what symbolic logic is all about and where to find a good introductory account of the subject.

Good question: What is symbolic logic? It is also called "mathematical logic," though the word "mathematics" here has a totally different meaning than the mathematics usually taught in the schools—arithmetic, algebra and geometry. Mathematical logic, instead of dealing specifically with things like numbers, lines and points, deals rather with *propositions* and *arguments*, but codified in symbolic forms to facilitate more precise reasoning.

As valuable as symbolic logic is, I have learned from much teaching experience that it is best to begin by devoting some time to *informal reasoning*. The logic of *lying* and *truth-telling* is admirably suited to this, being both entertaining and highly instructive; so this is how we begin in Part I of this book, in which you follow the remarkable journey of the anthropologist Abercrombie through successively more labyrinthine lands of more and more curious liars and truth-tellers of the most varied and sundry sort, culminating in one grand discovery that generalizes all the problems of the preceding adventures. This prepares the reader for Part II, which begins the formal study of symbolic logic. Part II begins with the subject known as *propositional logic*, shows how this logic formalizes the reasoning of the earlier chapters, and then advances to the main subject of this book, which is *first-order logic*, a subject that has many important applications in philosophy, mathematics and computer science.

In Part III, we journey through the amazing labyrinths of infinity, a concept that has stirred the imagination of mankind as much as, if not more than, any other subject.

In Part IV, we establish many of the most important results of first-order logic.

One important purpose of mathematical logic is to make precise the notion of a *proof*. Just what is meant by the word "proof"? One person suggested to me that a proof is simply an argument that convinces somebody. Not bad, but the trouble is that different people are convinced by different arguments, so the concept of *proof* in that sense is quite subjective. For example, someone once told me that he could *prove* that a certain political party was wrong. Well, I'm sure he could prove it to his own satisfaction, but hardly to the satisfaction of those who belong to that party! Can't one, then, find a more *objective* definition of the word "proof"? Well, so far no such definition has been found for the *general* notion of proof, but symbolic logic has done an admirable job in making absolutely precise the notion of "proof" for given axiom systems! The ancient Greeks had an excellent, albeit informal, concept of *proof*. Euclid starts out with certain basic propositions called *axioms*, which are sufficiently self-evident not to require any sort of proof, and then, by using logic, deduces from these axioms propositions that are far from self-evident. All this has been refined by modern symbolic logic in the study of formal *axiomatics*, which is the subject of Part V.

We then go on to further study of first-order logic, and, in the penultimate chapter of this book, we find one giant theorem that, in a truly remarkable manner, *generalizes* just about all the earlier results! Indeed, "generalization" is very much the hallmark of this book.

The final chapter is a guide to those readers who want to know more about mathematical logic and wish to find out what further literature exists.

This book can be used as a textbook for a one- or two-semester course in logic. A pre-publication copy has been successfully used in this manner at Harvard.

I wish to express my deepest thanks and appreciation to Dr. Malgosia Askanas, my former student, for her expert editing of my manuscript.

Elka Park, NY
2007

- Part I -

Be Wise, Generalize!

- Chapter 1 -

The Logic of Lying and Truth-Telling

I have found, over many years of teaching experience, that the logic of lying and truth-telling is something to which beginners at logic can easily relate, and is accordingly one of the best (and certainly one of the most entertaining) introductions to mathematical logic. The beauty is that the underlying reasoning behind these recreational puzzles is identical to the reasoning that underlies mathematics and computer science. We shall now pay a visit to one of the places I have written much about in the past—the Island of Knights and Knaves,[1] in which those called *knights* always tell the truth and *knaves* always lie. Furthermore, each inhabitant is either a knight or a knave.

Abercrombie Starts Out

Edgar Abercrombie was an anthropologist who was particularly interested in the logic and sociology of *lying* and *truth-telling*. One day he decided to visit a cluster of islands where a lot of lying and truth-telling activity was going on! The first island of his visit was the Island of Knights and Knaves.

[1]Several of the problems in this chapter have appeared in my earlier puzzle books. I include them here to make this book completely self-contained. Readers familiar with these puzzles can either review them, or skim them.

PROBLEM 1.1 (A CLASSIC CASE). On the day of his arrival, Abercrombie came across three inhabitants, whom we will call *A*, *B* and *C*. He asked *A*: "Are you a knight or a knave?" *A* answered, but so indistinctly that Abercrombie could not understand what he said. He then asked *B*: "What did he say?" *B* replied: "He said that he is a knave." At this point, *C* piped up and said: "Don't believe that; it's a lie!"

Was *C* a knight or a knave?

PROBLEM 1.2 (A VARIANT). According to another version of the story, Abercrombie didn't ask *A* whether he was a knight or a knave (because he would have known in advance what answer he would get), but instead asked *A* how many of the three were knaves. Again *A* answered indistinctly, so Abercrombie asked *B* what *A* had said. *B* then said that *A* had said that exactly two of them were knaves. Then, as before, *C* claimed that *B* was lying.

Is it now possible to determine whether *C* is a knight or a knave?

PROBLEM 1.3. Next, Abercrombie met just two inhabitants, *A* and *B*. *A* made the following statement: "Both of us are knaves." What is *A* and what is *B*?

REMARK. "But," some of you might say (after having read the solution to Problem 1.1), "how is this situation possible? You have already proved that no inhabitant of the island can claim to be a knave, so how can an inhabitant claim that *both* are knaves, when he can't even claim that *he* is a knave?"

This is an important point that is fully discussed in the solution.

PROBLEM 1.4. According to another version of the story, *A* didn't say "Both of us are knaves." All he said was "At least one of us is a knave."

If this version is correct, what are *A* and *B*?

PROBLEM 1.5. According to still another version, what *A* actually said was "We are of the same type—that is, we are either both knights or both knaves."

If this version is correct, then what can be deduced about *A* and *B*?

PROBLEM 1.6. On one occasion, Abercrombie came across two natives who were lazily lying in the sun. He asked one of them whether the other one was a knight and got an answer (*yes* or *no*). He then asked the other native whether the first one was a knight, and got an answer (*yes* or *no*). Were the two answers necessarily the same?

PROBLEM 1.7. On another occasion, Abercrombie came across just one native who was lazily lying in the sun. Abercrombie asked the native his name, and the native replied: "John." Was the native a knight or a knave?

PROBLEM 1.8. On another occasion, Abercrombie came across a native and remembered that his name was either Paul or Saul, but couldn't remember which. He asked him his name, and the native replied "Saul."

From this, it is not possible to tell whether the native was a knight or a knave, but one can tell with very high probability! How? (This is a genuine problem, not a monkey trick!)

PROBLEM 1.9. In the next incident, Abercrombie came across three natives, A, B, and C, who made the following statements:

> A: Exactly one of us is a knave.
> B: Exactly two of us are knaves.
> C: All of us are knaves.

What type is each?

PROBLEM 1.10 (WHO IS THE CHIEF?). Abercrombie knew that the island had a chief and was curious to find him. He finally narrowed his search down to two brothers named *Og* and *Bog*, and knew that one of the two was the chief, but didn't know which one until they made the following statements:

> Og: Bog is the chief and he is a knave!
> Bog: Og is not the chief, but he is a knight.

Which one is the chief?

THE NELSON GOODMAN PRINCIPLE AND SOME RELATED PROBLEMS

We shall leave Abercrombie for a while to discuss a theoretically vital principle that crops up again and again in various forms in the next several chapters.

PROBLEM 1.11 (INTRODUCING THE NELSON GOODMAN PRINCIPLE). Suppose that you visit the Island of Knights and Knaves because you have heard a rumor that there is gold buried there. You meet a native and you wish to find out from him whether there really is gold there, but you don't know whether he is a knight or a knave. You are allowed to ask him only one question answerable by *yes* or *no*.

What question would you ask? (The answer involves an important principle discovered by the philosopher Nelson Goodman.)

Problem 1.12 (A Neat Variant). (To be read after the solution of Problem 1.11.) There is an old problem in which it is said that the knights all live in one village and the knaves all live in another. You are standing at a fork in the road and one road leads to the village of knights, and the other to the village of knaves. You wish to go to the village of knights, but you don't know which road is the right one. A native is standing at the fork and you may ask him only one yes/no question. Of course, you could use the Nelson Goodman principle and ask: "Are you a knight if and only if the left road leads to the village of knights?" There is a much simpler and more natural-sounding question, however, that would do the job—a question using only eight words. Can you find such a question?

Problem 1.13 (Three Brothers). Here is another variant using a natural-sounding question: Three triplet brothers on this island are named Larry, Leon and Tim. They are indistinguishable in appearance, but Larry and Leon, whose names begin with "L," always lie and hence are knaves, whereas Tim, whose name begins with "T," is always truthful and hence is a knight.

One day you meet one of the three on the street and wish to know whether or not he is Larry, because Larry owes you money. You are allowed to ask him only one yes/no question, but to prevent you from using the Nelson Goodman principle, your question may not have more than three words! What question would work?

Problem 1.14 (Two Brothers). Arthur and Robert are twin brothers, indistinguishable in appearance. One of the two always lies and the other always tells the truth, but you are not told whether it is Arthur or Robert who is the liar. You meet one of the two one day and wish to find out whether he is Arthur or Robert. Again you may ask him only one yes/no question, and the question may not contain more than three words. What question would you ask?

Problem 1.15 (A Variant). Suppose that upon meeting one of the two brothers of the above problem, you are not interested in knowing whether he is Arthur or Robert, but rather whether it is Arthur or Robert who is the truthful one. What three-word yes/no question can determine this?

Problem 1.16 (A More Ambitious Task). What about a single yes/no question that would determine *both* whether he is Arthur or Robert and also whether he is truthful or lies?

Three Special Puzzles

Problem 1.17. One day I visited the Island of Knights and Knaves and met an unknown inhabitant who made a statement. I thought for a moment and then said: "You know, before you said that, I had no way of knowing whether it was true or false, but now that you have said it, I know that it must be false, and hence you must be a knave!" What did the inhabitant say?

Remark. A statement such as "Two plus two equals five" wouldn't satisfy the given conditions, because even before he spoke, I would certainly have known that it is false that two plus two is five. But, as I said, I didn't know the falsity of his statement before he made it.

Remark. I once presented this problem to a group of high school students. One of them suggested a solution that I thought was more clever and certainly more funny than mine. I give both in the solutions section.

Problem 1.18 (Another Special One). On another occasion during my visit, I met an inhabitant named Al who said: "My father once said that he and I are of different types—that one of us is a knight and the other a knave."

Is it possible that the father really said that? (The father is also an inhabitant of the island.)

Problem 1.19 (Enter a Spy!). One day a spy secretly entered the island. Now, the spy was neither a knight nor a knave; he could sometimes lie and sometimes tell the truth and would always do whatever most suited his convenience. It was known that a spy was on the island, but his identity was not known. After a fruitless hunt, the island police (who were all knights) gave up, hence Inspector Craig of Scotland Yard[2] had to be called in to catch the spy. Well, Craig finally found out that the spy was living with two friends, one of whom was a knight and the other a knave. The three were arrested and brought to trial. It was not known which of the three was the knight, which was the knave, and which was the spy. But Craig realized that by asking only two yes/no questions (not necessarily to the same one of the three) he could finger the spy.

What two questions would work?

[2]Inspector Craig is a character from several of my earlier puzzle books. His interest in logic is equal to his interest in criminal detection.

Three Metapuzzles

Problem 1.20. One day Abercrombie came across two brothers named Andrew and Bernard. Andrew said: "Both of us are knights." Abercrombie then asked Bernard: "Is that really true?" Bernard answered him (he either said *yes* or he said *no*), and Abercrombie then knew what type each was.

At this point, you have enough information to know the type of each. What type is each?

Problem 1.21 (Which Is Witch?). The island also has a witch doctor, and Abercrombie was curious to meet him. He finally found out that the witch doctor was one of two brothers named Jal and Tak. First Jal said: "I am a knight and my brother Tak is a knave." Abercrombie then asked: "Are you the witch doctor?" Jal replied *yes*. Then Abercrombie asked Tak: "Are *you* the witch doctor?" Tak answered (either *yes* or *no*), and Abercrombie then knew which one was the witch doctor. Which one was it?

Problem 1.22 (Innocent or Guilty?). Before leaving the island, Abercrombie attended the trial of a native named Snark who was suspected of having committed a robbery. In court were two witnesses named Ark and Bark. The judge (who was truthful) first asked Ark: "Is Snark innocent or guilty?" Ark replied: "He once claimed that he is innocent." Then Bark said: "He once claimed that he is guilty," at which point Snark said: "Bark is a liar!" The judge then asked Snark: "And what about Ark? Is he a liar?" Snark answered, either *yes* or *no*, and the judge (who was good at logic) then knew whether Snark was innocent or guilty.

Which was he?

Solutions

1.1. It is not possible for an inhabitant of this island to claim to be a knave, since a knight would never lie and claim to be a knave, and a knave would never truthfully admit to being a knave. Hence, no inhabitant of this island can ever claim to be a knave. Therefore, B was lying when he said that A claimed to be a knave. Thus also, C's statement was true. And so B is a knave and C is a knight. We have no idea what A is, since we don't know what he really said.

1.2. This is a more interesting version, because one aesthetic weakness of the above problem is that C doesn't really play a vital role, since before C spoke, we could tell that B was lying. The nice thing about the present version is that we cannot tell what type *B* is until we know what *C* said.

Since *C* said that *B* was lying, *C* must be the opposite of *B*. If *B* told the truth, then *C* lied in saying that *B* lied, but if *B* lied, then *C*'s claim was truthful. And so we see that of the two natives *B* and *C*, one is a knight and the other is a knave.

Now, could *A* really have said that exactly two of the three were knaves? Since we know that *B* and *C* are opposite, we see that *A* couldn't have said that, because if *A* is a knight, then there would be only one knave present (namely, *B* or *C*), and *A* wouldn't then have lied and said there were two! On the other hand, if *A* were a knave, then there really would be exactly two knaves present (namely *A* and one of *B* or *C*) and *A*, a knave, would then never have said that there were. Thus *A* never did say what *B* claimed he said, which makes *B* a knave and hence also *C* a knight. (Thus, like in the first problem, *B* is a knave and *C* is a knight, but the reasoning of the present problem is more elaborate.)

1.3. *A* is obviously a knave, for if he were a knight, he wouldn't have falsely claimed that he and *B* were both knaves. Since *A* is a knave, his statement was false, so *A* and *B* are not both knaves. Thus *B* is a knight. And so the answer is that *A* is a knave and *B* is a knight.

As for the objection to this problem, of course *A* couldn't claim to be a knave, yet since *B* is a knight, *A* can certainly make the false claim that *both* of them are knaves. Actually, *A* could say: "*B* is a knave and I am a knave," but he could not make the *separate* statements: "*B* is a knave," "I am a knave," since he couldn't make the second statement. And this illustrates a very interesting difference between liars and truth-tellers: When a truth-teller asserts the conjunction of two propositions (the statement that *both* propositions are true), then he can assert each one separately, but this is not the case for a constant liar! If one of the propositions is true and the other is false, then he could not assert the true one, but he could claim that both are true. For example, on this island, suppose *A* is a knave and *B* is married and *C* is single. If you ask *A* "Are *B* and *C* both married?" he will reply "Yes." Then if you ask "Is *B* married?" he will say "No." So he asserts that *B* and *C* are both married, yet he denies that *B* is married!

It is not surprising that a constant liar can be inconsistent!

1.4. If *A* were a knave, then it would be true that at least one of them was a knave, but knaves don't make true statements! Therefore, *A* must be a knight. Since he is a knight, his statement was true, hence at least one of them must be a knave, and it must then be B. Thus, A is a knight and B is a knave. (The answer is the opposite of that of Problem 1.3, in which A was a knave and B a knight.)

1.5. It is impossible to tell what A is, but B must be a knight, because A would never claim to be the same type as a knave, since this would be tantamount to claiming himself to be a knave.

Another way of looking at it is this: If A is a knight, then he really is the same type as B (as he said), and hence B is also a knight. On the other hand, if A is a knave, his statement is false, which means that the two are of different types, hence B must be a knight (unlike A). In either case, B is a knight.

Which of the two arguments do you like better? I prefer the first, since it is shorter and neater, but the second is also of interest.

1.6. Yes, because if they were both knights, then both would truthfully answer *yes*. If they were both knaves, then both would lie and again answer *yes*. If one were a knight and the other a knave, the knight would truthfully answer *no* and the knave would falsely answer *no*, and so again, they would give the same answer.

1.7. At first, it seems impossible to tell, doesn't it? This is what is known as a *monkey trick*, and the clue here is that he was lazily lying in the sun and, therefore, was lying in the sun, and hence he was lying; and since only knaves lie, he must have been a knave. (Sorry, but I feel entitled to a little monkey trick every now and then!)

1.8. If his name was not Saul, then the chances are extremely remote that in his lie, he would have happened to pick the very name that Abercrombie was thinking of! Hence, the probability is extremely high that his name really was Saul, and hence that he was a knight.

1.9. No two of them could both be right, hence at least two are knaves. Also C is obviously not a knight, and since he is a knave, his statement was false. Therefore, they are not all knaves, but at least two are, hence exactly two are. Thus, B was right, so he is a knight and the other two are knaves.

1.10. Og cannot be truthful, for if he were, then Bog would be the chief and a liar, hence Og would not be the chief, hence Bog's statement that Og is not the chief but is truthful, would be true, which is not possible if Bog is a liar. Thus, Og must be a liar. It then follows that Bog's statement was false, and so Bog is also a liar. Now, if

Bog were the chief, then he would be both the chief and a liar, hence Og's statement would be true, which cannot be, since Og is a liar. Therefore, it is Og who is the chief.

1.11. A question that works is: "Are you the type who could claim that there is gold on this island?" You are *not* asking him if there *is* gold on the island, but whether he is the type who could *claim* that there is. If he answers *yes*, then there must be gold on the island, regardless of whether he is a knight or knave.

Reason. Suppose he is a knight. Then he really is the type who could claim there is gold there, and since a knight could claim it, it must be true. On the other hand, suppose he is a knave. Then he is not really the type who could claim there is gold there. Well, if there were no gold there, then he *could* (falsely) claim that there is. Therefore, there must be gold there. And so, regardless of whether he is a knight or a knave, a *yes* answer indicates that there is gold on the island. A similar analysis, which I leave to the reader, reveals that a *no* answer indicates that there is no gold on the island, regardless of whether the native is a knight or a knave.

An alternative (and in some ways better) form of the question is this: "Are you a knight if and only if there is gold on the island?"

For those not familiar with the phrase "if and only if": Given any two propositions, to say that one *if and only if* the other means that if either one is true, so is the other—or, which is the same thing, that they are either both true or both false. Thus, the above question asks whether one of the following two alternatives holds:

(1) He is a knight and there is gold on the island.

(2) He is a knave and there is no gold on the island.

(Indeed, you could have asked him: "Is it the case that either you are a knight and there is gold on the island, or that you are a knave and there is no gold on the island?")

Let's see what our previous analysis looks like when applied to the question asked in this form. Well, suppose he answers *yes*. If he is a knight, his answer was correct, which means that one of the two alternatives (1), (2) really does hold. It can't be (2), hence it is (1), so there is gold there. On the other hand, if he is a knave, his answer was false, which means that neither alternative holds, hence (2), in particular, doesn't hold, hence there is gold on the island (because if there weren't, then (2) *would* hold, which it doesn't). So, in either case, there is gold on the island.

Remarks. Although the two questions use different words, they really ask the same thing. Why? Well, what is the type that could claim that there is gold on the island? The answer is a knight, if there is gold there, and a knave, if there is not. Thus, the following three propositions are equivalent:

(1) He can claim that there is gold on the island.

(2) He is a knight if and only if there is gold on the island.

(3) Either he is a knight and there is gold there, or he is a knave and there isn't gold there.

1.12. An eight-word question that works is: "Does the left road lead to your village?" Suppose he answers *yes*. If he is a knight, then the left road really does lead to his village, which is the village of knights. If he is a knave, then the left road doesn't lead to his village, which is the village of knaves, hence again, it leads to the village of knights. So in either case, a *yes* answer indicates that you should take the left road.

Suppose he answers *no*. If he is a knight, then the left road really doesn't lead to his village, so the right road leads to his village, which is the village of knights. If he is a knave, then the left road *does* lead to his village (since he indicates that it doesn't), which is the village of knaves, so again it is the right road that leads to the village of knights. Thus, a *no* answer indicates that you should take the right road.

1.13. You need merely ask: "Are you Leon?" Larry will lie and say *yes*. Leon will lie and say *no*. Tim will truthfully say *no*. Thus Larry will say *yes* and the other two will say *no*. So, if you get a *yes* answer, he is Larry, but if you get a *no* answer, he isn't Larry.

1.14. A three-word question that works is: "Is Arthur truthful?" Arthur will certainly say that Arthur is truthful, regardless of whether Arthur is truthful or not, and Robert will say that Arthur is not truthful, because if Robert is truthful, then Arthur is not truthful, hence Robert will truthfully say that he isn't. On the other hand, if Robert lies, then Arthur is truthful, hence Robert will falsely say that Arthur isn't truthful. Thus, if you get *yes* for an answer, he is Arthur, and if he answers *no*, then he is Robert.

1.15. To find out if Arthur is truthful, you need merely ask: "Are you Arthur?" Suppose he answers *yes*. If the answer is true, then he really is Arthur, hence Arthur answered truthfully, hence Arthur is truthful. On the other hand, if the answer is false, then he is not Arthur, so he is Robert, and hence Robert lied, so again Arthur is truthful. Thus, the answer *yes* implies that Arthur is truthful.

Suppose he answers *no*. If he told the truth, then he is Robert, hence Robert is truthful. If he lied, then he is Arthur, hence Arthur lies, hence again Robert is the truthful one. So the answer *no* would indicate that Robert is truthful.

Note. This problem and the last bear an interesting symmetrical relationship to each other: To find out whether he is Arthur, you ask whether Arthur is truthful, whereas to find out whether Arthur is truthful, you ask whether he is Arthur. Thus the two questions "Is Arthur truthful?" and "Are you Arthur?" are related in such a way that asking either one will give you the correct answer to the other!

1.16. The task is *overly* ambitious because no such question is possible!

Reason. There are four possibilities: (1) He is Arthur and truthful; (2) He is Arthur and lying; (3) He is Robert and truthful; (4) He is Robert and lying. But there are only two possible responses to your question, *yes* or *no*, and with only two responses you cannot possibly tell which of *four* possibilities holds! (This principle is basic to computer science.)

1.17. What he said was: "I am a married knave." A knight couldn't say that, and a married knave couldn't say that, but an *unmarried* knave could. Before he spoke, I had no idea whether he was a knight or a knave, nor whether or not he was married, but after he spoke, I knew that he must be an unmarried knave. (Note: Although a knave cannot claim to be a knave, he could claim to be a married knave, provided he is not married.)

The cute alternative solution suggested by the high school student was: "I am mute."

1.18. If Al hadn't *said* that his father had said that, then the father could have said it, but the fact that Al said that his father had said that implies that the father couldn't have said it, for the following reasons: The only way an inhabitant A can claim to be of a different type than an inhabitant B is that B is a knave (for if B were a knight, then claiming to be of a different type to B would be tantamount to claiming to be a knave, which an inhabitant cannot do). Therefore, if Al's father had really said that, then Al would be a knave and would never have made the true statement that his father had said that.

1.19. The idea is first to find, with one question, one of the three who is *not* the spy. Well, ask A: "Are you the type who could claim that B is the spy?" Suppose he answers *yes*. If he is not the spy, then the Nelson Goodman principle is applicable to him and hence B is the spy. In this case, C is not the spy. On the other hand, if A is the spy, C certainly is not. So whether A is the spy or not, if he answers *yes*, then C is definitely not the spy. By similar reasoning, if A answers *no*, then B is definitely not the spy. So in one question you have fingered someone who is not the spy—let's say it is B. Then the Nelson Goodman principle will definitely work with B, and so you can ask him: "Are you the type who could claim that A is the spy?" If B answers *yes*, then A is the spy, and if he answers *no*, then C is the spy. (Of course, if after the first question you found out that C is not the spy, then your second question is as before, and if C answers *yes*, then A is the spy; otherwise, B is the spy.)

1.20. This is called a *metapuzzle* because we are not given complete information as to what happened: We are not told what answer Bernard gave. The only way we can solve this is by using the fact that Abercrombie could solve it. If we had not been told that Abercrombie knew what they were, then there would be no way that *we* could know!

Well, had Bernard answered *yes*, then the two could either be both knights or both knaves, and Abercrombie could not have known which. But Abercrombie *did* know, and the only way he could know is that Bernard answered *no*, thus contradicting Andrew and hence indicating that the two couldn't both be knights. And so Bernard was right, hence he is a knight and Andrew is a knave.

1.21. Step 1. Jal said that he is a knight and Tak is a knave. Well, if Jal is really a knight, then, of course, Tak is a knave. If Jal is a knave, then Tak could be either a knight or a knave (in both cases, Jal's statement would be false). So all that can be inferred is that at least one of the two is a knave (and maybe both of them). At this point Abercrombie knew that they were not both knights.

Step 2. If Tak had also answered *yes*, then it would follow that one of the two must be lying and the other telling the truth, but Abercrombie would have no way of knowing which. On the other hand, if Tak answered *no*, then it would follow that they were either both lying or both telling the truth, but Abercrombie already knew they were not both knights, hence they must both be lying, and therefore Tak was the witch doctor.

1.22. Even though we are not told what Snark's answer was, we can solve this puzzle because, as we will see, if Snark answered *yes*, the judge would have no way of knowing whether he was innocent or guilty; whereas if he answered *no*, the judge could tell.

Well, suppose he answered *yes*. Then he is denying Ark's claim, so is in effect saying that he has never claimed to be innocent. Also, since he has denied Bark's claim, he in effect has said that he has never claimed to be guilty. Thus, in effect, he has said that he never claimed to be innocent and he has said that he never claimed to be guilty. If he is a liar, then he did once claim to be innocent and did once claim to be guilty, which is not possible, since the two claims couldn't both be false. Therefore, he is truthful and he has never made either claim. Then the judge would have no way of knowing whether Snark was innocent or guilty, even though he would know whether or not Snark was truthful. But the judge *did* know the innocence or guilt of Snark, and so Snark couldn't have answered *yes*; he must have answered *no*.

By answering *no*, Snark is in effect saying that he did once claim to be innocent. Thus, Snark has affirmed two things: (1) He once claimed to be innocent, and (2) he never claimed to be guilty. Now, Snark is either truthful or a liar. Suppose he is truthful. Then he really did once claim to be innocent—and, being truthful, he really is innocent. On the other hand, suppose he is a liar. Then (2) is false, which means that he did once claim to be guilty—but being a liar, he is really innocent! Thus, regardless whether he is truthful or a liar, he must be innocent. (His veracity, however, cannot be determined.)

- Chapter 2 -

Male or Female?

The next island visited by Abercrombie was a curious one in which women were also classified as knights or knaves! And the most curious part of all is that while male knights told the truth and male knaves lied, the females did the opposite: The female *knaves* told the truth and the female *knights* lied! Visitors to the island are required to take a series of logic tests. Well, here are some of the logic tests that Abercrombie took.

Problem 2.1 (The First Test). Abercrombie was led into an auditorium, and an inhabitant of the island came on the stage wearing a mask. Abercrombie was to determine whether the inhabitant was male or female. He was allowed to ask only one yes/no question, and the inhabitant would then write his or her answer on the blackboard (so as not to be betrayed by his or her voice). At first, Abercrombie thought of using a modification of the Nelson Goodman principle, but suddenly realized that he could ask a much simpler and less convoluted question that would work. He asked a direct yes/no question that involved only four words. What question would work?

Problem 2.2. The inhabitant then left the stage and a new masked inhabitant appeared. This time Abercrombie's task was to find out, not the sex of the inhabitant, but whether the inhabitant was a knight or a knave. Abercrombie then asked a three-word yes/no question that accomplished this. What question would do this?

PROBLEM 2.3. Next, a new inhabitant appeared on the stage, and Abercrombie was to determine whether or not the inhabitant was married, but he could ask only one yes/no question. What question should he ask? (The solution involves but a minor modification of the Nelson Goodman principle. Indeed, Abercrombie could find out any information he wanted by asking just one yes/no question.)

PROBLEM 2.4. In the next test, an inhabitant wrote a sentence from which Abercrombie could deduce that the inhabitant must be a female knight. What sentence could do this?

PROBLEM 2.5. Next, a sentence was written from which Abercrombie could deduce that the writer must be a male knight. What could this sentence have been?

PROBLEM 2.6. Then an inhabitant wrote a sentence from which it could be deduced that the writer must be either a female or a knight (or maybe both), but there was no way to tell which. What sentence would work?

PROBLEM 2.7. When Abercrombie returned from the island, he told his friend, a lawyer, that the day before he left, he had witnessed an extremely unusual trial! A valuable diamond had been stolen, and it was not known whether the thief was male or female, or whether a knight or a knave. Three suspects were tried in court, and it was known that one of the three was the thief, but it was not known which one. The strangest thing about the trial was that all three appeared masked, since they did not want their sex to be known. [They had the legal right to do this!] They also refused to speak, since they did not want their voices to betray their sex, but they were willing to write down answers to any questions asked of them. The judge was a truthful person. We will call the three defendants *A*, *B* and *C*, and here are the questions asked by the judge and the answers they wrote:

Judge (*to A*):	What do you know about all this?
A:	The thief is male.
Judge:	Is he a knight or a knave?
A:	He is a knave.
Judge (*to B*):	What do you know about *A*?
B:	*A* is female.
Judge:	Is she a knight or a knave?
B:	She is a knave.

The judge thought for a while and then asked *C*: "Are you, by any chance, the thief?" *C* then wrote down his or her answer, and the judge then knew which one was the thief.

At this point of Abercrombie's account of the trial, his friend asked: "What answer did C write?"

"Unfortunately, I don't remember," replied Abercrombie. "C either wrote *yes* or wrote *no*, but I don't remember which. At any rate, at this point the judge knew which one was the thief."

Well, at this point, you, the reader, have enough information to solve the case! Which of the three was the thief?

Solutions

2.1. The question Abercrombie asked was "Are you a knight?" Any male would answer *yes* (a male knight, truthfully so, and a male knave, falsely), whereas any female would answer *no* (a female knave truthfully, and a female knight, falsely). Thus, the answer *yes* would indicate that the inhabitant was male, and *no* would indicate that the inhabitant was female.

2.2. A question that works is "Are you male?" Knights will answer *yes* and knaves will answer *no*, as the reader can verify by a case analysis similar to that of the last problem.

Again we have a pretty symmetry: To find out if the inhabitant is male, you ask "Are you a knight?" and to find out if the inhabitant is a knight, you ask "Are you male?" (Of course, the questions "Are you a knave?" and "Are you female?" would work just as well.)

2.3. Instead of asking "Are you a knight if and only if you are married?" which would work fine for the island of the last chapter, you should now ask "Are you truthful if and only if you are married?" (or, alternatively, "Are you married if and only if you are either a male knight or a female knave?") Also, the question "Are you the type who could claim to be married?" would work just as well for this island as for the last island.

2.4. A sentence that works is "I am a male knave." No truthful person (male knight or female knave) could write that, and a male knave wouldn't truthfully write that. Only a female knight could write that.

2.5. A sentence that works is "I am not a female knave." Neither a female knight nor a male knave would write such a true statement, and a female knave wouldn't write such a false statement. Only a male knight could write that.

2.6. A sentence that works is "I am not a female knight." This could be written (truthfully) by a male knight or a female knave, or (falsely) by a female knight. The only type who couldn't write it is a male knave. Thus, the writer is either female or a knight, and possibly both.

2.7. B's answers were either both true or both false. If the former, then A is a female knave, and if the latter, then A is a male knight. In either case, A is truthful. Since A was telling the truth, the thief is really a male knave, and hence lies. Since A is truthful, A is not the thief.

At this point, the judge knew that the thief was either B or C and that he lied. The judge then asked C whether C was the thief. Now, if C had answered *no*, then the judge would have had no way of knowing whether B or C was the thief, since it could be that C answered truthfully, in which case B would be the thief, or that C answered falsely, in which case C would be the thief, which is consistent with the thief's being a liar. On the other hand, if C answered *yes*, then the judge would have known that B was the thief, because if C were the thief, C would have answered truthfully, contrary to the fact that C is a liar! So, in short, if C answered *no*, the judge would be in the dark, whereas if C answered *yes*, then the judge would know that B was the thief. Since we are given that the judge did know, then it must have been that C answered *yes*, and the judge then knew that the thief was B.

- Chapter 3 -

Silent Knights and Knaves

We now visit another knight/knave island on which, like on the first one, all knights tell the truth and all knaves lie. But now there is another complication! For some reason, the natives refuse to speak to strangers, but they are willing to answer yes/no questions using a *secret* sign language that works like this:

Each native carries two cards on his person; one is red and the other is black. One of them means *yes* and the other means *no*, but you are not told which color means what. If you ask a yes/no question, the native will flash one of the two cards, but unfortunately, you will not know whether the card means *yes* or *no*!

Problem 3.1. Abercrombie, who knew the rules of this island, decided to pay it a visit. He met a native and asked him: "Does a red card signify *yes*?" The native then showed him a red card.

From this, is it possible to deduce what a red card signifies? Is it possible to deduce whether the native was a knight or a knave?

Problem 3.2. Suppose one wishes to find out whether it is a red card or a black card that signifies *yes*. What simple yes/no question should one ask?

Problem 3.3. Suppose, instead, one wishes to find out whether the native is a knight or a knave. What yes/no question should one ask?

Remarks. Actually, there is a Nelson Goodman-type principle for this island; one can, with just a single yes/no question, find out any information one wants—such as whether there is gold on the island. In Chapter 8 we will provide such a principle in a far more general form that works for a whole host of islands, including this one.

PROBLEM 3.4. Without knowing what red and black signify, what yes/no question is such that the native would be sure to flash a red card?

PROBLEM 3.5. While Abercrombie was on the island, he attended a curious trial: A valuable diamond had been stolen. A suspect was tried, and three witnesses *A*, *B* and *C* were questioned. The presiding judge was from another land and didn't know what the colors red and black signified. Since non-inhabitants were present at the trial, the three witnesses were willing to answer questions only in their sign language of red and black.

First, the judge asked *A* whether the defendant was innocent. *A* responded by flashing a red card.

Then, the judge asked the same question of *B*, who then flashed a black card.

Then, the judge asked *B* a second question: "Are *A* and *C* of the same type?" (meaning both knights or both knaves). *B* flashed a red card.

Finally, the judge asked *C* a curious question: "Will you flash a red card in answer to this question?" *C* then flashed a red card.

Is the defendant innocent or guilty?

SOLUTIONS

3.1. It is not possible to determine which color means what, but the native must be a knight for the following reasons:

Suppose red means *yes*. Then by flashing a red card, the native is affirming that red does mean *yes*, and the affirmation is truthful, and so in this case, the native is a knight.

On the other hand, suppose red means *no*. Then by flashing a red card, the native is denying that red means *yes*, and the denial is correct, hence in this case the native is again a knight. Thus regardless of whether red means *yes* or *no*, the native must be a knight.

3.2. All you need ask is: "Are you a knight?" Both knights and knaves answer this question affirmatively, hence whatever color is flashed must mean *yes*.

3.3. The question of Problem 3.1 works; just ask him whether red means *yes*. We have already seen that if he flashes a red card, then he is a knight. A similar analysis, which we leave to the reader, reveals that if he flashes a black card, he is a knave.

3.4. There are several ways to do this. One way is to ask: "Is it the case that either you are a knight and red means *yes*, or you are a knave and red means *no*?"

You are asking whether one of the following two alternatives holds:

(1) You are a knight and red means *yes*.

(2) You are a knave and red means *no*.

Suppose red does mean *yes*. If he is a knight, then one of the two alternatives does hold (namely, (1)), hence the knight will correctly affirm this by flashing red (which means *yes*). On the other hand, if he is a knave, then neither alternative holds, so the knave will falsely answer *yes* by flashing red. Thus if red means *yes*, then both a knight and a knave will flash red.

Now, suppose red means *no*. If he is a knight, then neither alternative holds, hence he will honestly signify *no* by flashing red. If he is a knave, then (2) holds, hence one of the alternatives *does* hold, hence the knave will falsely signify *no* by flashing red.

Thus, regardless of what red really means, and regardless of whether the native is a knight or a knave, he will flash a red card.

3.5. Step 1. It follows from C's response that if C is a knight, then red means yes, and if C is a knave, then red means no.

Reason. Since C did flash red, the correct answer to the judge's question is *yes*. If C is a knight, then he answered truthfully, hence red then means *yes*. If C is a knave, then he lied, hence intended to answer *no*, in which case red means *no*.

Step 2. Since B flashed two different colors to the judge's two questions to him, the correct answers to the two questions must be different. Now, suppose the defendant is guilty. Then the correct answer to the judge's first question to B is *no*, hence the correct answer to the judge's second question to B is *yes*, which means that A and C really are of the same type. If A and C are knights, then red does mean *yes* (by Step 1, since C is then a knight), hence A meant *yes* by his response to the judge's question, and since A is a knight, *yes* was the correct answer, which means that the defendant is innocent, contrary to our assumption that the defendant is guilty. On the other hand (still assuming that the defendant is guilty), if A and C are knaves, we also get a contradiction, for then red means *no* (by Step 1, since C is a knave), hence A, a knave, meant *no* by flashing red, and since he lied, the correct answer to the judge's question is again *yes*, which means the defendant is innocent, contrary to our supposition that the defendant is guilty. Thus, the supposition that the defendant is guilty leads to a contradiction, so the defendant must be innocent.

- Chapter 4 -

Mad or Sane?

When Abercrombie reached his destination, he found himself on a very strange island, indeed! All the inhabitants of this island are completely truthful—they always tell you honestly what they believe, but the trouble is that half the inhabitants are totally mad, and all their beliefs are wrong! The other half are totally sane and accurate in their judgments; all their beliefs are correct.

Problem 4.1 (Let's Be Careful!). We shall start with a very tricky puzzle, but also one that illustrates a basic principle. Is it possible for an inhabitant of this island to say: "I believe I am mad"?

Problem 4.2 (A Simple Form of the Nelson Goodman Principle). The Nelson Goodman principles for the islands of Chapters 1 and 2 involve questions that are rather convoluted and unnatural. Well, on the present island, there is a much more natural-sounding yes/no question you could ask to ascertain any information you want. For example, if you want to find out whether a given native is married or not, there is a relatively natural-sounding question you can ask—one, in fact, having only six words. What question would work?

Problem 4.3. When Abercrombie got settled on this island, he first interviewed three siblings named Henry, Dianne, and Maxwell. Henry and Dianne made the following statements:

Henry: Maxwell believes that at least one of us is mad.
Dianne: Maxwell is sane.

What type is each?

Problem 4.4. Next, Abercrombie interviewed Mary and Gerald, a married couple, together with their only child, Lenore. Here is the dialogue that took place.

Abercrombie (*to Gerald*):	I heard that your wife once said that all three of you are mad. Is that true?
Gerald:	No, my wife never said that.
Abercrombie (*to Lenore*):	Did your father once say that exactly one of you three is sane?
Lenore:	Yes, he once said that.
Abercrombie (*to Mary*):	Is your husband sane?
Mary:	Yes.

What type is each?

Problem 4.5. Abercrombie's next interview was a bit more puzzling. He met a married couple, Arthur and Lillian Smith. Arthur was the only one who said anything, and what he said was: "My wife once said that I believe that she believes I am mad."

What can be deduced about either one?

Problem 4.6. Abercrombie next interviewed eight brothers named Arthur, Bernard, Charles, David, Ernest, Frank, Harold, and Peter. They made the following statements:

Charles:	Arthur is mad.
David:	Bernard and Charles are not both mad.
Ernest:	Arthur and David are alike, as far as sanity goes.
Frank:	Arthur and Bernard are both sane.
Harold:	Ernest is mad.
Peter:	Frank and Harold are not both mad.

From this confusing tangle, it is possible to determine the madness or sanity of one of the eight. Which one, and what is he?

Problem 4.7 (A Metapuzzle). Before Abercrombie left the island, one of the sane inhabitants, whose name was David, told him of a trial he had attended some time ago. The defendant was suspected of having stolen a watch. First, the judge (who was sane) said to the defendant: "I have heard that you once claimed that you stole the watch. Is that true?" The defendant answered (either *yes* or *no*). Then the judge asked: "Did you steal the watch?" The defendant then answered (either *yes* or *no*) and the judge then knew whether he was innocent or guilty.

"What answers did the defendant give?" asked Abercrombie.

"I don't quite remember," replied David. "It was quite some time ago. I do, however, remember that he didn't answer *no* both times."

Was the defendant innocent or guilty?

Solutions

4.1. Many of you will say that it is *not* possible. You will reason that a sane person knows he is sane, hence does not believe he is mad, and a mad person erroneously believes that he is sane, hence does not believe the true fact that he is mad. Thus, no inhabitant can believe he is mad.

Well, so far, so good. It is indeed true that no inhabitant can believe he is mad. And since the inhabitants honestly state what they believe, then no inhabitant can say that he is mad. But I didn't ask you whether an inhabitant can say that he is mad, nor did I ask whether an inhabitant can believe that he is mad; what I asked you was whether an inhabitant can say that he believes that he is mad, and that's a different story!

Look, a mad person doesn't believe that he is mad, and so it is *false* that he believes he is mad, but since he believes false propositions, then he also believes that one—he believes that he believes he is mad!! Thus, he doesn't believe he is mad, yet he believes that he *does* believe he is mad. And, being honest, he could indeed say that he believes he is mad. Indeed, if you ask a mad inhabitant: "Are you mad?" he will say *no*, but if you ask him: "Do you *believe* you are mad?" he will answer *yes* (since he doesn't really believe he is mad).

The fact of the matter is that given any true proposition, he will disbelieve the proposition, but also believe that he believes the proposition! Conversely, whatever a mad person believes that he believes must be true. (Also, of course, whatever a sane person believes that he believes must be true.) Thus, whatever any inhabitant, mad or sane, believes that he believes must be true. Also, if an inhabitant believes that he *doesn't* believe a certain proposition, then the proposition must be false. (This is obvious for a sane inhabitant, but if the inhabitant is mad, then it is false that he doesn't believe the proposition (since he erroneously believes that he doesn't believe it), which means that he does believe it, and hence it is false.)

Let us record two of the things we have just learned:

Fact 1. When an inhabitant believes that he believes something (whatever that *something* is), that *something* must be true.

Fact 2. When an inhabitant believes that he doesn't believe something, that *something* must be false.

4.2. For the Island of Knights and Knaves of Chapter 2, to find out if an inhabitant is married, you can ask: "Are you a knight if and only if you are married?" For the island of the present chapter, you can ask instead: "Are you sane if and only if you are married?" Also, the question "Are you the type who can claim that you are married?" would work for this island as well as the islands of Chapters 2 and 3. For the present island, however, a much more economical and natural-sounding question is possible. All you need ask is "Do you *believe* you are married?" If he answers *yes*, then he believes that he believes he is married, hence he really is married (by Fact 1, stated and proved in the solution of the last problem). If he answers *no*, then he believes that he doesn't believe that he is married, hence he is not married (by Fact 2).

Thus, in general, to find out if something is the case, you ask the inhabitant if he *believes* that the something is the case. For example, if you want to know if there is gold on the island, all you need to ask is "Do you believe there is gold on the island?"

Neat, eh?

4.3. Suppose Henry is sane. Then his statement is true; hence Maxwell really does believe that at least one of the three is mad. If Maxwell were mad, then it would be true that at least one is mad, and so mad Maxwell would have a true belief, which is not possible. Hence, Maxwell is sane (still under the assumption that Henry is sane). Then, also, Dianne is sane (since she correctly believes that Maxwell is sane); hence all three are sane, contrary to Maxwell's sane belief that at least one is mad! Thus, it is contradictory to assume that Henry is sane. Thus Henry is mad.

Since Henry is mad, what he said is false, so Maxwell doesn't really believe that at least one of the three is mad; he believes that all three are sane. But his belief is wrong (since Henry is mad), and so, Maxwell is mad. Hence Dianne is also mad (since she believes that Maxwell is sane). Thus, all three are mad.

4.4. Step 1. Gerald and Mary are alike, as far as their sanity goes.

Reason. Mary believes that Gerald is sane. If Mary is sane, her belief is correct, hence Gerald is also sane. If Mary is mad, her belief is wrong, which means that Gerald is not sane, but mad.

Step 2. Lenore must be mad.

Reason. Suppose Lenore were sane. Then her statement would be true, hence Gerald did once say that exactly one of the three is sane, but this leads to a contradiction because:

(1) If Gerald is sane, so is Mary (by Step 1); hence all three are sane, so it is false that exactly one is sane, but sane people here don't make false statements.

(2) On the other hand, if Gerald is mad, then so is Mary (Step 1) and Lenore is then the only sane one, so it is true that exactly one is sane, but mad inhabitants don't make true statements.

Thus Lenore can't be sane: she is mad.

Step 3. Suppose Gerald is mad. Then, so is Mary (by Step 1); hence all three are mad. Then Mary, who is mad, never did make the true statement that all three are mad; hence Gerald was right when he denied that Mary did, but mad people here don't make true statements! Thus, it is contradictory to assume that Gerald is mad. Hence Gerald is sane and so is his wife (by Step 1). Thus, the mother and father are both sane, but their daughter Lenore is mad.

4.5. Suppose Arthur is sane. Then Lillian did once claim that Arthur believes that Lillian believes that Arthur is mad. Suppose Lillian is sane. Then Arthur believes that Lillian believes that Arthur is mad. Since Arthur is sane, then Lillian does believe that Arthur is mad, and since Lillian is sane, then Arthur is mad, contrary to the assumption that Arthur is sane.

Suppose Lillian is mad. Then Arthur doesn't really believe that Lillian believes that Arthur is mad. Since Arthur is sane (by assumption), then it is false that Lillian believes that Arthur is mad. But Lillian is mad, and since she doesn't believe that Arthur is mad, then Arthur is mad, again contrary to the assumption that Arthur is sane. Thus, Arthur must be mad. Hence Lillian never did say what Arthur said she said, and so nothing can be deduced about Lillian.

4.6. I will prove that Peter must be sane.

Step 1. Arthur and David cannot both be mad.

Reason. Suppose David is mad. Then, contrary to what he said, Bernard and Charles are both mad. Since Charles is mad, then, contrary to what he said, Arthur must be sane. This proves that if David is mad, Arthur is sane; hence Arthur and David cannot both be mad.

Step 2. Frank and Harold cannot both be mad.

Reason. Suppose Harold is mad. Then Ernest must be sane; hence Arthur and David are really alike, as far as their sanity goes. But Arthur and David are not both mad (as we proved in Step 1), so they are both sane. Hence, Frank's statement was true, so Frank is sane. This proves that if Harold is mad, then Frank is sane, so Harold and Frank are not both mad.

Step 3. Therefore, Peter's statement was true, so Peter is sane.

4.7. Since we know that the defendant didn't answer *no* both times, there are only three cases to consider.

Case 1: He Answered *Yes* Both Times. It could be that he is sane and stole the watch and did once claim that he stole it. But it is also possible that he is mad and never stole it and also never claimed that he did. The judge would then have no way of knowing whether the defendant was innocent or guilty.

Case 2: His First Answer Was *No* and His Second Was *Yes*. Then it could be that he is sane and guilty but never claimed that he had stolen the watch, but it is also possible that he is mad and never stole the watch, but once claimed that he had. Again, the judge would have no way of knowing which of these possibilities held.

Case 3: His First Answer Was *Yes* and His Second Was *No*. In this case, could he be sane? No, because then he would be innocent of the theft (by virtue of his second answer) but also would have once claimed that he had stolen the watch, which would be a false claim, hence not possible for a sane inhabitant. Therefore, he must be mad, hence he did steal the watch (since he indicated that he hadn't) but also never claimed that he had stolen it (since he indicated that he did make such a claim). This is the only possibility, and the judge would then know that the defendant was guilty.

Since the judge *did* know, Cases 1 and 2 are out; hence it must be Case 3 that actually held. Thus the defendant was guilty (and also mad).

- Chapter 5 -

The Difficulties Double!

The next island visited by Abercrombie was more baffling yet! It compounded the difficulties of a knight/knave island with those of the island of the sane and mad.

On this island, each inhabitant was either a knight or a knave; knights were truthful and knaves were liars. But half of all the inhabitants were mad and had only false beliefs, and the other half were sane and had only correct beliefs. Thus each inhabitant was one of the following four types:

(1) Sane knight

(2) Mad knight

(3) Sane knave

(4) Mad knave

We note the following facts:

Fact 1. Anything a sane knight says is true.

Fact 2. Anything a mad knight says is false. (He tries to make true statements, but cannot.)

Fact 3. Anything a sane knave says is false.

Fact 4. Anything a mad knave says is true. (He tries to deceive you, but is unable to do so.)

For example, suppose you ask an inhabitant whether two plus two is four. A sane knight knows that this is true and honestly says *yes*. A mad knight believes it isn't, so, true to his belief, he says *no*. A sane knave knows that two plus two is four, and then lies and says it isn't. A mad knave believes that two plus two *doesn't* equal four, and then lies and says it *does*! Thus a mad knight answers *no*, but a mad knave answers *yes*.

Problem 5.1. Suppose you meet a native of this island and want to know whether he is sane or mad. What single yes/no question could determine this?

Problem 5.2. Suppose, instead, you wanted to find out whether he is a knight or a knave?

Problem 5.3. What yes/no question could you ask that would ensure that he will answer *yes*?

Problem 5.4. There is a Nelson Goodman principle for this island; one can find out any information one wants with just one yes/no question. For example, suppose you wanted to know whether there is gold on this island. What single yes/no question would you ask?

Problem 5.5. When Abercrombie arrived on this island, he met a native named Hal who made a statement from which Abercrombie could deduce that he must be a sane knight. What statement would work?

Problem 5.6. Abercrombie and Hal became fast friends. They often went on walks together, and Hal was sometimes quite useful in helping Abercrombie in his investigations. On one occasion, they spied two inhabitants walking toward them.

"I know them!" said Hal. "They are Auk and Bog. I know that one of them is sane and the other is mad, but I don't remember which one is which."

When the two came up to them, Abercrombie asked them to tell him something about themselves. Here is what they said:

Auk: Both of us are knaves.
Bog: That is not true!

Which of the two is mad?

Problem 5.7. On another occasion, they came across two other natives named Ceg and Fek. Hal told Abercrombie that he remembered that one was sane and one was mad, but wasn't sure which was which. He also remembered that one was a knight and the other a knave, but again was not sure which was which.

"As a matter of fact," said Abercrombie, "I came across these same two natives a couple of days ago and Ceg said that Fek is mad and Fek said that Ceg was a knave."

"Ah, that settles it!" said Hal.

Hal was right. What type is each?

Problem 5.8. On another occasion, Abercrombie and Hal came across two natives named Bek and Drog, who made the following statements:

Bek: Drog is mad.
Drog: Bek is sane.
Bek: Drog is a knight.
Drog: Bek is a knave.

What type is each?

Problem 5.9 (A Metapuzzle). Several days later, after Abercrombie learned more about some of the natives, he and Hal were walking along one late afternoon and spied a native mumbling something to himself.

"I know something about him," said Abercrombie. "I know whether he is a knight or a knave, but I don't know whether he is sane or mad."

"That's interesting!" said Hal. "I, on the other hand, happen to know whether he is sane or mad, but I don't know whether he is a knight or a knave."

When the two got closer, they heard what the native was mumbling, which was "I am not a sane knave." The two both thought for a while, but Abercrombie still didn't have enough information to determine whether the native was sane or mad, nor did Hal have enough information to determine whether the native was a knight or a knave.

At this point, *you* have enough information to determine the type of the native. Was he sane or mad, and was he a knight or a knave?

Solutions

5.1. A particularly simple question is "Are you a knight?" A sane knight will correctly say *yes*; a sane knave will falsely say *yes*; a mad knight will incorrectly say *no*; and a mad knave will correctly say *no*. Thus, a sane inhabitant will say *yes* and a mad inhabitant will say *no*.

5.2. The question to ask is "Are you sane?" By a case analysis similar to that of Problem 5.1, you will see that a knight (whether sane or mad) will say *yes* and a knave will say *no*.

Again, we have this nice symmetry: To find out if he is sane, you ask whether he is a knight, and to find out if he is a knight, you ask whether he is sane.

5.3. Call a native *reliable* if he makes true statements and answers questions correctly and *unreliable* otherwise. Reliable natives are sane knights and mad knaves; unreliable natives are mad knights and sane knaves.

A question that guarantees the answer *yes* is "Are you reliable?" or "Are you either a sane knight or a mad knave?" If he is reliable, he will answer correctly and say *yes*. If he is unreliable, he will answer incorrectly and say *yes*. In either case he will say *yes*.

5.4. A question that works is "Are you the type who could claim that you *believe* there is gold on this island?" Another is "Do you believe that you are the type who could claim that there is gold on this island?"

But a much neater and simpler one is "Are you reliable if and only if there is gold on this island?"

Actually, in the next chapter, we will present an extremely general Nelson Goodman-type principle that works simultaneously for all the islands considered up to now (even the one where natives answer by flashing red or black cards) as well as the more bizarre island of the next chapter, and we will prove that it always works.

5.5. A statement that works is: "I am not a mad knave." A sane knight could (correctly) say that; a mad knight could not (correctly) say it; and a mad knave wouldn't (correctly) say it. Thus, only a sane knight could say it.

5.6. We are given that one and only one of the two is sane. Suppose Auk is sane. Then he couldn't be a knight, for then his statement would be true, which would mean that both are knaves, which is impossible if he is a knight. Therefore (assuming Auk is sane), he must be a knave. Since he is a sane knave, his statement is false, so it is not really true that both are knaves, and hence Bog must be a knight. Also Bog is mad (since Auk is sane), so Bog is a mad knight, hence his statement is false, which would mean Auk's statement is true, which it isn't, since they are not both knaves. Thus, the assumption that Auk is sane leads to a contradiction. Therefore, Auk is mad.

5.7. Step 1. Suppose Ceg is the knave. Then Fek is the knight and also has made a true statement, hence Fek is a sane knight, and, therefore, Ceg is a mad knave. But then, a mad knave wouldn't have made the false statement that Fek is a knave! Thus, it is contradictory to assume that Ceg is the knave, so Ceg is the knight and Fek is the knave.

Step 2. Fek's statement was false (since Ceg is actually a knight), and since Fek is a knave, Fek must be a sane knave. Therefore, Ceg must be a mad knight.

Thus, Ceg is a mad knight and Fek is a sane knave. (What a pair!)

5.8. Bek's statements are either both true or both false. If both true, then Drog is a mad knight; if both false, Drog is a sane knave. In both cases, Drog makes wrong statements. Since both of Drog's statements are wrong, Bek is a mad knight. Then Bek's statements are also both false, and so Drog is a sane knave.

5.9. Step 1. All that follows from the native's statement is that he is not a mad knight, because a sane knight could correctly say that he is not a sane knave, and a mad knave could correctly say that, and a sane knave could falsely say that he is not a sane knave, but a mad knight could not make the correct statement that he is not a sane knave.

Step 2. Abercrombie had already known whether the native was a knight or a knave. Had he known that the native was a knight, he would then have learned from the native's statement that he wasn't a mad knight, and hence that he must be a sane knight. Therefore, he would have had enough information to know that he was sane. But he didn't have enough information, so it must be that he had previously known that the native was a knave, and hence got no additional information from knowing that he was not a mad knight. Thus the native must be a knave.

Hal, on the other hand, had already known whether the native was sane or mad, but not whether he was a knight or knave. Had he previously known that the native was mad, then from his later knowledge that the native was not a mad knight, he would have had enough information to know that the native was a knave. But since he didn't have enough information, then it must be that he had previously known that the native was sane, and already knew that he couldn't be a mad knight.

Thus, Abercrombie had previously known that the native was a knave, and Hal had previously known that he was sane. Consequently, the native was a sane knave.

- Chapter 6 -

A Unification

Oh, No!

When Abercrombie left the last island, he visited another one which was by far the most bizarre of all! It combined all the difficulties of all the islands previously visited. This island had the following features:

(1) Every inhabitant was classified as a knight or a knave.

(2) Male knights were truthful and male knaves were liars, but female knights lied and female knaves were truthful.

(3) Half of the inhabitants were mad and had only false beliefs, whereas the other half were sane and had only correct beliefs.

(4) When you asked a native a yes/no question, instead of answering *yes* or *no*, he or she would show you either a red card or a black card, one of which signified *yes* and the other, *no*.

(5) But different inhabitants might have meant different things by the two colors: Some of them would show a red card to signify *yes* and a black card to signify *no*, whereas some others would do the opposite!

Problem 6.1. Is there a Nelson Goodman-type principle for this crazy island? That is, can one find out any information one wants by asking just one yes/no question? For example, suppose you visit the island and want to know if there is gold on it. You meet a masked native and don't know the sex of the native, nor whether he or she is a knight or a knave, nor whether mad or sane, nor what the colors red and black signify to him or her. Is there a single yes/no question that you could ask to determine whether there is gold on the island, or is no such question possible?

Problem 6.2. A related problem is this: Is there a yes/no question that will ensure that the native addressed will respond by flashing a red card?

In Quest of a Unifying Principle

It has been wisely remarked that the existence of similarities among the central features of various theories implies the existence of a general theory that underlies the particular theories and unifies them with respect to those central features.

Problem 6.3. We have considered the Nelson Goodman principle in six different situations: the island of knights and knaves, the knight/knave island in which males and females respond differently, the island of knights and knaves where natives respond by flashing red or black cards instead of saying *yes* or *no*, the island of the sane and the mad, the island that combines knights and knaves with sane and mad, and, finally, the crazy island of this chapter. We have given different Nelson Goodman principles for four of them and stated that there is also one for the island of Chapter 3. Is it not possible to unify them into one principle that is simultaneously applicable to all six cases, and possibly others?

Yes, there is! Can the reader find one? (There is not just one that works; different readers may well find different general principles, all of which fit the bill. We provide one in the Solutions.)

Solutions

6.1. Such a question is possible. There is a suitable modification of Nelson Goodman's principle for this island.

Let me begin by saying that what is really important about an inhabitant is *not* the inner workings of his or her mind, whether he or she is truthful, or sane, or what the two colors mean to the person. The important thing is the *responses* given to yes/no questions. Let us define an inhabitant to be of Type 1 if he or she flashes red to questions whose correct answer is *yes* (and hence flashes black to questions whose correct answer is *no*). Inhabitants not of Type 1, whom we will define to be of Type 2, will flash black to questions whose correct answer is *yes* and red to questions whose correct answer is *no*.

To find out if there is gold on the island, instead of asking "Are you a knight if and only if there is gold on the island?" which would work fine for the island of Chapter 1, you now ask: "Are you of Type 1 if and only if there is gold on the island?" We will see that if the native flashes red, then there is gold, and if he flashes black, there isn't. Rather than prove this directly, we will derive it as a special case of a solution to Problem 6.3.

6.2. You need simply ask: "Are you of Type 1?" That this works is but a special case of a more general result to be established below.

6.3. Now we shall generalize all the previous Nelson Goodman principles. We consider a very general-type situation in which we visit a land where each inhabitant responds to a yes/no question in one of two ways, which we will call *Response 1* and *Response 2*. (In application to the islands already considered, for the islands of Chapters 1, 2, 4, and 5, we will take Response 1 to be the act of replying *yes* and Response 2 to be the act of saying *no*. In application to the islands of Chapter 3 and the present chapter, we will take Response 1 to be the act of flashing a red card and Response 2 to be the act of flashing a black card. Other applications are possible; for example, natives might respond by saying *yes* or *no*, but in a foreign language that you don't understand.) It is understood that a native will give the same response to all questions whose correct answer is *yes*, and the other response to all questions whose correct answer is *no*. We then define a native to be of *Type 1* if he gives Response 1 to yes/no questions whose correct answer is *yes* (and hence Response 2 to questions whose correct answer is *no*). The other type we will call *Type 2*—this is the type who gives Response 2 to questions whose correct answer is *yes*, and Response 1 to questions whose correct answer is *no*. (For the islands of Chapters 1, 2, 4 and 5, Type 1 inhabitants are those who answer *yes* to questions whose correct answer is *yes*—in other words, those inhabitants who answer yes/no questions correctly. For the islands of Chapter 3 and this chapter, Type 1 inhabitants are those who flash a red card in response to questions whose correct answer is *yes*.)

First, let us consider the simpler problem of designing a question that will force the native to give Response 1. Well, such a question is simply "Are you of Type 1?"

CASE 1: THE NATIVE IS OF TYPE 1. Then the correct answer to the question is *yes*; hence, since he is of Type 1, the native will give Response 1.

Case 2: The Native Is of Type 2. Then the correct answer is *no*; hence, since the native is of Type 2, he will give Response 1.

Thus, in both cases, the native will give Response 1. (Of course, if you wished the native to give Response 2 instead, you would ask: "Are you of Type 2?")

Now we give a generalized Nelson Goodman principle. Suppose you want to find out, say, whether there is gold on the island. What you ask is the same question as in Problem 6.1 (and it works for *all* the islands so far considered). You ask: "Are you of Type 1 if and only if there is gold on the island?" Thus, you are asking whether one of the following two alternatives holds:

(1) The native is of Type 1 and there is gold on the island.

(2) The native is of Type 2 and there is no gold on the island.

Suppose the native gives Response 1.

Case 1: The Native Is of Type 1. Then, since he gave Response 1, the correct answer to the question is *yes*. Thus, one of the two alternatives (1), (2) *does* hold. It cannot be (2), since the native is not of Type 2, so it must be (1). And so, in this case, there is gold on the island.

Case 2: The Native Is of Type 2. Then, since he gave Response 1 and he is of Type 2, the correct answer to the question is *no*. Thus, neither of the two alternatives holds—in particular, (2) doesn't hold. Hence, there must be gold on the island (because if there weren't, then (2) *would* hold, which it doesn't).

This proves that if the native gives Response 1, there is gold on the island.

Now, what if he gives Response 2? Well, to be brief, if he is of Type 2, then the correct answer to the question is *yes*; hence one of the alternatives *does* hold, and it must be (2); hence there is no gold; whereas if the native is of Type 1, then the correct answer to the question is *no*; hence, neither alternative holds; hence (1) doesn't hold; hence there is no gold on the island. This proves that Response 2 implies that there is no gold on the island.

Pretty neat, huh?

- Part II -

Be Wise, Symbolize!

- Chapter 7 -

Beginning Propositional Logic

Propositional logic, the logic of propositions, is something we have been doing all along, albeit on a purely informal level. Now we shall approach this basic field—basic for all of formal mathematics as well as computer science—on a more formal symbolic level, and in the next chapter we will see how it is beautifully applicable to the logic of lying and truth-telling.

The Logical Connectives

Just as in ordinary algebra we use letters x, y, z, with or without subscripts, to stand for arbitrary numbers, so in propositional logic we use letters p, q, r, with or without subscripts, to stand for arbitrary propositions.

Propositions can be combined by using *logical connectives.* The principal ones are

(1) negation (not) $\sim$,

(2) conjunction (and) $\wedge$,

(3) disjunction (or) $\vee$,

(4) if-then $\Rightarrow$,

(5) if and only if $\equiv$.

Here is what they mean.

NEGATION

For any proposition p, by $\sim p$ (sometimes written $\neg p$) is meant the *opposite* or *contrary* of p. For example, if p is the proposition that Jack is guilty, then $\sim p$ is the proposition that Jack is *not* guilty. The proposition $\sim p$ is read "it is not the case that p," or, more briefly, "not p." The proposition $\sim p$ is called the *negation* of p, and is true if p is false, and false if p is true. These two facts are summarized in the following table, which is called the *truth table for negation*. In this table, as in all the truth tables that follow, we shall use the letter "T" to stand for *truth* and "F" to stand for *falsehood*.

p	$\sim p$
T	F
F	T

The first line of this truth table says that if p has the value T (in other words, if p is true), then $\sim p$ has the value F ($\sim p$ is false). The second line says that if p has the value F, then $\sim p$ has the value T. We can also express this by the following equations:

$$\sim \mathrm{T} = \mathrm{F},$$
$$\sim \mathrm{F} = \mathrm{T}.$$

CONJUNCTION

For any propositions p and q, the proposition that p *and* q are both true is written "$p \wedge q$" (sometimes "$p \& q$"). We call $p \wedge q$ the *conjunction* of p and q, and it is read "p and q are both true," or more briefly "p and q." For example, if p is the proposition that Jack is guilty and q is the proposition that Jill is guilty, then $p \wedge q$ is the proposition that Jack and Jill are both guilty.

The proposition $p \wedge q$ is true if p and q are both true and is false if at least one of them is false. We thus have the following four laws of conjunction:

$$\mathrm{T} \wedge \mathrm{T} = \mathrm{T},$$
$$\mathrm{T} \wedge \mathrm{F} = \mathrm{F},$$
$$F \wedge \mathrm{T} = \mathrm{F},$$
$$F \wedge \mathrm{F} = \mathrm{F}.$$

This is also expressed by the following table, the truth table for *conjunction*:

p	q	$p \wedge q$
T	T	T
T	F	F
F	T	F
F	F	F

Disjunction

We write "$p \vee q$" to mean that at least one of the propositions p, q is true (and maybe both). We read "$p \vee q$" as "either p or q," or, more briefly, "p or q." It is true if at least one of the propositions p, q is true, and false only if p and q are both false.

For example, if p is the proposition the Jack is guilty and q is the proposition that Jill is guilty, then $p \vee q$ is the proposition that at least one of the two persons, Jack or Jill, is guilty (and maybe both).

It should be pointed out that in ordinary English, the phrase "either-or" is used in two senses: the *strict* or *exclusive* sense, meaning *exactly one*, and the *loose* or *inclusive* sense, meaning *at least one*. For example, if I say that tomorrow I will marry either Betty or Jane, I of course mean that I will marry one *and only one* of the two ladies, so I am using "either-or" in the exclusive sense. On the other hand, if an advertisement for a secretary requires that the applicant know either French or German, an applicant is certainly not going to be rejected because he or she happens to know both! So in this case, "either-or" is used in the *inclusive* sense. Now, in formal logic, mathematics and computer science, we always use "either-or" in the *inclusive* sense, and so $p \vee q$ means *at least one* of p, q is true.

I might also point out that, in Latin, there are two different words for the two different senses: "aut" is used for the exclusive sense, and "vel" for the inclusive sense. In fact, the logical symbol "$\vee$" for "or" actually comes from the Latin word "vel."

The proposition $p \vee q$ is called the *disjunction* of p and q and has the following truth table:

p	q	$p \vee q$
T	T	T
T	F	T
F	T	T
F	F	F

We note that we have an F for $p \vee q$ only in the last row (in which p and q are both F). This table can also be expressed by the following equations:

$$\begin{aligned} T \vee T &= T, \\ T \vee F &= T, \\ F \vee T &= T, \\ F \vee F &= F. \end{aligned}$$

IF-THEN

The if-then symbol "$\Rightarrow$" is particularly troublesome to those in first contact with symbolic logic, since it is questionable whether the meaning of "if-then" as used technically by logicians is quite the same as that of common use.

For any propositions p and q, we write "$p \Rightarrow q$" to mean "if p, then q"; also read "p *implies* q," or "it is not the case that p is true and q is false," or "either p is false, or p and q are both true." Thus, $p \Rightarrow q$ means that you can't have p without also having q, or in other words that either p is false, or p and q are both true.

How are we to evaluate the truth or falsity of $p \Rightarrow q$, given the truth or falsity of each of p and q? Well, there are four cases to consider. Either p and q are both true, or p is true and q is false, or p is false and q is true, or p and q are both false. In the first case, since q is true, it is certainly the case that *if* p, *then* q (because, for that matter, if *not* p, then q would *also* hold; if q is true absolutely, then "if p then q" is quite independent of p). And so we clearly have

$$T \Rightarrow T = T.$$

Next, if p is true and q is false, then $p \Rightarrow q$ must be false (because a true proposition can never imply a false proposition), so we have

$$T \Rightarrow F = F.$$

In the third case, since q is true, it is again the case that $p \Rightarrow q$ is true regardless of whether p is true or false, so we have

$$F \Rightarrow T = T.$$

The fourth case is the puzzling one: Suppose p and q are both false; what should I make of "*if* p, *then* q"? Some might guess that it should be false, others that it is true, and others that it is inapplicable. At that point, a decision must be made once and for all, and the decision that *has* been made by logicians and computer scientists is that in this case, $p \Rightarrow q$ should be declared *true*. Let me give you what I believe is a good argument why this decision is a wise one.

Suppose I put a card face down on the table, without first letting you see the face. I then say: "If this card is the queen of spades, then it is black." Surely, you will agree! Thus, letting p be the proposition that the card is the queen of spades and q the proposition that the card is black, I am asserting $p \Rightarrow q$, and you agree. Now, suppose I turn this card over, and it turns out to be the five of diamonds, am I then to retract my statement? You originally agreed that $p \Rightarrow q$ is true, but you have subsequently seen that p and q are both false (the card is neither the queen of spades nor black), but isn't it still true that *if* the card had been the queen of spades, *then* it would have been black? And so, here is a perfect example of a case where $p \Rightarrow q$ is true even though p and q are themselves both false. Hence, we have

$$\mathrm{F} \Rightarrow \mathrm{F} = \mathrm{T}.$$

Thus, the truth table for $\Rightarrow$ is the following:

p	q	$p \Rightarrow q$
T	T	T
T	F	F
F	T	T
F	F	T

It must be emphasized that $p \Rightarrow q$ is false *only* in the case that p is true and q is false—in the other three cases, it is true.

IF AND ONLY IF

We write $p \equiv q$ to mean that p and q are either both true or both false, or what is the same thing: if either is true, so is the other. We read $p \equiv q$ as "p if and only if q" or "p and q are equivalent" (as far as their truth or falsity are concerned).

Since $p \equiv q$ is true when and only when p and q are either both true or both false, $p \equiv q$ is false when and only when one of p, q is true and the other false (either p true and q false, or p false and q true), so here is the truth table for $\equiv$.

p	q	$p \equiv q$
T	T	T
T	F	F
F	T	F
F	F	T

Or, as equations,

$$\begin{aligned} \mathrm{T} \equiv \mathrm{T} &= \mathrm{T}, \\ \mathrm{T} \equiv \mathrm{F} &= \mathrm{F}, \\ \mathrm{F} \equiv \mathrm{T} &= \mathrm{F}, \\ \mathrm{F} \equiv \mathrm{F} &= \mathrm{T}. \end{aligned}$$

We note that $p \equiv q$ holds if and only if $p \Rightarrow q$ and $q \Rightarrow p$ both hold. Thus, $p \equiv q$ could be regarded as shorthand for $(p \Rightarrow q) \wedge (q \Rightarrow p)$. We remark that the operation $\Rightarrow$ is sometimes called the *conditional* and $\equiv$, the *bi-conditional*.

Note. The "if and only if" construct plays an important role in mathematical definitions. For example, to define the concept "even number," we might write "We say that X is an *even* number if and only if X is a whole number and is divisible by 2." Because of the ubiquity of its usage in mathematics, the phrase "if and only if" is frequently abbreviated "*iff*." We will make liberal use of this abbreviation throughout the book.

The operations $\sim$, $\wedge$, $\vee$, $\Rightarrow$, $\equiv$ are examples of what are called *logical connectives*.

Parentheses

One can combine simple propositions into compound ones in many ways by using the logical connectives. We usually need *parentheses* to avoid ambiguity. For example, if we write $p \wedge q \vee r$ without parentheses, we cannot tell which of the following is meant:

(1) Either $p \wedge q$ is true, or r is true.

(2) p is true and $q \vee r$ is true.

If we mean (1), we should write $(p \wedge q) \vee r$, whereas if we mean (2), we should write $p \wedge (q \vee r)$. (The situation is analogous to algebra: $(x+y) \times z$ has a different meaning from $x + (y \times z)$—for example, $(2+3) \times 4 = 20$, whereas $2 + (3 \times 4) = 14$.) We thus need parentheses in propositional logic for punctuation.

Compound Truth Tables

By the *truth value* of a proposition p is meant its truth or falsity—that is, T if p is true, and F if p is false. Thus the propositions "$2+3=5$" and "Paris is the capital of France," though different propositions, have the same truth value, namely T.

Consider now two propositions p and q. If we know the truth value of p and the truth value of q, then, by the simple truth tables already constructed, we can determine the truth values of $\sim p$, $p \wedge q$, $p \vee q$, $p \Rightarrow q$, and $p \equiv q$. It therefore follows that given any combination of p and q—that is, any proposition expressible in terms of p and q using the logical connectives—we can determine the truth value of the combination given the truth values of p and q. For example, suppose X is the combination

$(p{\equiv}(q{\wedge}p)){\Rightarrow}({\sim}p{\Rightarrow}q)$. Given the truth values of p and q, we can successively find the truth values of $q{\wedge}p$, $p{\equiv}(q{\wedge}p)$, ${\sim}p$, ${\sim}p{\Rightarrow}q$, and, finally, $(p{\equiv}(q{\wedge}p)){\Rightarrow}({\sim}p{\Rightarrow}q)$. There are four possible distributions of truth values for p and q (p true, q true; p true, q false; p false, q true; and p false, q false), and in each of the four cases, we can determine the truth value of X. We can do this systematically by constructing the following table (an example of a *compound* truth table):

p	q	$q{\wedge}p$	$p{\equiv}(q{\wedge}p)$	${\sim}p$	${\sim}p{\Rightarrow}q$	$(p{\equiv}(q{\wedge}p)){\Rightarrow}({\sim}p{\Rightarrow}q)$
T	T	T	T	F	T	T
T	F	F	F	F	T	T
F	T	F	T	T	T	T
F	F	F	T	T	F	F

We see that X is true in the first three cases and false in the fourth.

We can also construct a truth table for a combination of three propositional unknowns (p, q and r) but now there are eight cases to consider (because there are four distributions of T's and F's to p and q, and with each of these four distributions there are two possibilities for r). For example, suppose X is the combination $(p{\wedge}q){\equiv}({\sim}p{\Rightarrow}r)$. Here is the truth table for X.

p	q	r	$p{\wedge}q$	${\sim}p$	${\sim}p{\Rightarrow}r$	$(p{\wedge}q){\equiv}({\sim}p{\Rightarrow}r)$
T	T	T	T	F	T	T
T	T	F	T	F	T	T
T	F	T	F	F	T	F
T	F	F	F	F	T	F
F	T	T	F	T	T	F
F	T	F	F	T	F	T
F	F	T	F	T	T	F
F	F	F	F	T	F	T

We see that X is true in cases 1, 2, 6, and 8.

TAUTOLOGIES

Consider the expression

$$(p{\Rightarrow}q){\equiv}({\sim}q{\Rightarrow}{\sim}p).$$

Its truth table is the following:

p	q	${\sim}p$	${\sim}q$	$p{\Rightarrow}q$	${\sim}q{\Rightarrow}{\sim}p$	$(p{\Rightarrow}q){\equiv}({\sim}q{\Rightarrow}{\sim}p)$
T	T	F	F	T	T	T
T	F	F	T	F	F	T
F	T	T	F	T	T	T
F	F	T	T	T	T	T

We notice that the last column contains all T's. Thus $(p \Rightarrow q) \equiv (\sim q \Rightarrow \sim p)$ is true in all four cases. For *any* propositions p and q, the proposition $(p \Rightarrow q) \equiv (\sim q \Rightarrow \sim p)$ is true. Such a proposition is known as a *tautology*. Tautologies are true in all possible cases. The purpose of propositional logic is to provide methods for determining which propositions are tautologies. Truth tables constitute one sure-fire method. Other methods are provided in later chapters.

Formulas

To approach our subject more rigorously, we need to define a *formula*. The letters p, q, r, with or without subscripts, are called *propositional variables*; these are the simplest possible formulas, as they stand for unknown propositions (just as in algebra the letters x, y, z, with or without subscripts, stand for unknown numbers). By a *formula* we mean any expression constructed according to the following rules:

(1) Each propositional variable is a formula.

(2) Given any formulas X and Y already constructed, the expressions $\sim X$, $(X \wedge Y)$, $(X \vee Y)$, $(X \Rightarrow Y)$, $(X \equiv Y)$ are also formulas.

It is to be understood that no expression is a formula unless it is constructed according to rules (1) and (2) above.

When displaying a formula standing alone, we can dispense with outer parentheses without incurring any ambiguity. For example, when we say "the formula $p \Rightarrow \sim\sim q$," we mean "the formula $(p \Rightarrow \sim\sim q)$."

A formula in itself is neither true nor false, but only becomes true or false when we *interpret* the propositional variables as standing for specific propositions. We can, however, say that a formula is *always true*, *never true* or *sometimes true and sometimes false*, if it is, respectively, true in all cases, true in no cases, or true in some cases and false in others. For example, $p \vee \sim p$ is always true (it is a tautology); $p \wedge \sim p$ is always false, whereas $(p \vee q) \Rightarrow (p \wedge q)$ is true in some cases (the cases when p and q are both true, or both false) and false in the other cases. Formulas that are always false are called *contradictory* formulas, or *contradictions*. Formulas that are always true are called *tautologies* (as we have already indicated), and formulas that are true in some cases and false in others are sometimes called *contingent*. Formulas that are true in at least one case are called *satisfiable*.

Some Tautologies

The truth table is a systematic method of verifying tautologies, but some tautologies are so obvious that they can be immediately perceived as such. Here are some examples:

(1) $((p \Rightarrow q) \wedge (q \Rightarrow r)) \Rightarrow (p \Rightarrow r)$: This says that if p implies q and q implies r, then p implies r. This is surely self-evident, although, of course, verifiable by a truth table. This tautology has a name—it is called the *syllogism*.

(2) $(p \wedge (p \Rightarrow q)) \Rightarrow q$: This says that if p and $p \Rightarrow q$ are both true, so is q. This is sometimes paraphrased as "Anything implied by a true proposition is true."

(3) $((p \Rightarrow q) \wedge \sim q) \Rightarrow \sim p$: Thus, if p implies q and q is false, then p must be false. More briefly, "Any proposition implying a false proposition must be false." Thus, a true proposition can never imply a false one, so we could write (3) in the equivalent form $(p \wedge \sim q) \Rightarrow \sim (p \Rightarrow q)$

(4) $((\sim p \Rightarrow q) \wedge (\sim p \Rightarrow \sim q)) \Rightarrow p$: This principle is known as *reductio ad absurdum*. To show that p is true, it is sufficient to show that $\sim p$ implies some proposition q as well as its negation $\sim q$. No true proposition could imply both q and $\sim q$, so if $\sim p$ implies them both, then $\sim p$ must be false, which means that p must be true. (Symbolic logic is, in the last analysis, merely a systematization of common sense.)

(5) $((p \Rightarrow q) \wedge (p \Rightarrow r) \Rightarrow (p \Rightarrow (q \wedge r))$: Of course, if p implies q and p implies r, then p must imply both q and r.

(6) $((p \vee q) \wedge (p \Rightarrow r) \wedge (q \Rightarrow r))) \Rightarrow r$: This principle is known as *proof by cases*. Suppose $p \vee q$ is true. Suppose also that p implies r and q implies r. Then r must be true, regardless of whether it is p or q that is true (or both).

The reader with little experience in propositional logic should benefit from the following exercise.

EXERCISE 7.1. State which of the following are tautologies, which are contradictions and which are contingent (sometimes true, sometimes false).

(a) $(p \Rightarrow q) \Rightarrow (q \Rightarrow p)$.

(b) $(p \Rightarrow q) \Rightarrow (\sim p \Rightarrow \sim q)$.

(c) $(p \Rightarrow q) \Rightarrow (\sim q \Rightarrow \sim p)$.

(d) $(p \equiv q) \equiv (\sim p \equiv \sim q)$.

(e) $(p \Rightarrow \sim p)$.

(f) $(p \equiv \sim p)$.

(g) $\sim(p \wedge q) \equiv (\sim p \wedge \sim q)$.

(h) $\sim(p \wedge q) \equiv (\sim p \vee \sim q)$.

(i) $(\sim p \vee \sim q) \Rightarrow \sim(p \vee q)$.

(j) $\sim(p \vee q) \Rightarrow (\sim p \wedge \sim q)$.

(k) $(\sim p \vee \sim q) \wedge (p \equiv (p \Rightarrow q))$.

(l) $(p \equiv (p \wedge q)) \equiv (q \equiv (p \vee q))$.

Answers.

(a) Contingent.

(b) Contingent.

(c) Tautology.

(d) Tautology.

(e) Contingent (see remarks below).

(f) Contradiction.

(g) Contingent.

(h) Tautology.

(i) Contingent.

(j) Tautology.

(k) Contradiction.

(l) Tautology (see remarks below).

Remark. Concerning (e), many beginners fall into the trap of thinking that (e) is a contradiction. They think that no proposition can imply its own negation. This is not so; if p is *false*, then $\sim p$ is true, hence $p \Rightarrow \sim p$ is then true (F⇒T=T). Thus, when p is *true*, then $(p \Rightarrow \sim p)$ is false, but when p is false, then $(p \Rightarrow \sim p)$ is true. So $(p \Rightarrow \sim p)$ is true in one case and false in the other.

Remark. Concerning (l), both $p \equiv (p \wedge q)$ and $q \equiv (p \vee q)$ have the same truth tables as $p \Rightarrow q$.

Logical Implication and Equivalence

A formula X is said to *imply* a formula Y if Y is true in all cases in which X is true, or, what is the same thing, if $X \Rightarrow Y$ is a tautology. Formulas X and Y are said to be *equivalent* if they are true in exactly the same cases, or, what is the same thing, if $X \equiv Y$ is a tautology, or, what is again the same thing, if the truth tables for X and Y are the same (in their last columns).

Finding a Formula, Given Its Truth Table

Suppose I tell you the distribution of T's and F's in the last column of a truth table; can you find a formula having that as its truth table? For example, suppose I consider a case of three variables p, q, and r, and I write down at random T's and F's in the last column, thus:

p	q	r	?
T	T	T	F
T	T	F	F
T	F	T	T
T	F	F	F
F	T	T	F
F	T	F	T
F	F	T	F
F	F	F	T

The problem is to find a formula for which the last column of its truth table is the column under the question mark.

Do you think that cleverness and ingenuity are required? Well, it so happens that there is a ridiculously simple mechanical method that solves all problems of this type! Once you realize the method, then regardless of the distribution of T's and F's in the last column, you can instantly write down the required formula.

Problem 7.1. What is the method?

Formulas Involving t and f

For certain purposes (see, e.g., Chapters 10 and 23), it is desirable to add the symbols t and f to the language of propositional logic and extend the notion of *formula* by replacing (1), in our definition of this notion, by "Each propositional variable is a formula, and so are t and f." Thus, for example, $(p \Rightarrow t) \vee (f \wedge q)$ is a formula. The symbols t and f are called *propositional constants* and stand for *truth* and *falsity*, respectively. That is, under any interpretation, t is given the value *truth* and f the value *falsehood*. (Thus the formula consisting of t alone is a tautology, and of f alone, a contradiction.) Also, under any interpretation, $t \Rightarrow X$ has the same truth value as X (i.e., both are true or both false; $t \Rightarrow X$ is true if and only if X is true). Also, $X \Rightarrow f$ has the opposite truth value to X—i.e., $X \Rightarrow f$ is true if and only if X is false.

Now, any formula involving t and/or f is equivalent either to a formula involving neither t nor f, or to t itself or to f itself. This is easily established by virtue of the following equivalences (we abbreviate "is equivalent to" by "equ"):

$$\begin{array}{rcrrcr}
X \wedge t & \text{equ} & X, & t \wedge X & \text{equ} & X, \\
X \wedge f & \text{equ} & f, & f \wedge X & \text{equ} & f, \\
X \vee t & \text{equ} & t, & t \vee X & \text{equ} & t, \\
\\
X \vee f & \text{equ} & X, & f \vee X & \text{equ} & X, \\
X \Rightarrow t & \text{equ} & t, & t \Rightarrow X & \text{equ} & X, \\
X \Rightarrow f & \text{equ} & \sim X, & f \Rightarrow X & \text{equ} & t, \\
\sim t & \text{equ} & f, & \sim f & \text{equ} & t.
\end{array}$$

For example, consider the formula $((t \Rightarrow p) \wedge (q \vee f)) \Rightarrow ((q \Rightarrow f) \vee (r \wedge t))$. We can respectively replace the parts $(t \Rightarrow p)$, $(q \vee f)$, $(q \Rightarrow f)$, $(r \wedge t)$ by p, q, $\sim q$, r, thus obtaining the equivalent formula $(p \wedge q) \Rightarrow (\sim q \vee r)$.

Another example: Consider the formula $(p \vee t) \Rightarrow r$. We first replace $(p \vee t)$ by t, thus obtaining $t \Rightarrow r$, which in turn is equivalent to r itself. Thus $(p \vee t) \Rightarrow r$ is equivalent simply to r.

SOLUTIONS

7.1. This particular case will illustrate the general method perfectly!

In this case, the formula is to come out T in the third, sixth, and eighth rows. Well, the third row is the case when p is true, q is false, and r is true—in other words, when $(p \wedge \sim q \wedge r)$ is true. The sixth row is the case when $(\sim p \wedge q \wedge \sim r)$ is true, and the eighth is the case when $(\sim p \wedge \sim q \wedge \sim r)$ is true. Thus, the formula is to be true when and only when *at least one* of those three cases holds, so the solution is simply

$$(p \wedge \sim q \wedge r) \vee (\sim p \wedge q \wedge \sim r) \vee (\sim p \wedge \sim q \wedge \sim r).$$

- Chapter 8 -

Liars, Truth-Tellers, and Propositional Logic

We shall now look at the logic of lying and truth-telling from the viewpoint of propositional logic.

Knights and Knaves Revisited

Let us revisit the island of knights and knaves of Chapter 1. Let A be a native of the island, and let k be the proposition that A is a knight. Then the proposition that A is a knave can be simply written as $\sim k$ (A is *not* a knight). Now, suppose that A asserts a proposition $\mathcal{P}$. We don't know whether or not A is a knight, nor whether or not $\mathcal{P}$ is true, but this much we *do* know: If A is a knight, then $\mathcal{P}$ is true, and conversely, if $\mathcal{P}$ is true, then A is a knight. Thus A is a knight *if and only if* $\mathcal{P}$ is true—symbolically, $k \equiv \mathcal{P}$. And so when A asserts $\mathcal{P}$, the *reality* of the situation is

$$k \equiv \mathcal{P}.$$

Now, we usually deal with more than one native at a time, so we let $A_1, A_2, A_3, \ldots$, etc., stand for natives, and we let $k_1, k_2, k_3, \ldots$, etc., stand for the respective propositions that A_1 is a knight, A_2 is a knight, A_3 is a knight, and so forth. Thus if A_1 asserts a proposition $\mathcal{P}$, we know $k_1 \equiv \mathcal{P}$; if A_2 asserts $\mathcal{P}$, we know $k_2 \equiv \mathcal{P}$, and so forth. We thus can translate knight-knave problems into problems of propositional logic, as the following examples will show.

Let us look at Problem 1.3, in which we have natives A_1 and A_2, and A_1 asserts that A_1 and A_2 are both knaves. Now, the statement that A_1 and A_2 are both knaves is rendered $\sim k_1 \wedge \sim k_2$. Since A_1 asserts this, the reality of the situation is that $k_1 \equiv (\sim k_1 \wedge \sim k_2)$. From this we are to determine the truth value of k_1 and of k_2. We could solve this by a truth table for $k_1 \equiv (\sim k_1 \wedge \sim k_2)$.

k_1	k_2	$\sim k_1$	$\sim k_2$	$\sim k_1 \wedge \sim k_2$	$k_1 \equiv (\sim k_1 \wedge \sim k_2)$
T	T	F	F	F	F
T	F	F	T	F	F
F	T	T	F	F	T
F	F	T	T	T	F

We thus see that the only case in which $k_1 \equiv (\sim k_1 \wedge \sim k_2)$ comes out *true* is when k_1 is false and k_2 is true—in other words, when A_1 is a knave and A_2 is a knight (which we also saw by informal reasoning in Chapter 1). The upshot of the problem, then, is that the following is a tautology:

$$(k_1 \equiv (\sim k_1 \wedge \sim k_2)) \Rightarrow (\sim k_1 \wedge k_2).$$

Another way of putting it is that both $\sim k_1$ and k_2 are logical consequences of $k_1 \equiv (\sim k_1 \wedge \sim k_2)$.

Next, let us look at Problem 1.4, in which A_1 claims that at least one of A_1, A_2 is a knave, from which we deduced that A_1 must be a knight and A_2 a knave, and we thus have the tautology $(k_1 \equiv (\sim k_1 \vee \sim k_2)) \Rightarrow (k_1 \wedge \sim k_2)$. Again, we could have solved this by making a truth table for $k_1 \equiv (\sim k_1 \vee \sim k_2)$, and we would have seen that the only case in which it comes out true is when k_1 is true and k_2 is false.

In the next problem (Problem 8.5), A_1 claims to be of the same type as A_2—in other words, he claims that $k_1 \equiv k_2$. Thus we have $k_1 \equiv (k_1 \equiv k_2)$, from which we deduced that k_2 is true, regardless of whether k_1 is true or false, and we have the tautology $(k_1 \equiv (k_1 \equiv k_2)) \Rightarrow k_2$, which can also be verified by a truth table.

PROBLEM 8.1 (AN IF-THEN PROBLEM). Suppose you visit the island of knights and knaves and wish to find out some information such as whether there is gold on the island. You meet a native and ask him whether there is gold on the island. All he replies is: "If I am a knight, then there is gold on the island." From this, is it possible to tell whether there is gold, and whether he is a knight or a knave? Yes, it is possible to determine both! How? (The solution is important!)

PROBLEM 8.2 (A VARIANT). Suppose that the native had instead said: "Either I am a knave or there is gold on the island." Is it now possible to determine whether there is gold there, or the type of the native?

PROBLEM 8.3 (ANOTHER VARIANT). Suppose that the native had instead said: "I am a knave and there is no gold on the island." What can be deduced?

PROBLEM 8.4 (GOLD OR SILVER?). Suppose that a native makes the following statement: "There is gold on this island, and if there is gold, then there also is silver." Can it be deduced whether he is a knight or a knave? Can it be determined whether there is gold? What about silver?

PROBLEM 8.5 (GOLD OR SILVER?). Suppose that he instead makes the following two separate statements:

(1) There is gold on this island.

(2) If there is gold here, then there also is silver.

Is the solution the same as that of the last problem?

PROBLEM 8.6 (A METAPUZZLE). One day a man was tried for a crime on this island. First the prosecutor pointed to the defendant and said: "If he is guilty, then he had an accomplice." Then the defense attorney said: "That's not true!" Now, the judge didn't know whether the prosecutor was a knight or a knave, nor did he know the type of the defense attorney. Later, he found out whether the defense attorney was a knight or a knave and was then able to convict or acquit the defendant. Which did he do?

THE NELSON GOODMAN PRINCIPLE REVISITED

We recall the Nelson Goodman principle for the island of knights and knaves: To ascertain whether a given proposition is true or not, you ask a native: "Is it the case that you are a knight of and only if $\mathcal{P}$?" Thus you are asking whether $k \equiv \mathcal{P}$ holds. We informally proved that if he answers *yes*, then $\mathcal{P}$ is true, and if he answers *no*, then $\mathcal{P}$ is not true—thus he answers *yes* if and only if $\mathcal{P}$ holds. His answering *yes* is *equivalent* to his asserting that $k \equiv \mathcal{P}$ holds, so the reality of the situation is then $k \equiv (k \equiv \mathcal{P})$. But, as we have seen at the end of the last chapter, this is equivalent to $(k{\equiv}k) \equiv \mathcal{P}$, which in turn is equivalent to $\mathcal{P}$. And so by the Nelson Goodman tautology $(k \equiv (k \equiv \mathcal{P})) \equiv \mathcal{P}$, we see that $\mathcal{P}$ must be true (assuming he answers *yes*, and thus asserts $k \equiv \mathcal{P}$). If he answers *no*, then he is asserting $\sim(k \equiv \mathcal{P})$, and hence the reality is $k \equiv \sim(k \equiv \mathcal{P})$, which is equivalent to $\sim\mathcal{P}$ (as can be verified by a truth table).

PROBLEM 8.7. What question could you ask a native such that it is impossible for him to answer either *yes* or *no* without violating his type?

Problem 8.8. Suppose there is a *spy* on the island (who is neither a knight or a knave) and you ask him the question given in the solution to the last problem. Could he answer it truthfully? Could he answer it falsely?

Some Problems Involving More Than One Inhabitant

Suppose there is a trial on this island, and two witnesses A_1 and A_2 make the following statements:

A_1: If A_2 and I are both knights, then the defendant is guilty.
A_2: A_1 is a knight.

Is it possible to determine the types of A_1 and A_2 and whether the defendant is guilty? Yes it is, and one can solve this either informally or systematically by truth tables, using our translation device.

Let $\mathcal{P}$ be the proposition that the defendant is guilty. Then A_1 has asserted $(k_2 \wedge k_1) \Rightarrow \mathcal{P}$, so we know that $k_1 \equiv ((k_2 \wedge k_1) \Rightarrow \mathcal{P})$. By A_2's assertion, we know that $k_2 \equiv k_1$. If we make a truth table for $(k_1 \equiv ((k_2 \wedge k_1) \Rightarrow \mathcal{P})) \wedge (k_2 \equiv k_1)$, we see that the only case in which it comes out true is when k_1, k_2 and $\mathcal{P}$ are all true (and thus the defendant is guilty, and both witnesses are knights).

A little ingenuity provides a shortcut to the truth table: Since $k_2 \equiv k_1$ holds, k_2 and k_1 are interchangeable, so $k_1 \equiv ((k_2 \wedge k_1) \Rightarrow \mathcal{P})$ reduces to $k_1 \equiv ((k_1 \wedge k_1) \Rightarrow \mathcal{P})$. But $k_1 \wedge k_1$ reduces to k_1, so $k_1 \equiv ((k_2 \wedge k_1) \Rightarrow \mathcal{P})$ reduces to $k_1 \equiv (k_1 \Rightarrow \mathcal{P})$, which we have already seen to be equivalent to $k_1 \wedge \mathcal{P}$. Thus we have k_1 (and hence also k_2) and $\mathcal{P}$.

The next three problems can be solved similarly (using truth tables or, better still, by a little ingenuity).

Problem 8.9. In the above trial, suppose A_1 and A_2 had made the following statements instead:

A_1: If A_2 is a knight, then the defendant is innocent.
A_2: If A_1 is a knave, then the defendant is innocent.

What can be deduced?

Problem 8.10. Suppose that A_1 and A_2 had made the following statements instead:

A_1: If either of us is a knight, then the defendant is guilty.
A_2: If either of us is a knave, then the defendant is guilty.

What is the solution?

Problem 8.11. Suppose that A_1 and A_2 had made the following statements instead:

A_1: If I am a knight and A_2 is a knave, then the defendant is guilty.
A_2: That is not true!

Is the defendant guilty or not?

Some of the Other Islands

Male and Female Knights and Knaves

Let us now revisit the island of Chapter 2, in which female knights lie and female knaves tell the truth, whereas male knights tell the truth and male knaves lie.

How do we translate problems about this island into problems in propositional logic? Well, let A be an inhabitant, k be the proposition that A is a knight, and m be the proposition that A is male. Then "A is a knave" is equivalent to $\sim k$, and "A is female" is equivalent to $\sim m$. Now, A is truthful if and only if A is a male knight or a female knave—thus if and only if $(k \wedge m) \vee (\sim k \wedge \sim m)$, which can be more conveniently written as $k \equiv m$. Thus "A is truthful" can be translated as $k \equiv m$, so when A asserts a proposition $\mathcal{P}$, the reality is now $(k \equiv m) \equiv \mathcal{P}$ (compared with $k \equiv \mathcal{P}$ for the island of Chapter 2).

We saw in Chapter 3 that the only way an inhabitant A can assert that he or she is a knight is if A is male. Looked at symbolically, if A asserts that A is a knight, then the reality is $(k \equiv m) \equiv k$, which is logically equivalent to m! Likewise, if A asserts that A is male, then $(k \equiv m) \equiv m$, which is logically equivalent to k (and thus A is a knight).

Symbolically, the Nelson Goodman principle for this island is that to find out whether a given proposition $\mathcal{P}$ is true or not, the question you ask is "$(k \equiv m) \equiv \mathcal{P}$?" If you get the answer *yes*, then $(k \equiv m) \equiv ((k \equiv m) \equiv \mathcal{P})$ holds, which is a special case of $q \equiv (q \equiv \mathcal{P})$ (taking $k \equiv m$ for q), which reduces to $\mathcal{P}$. (If you get the answer *no*, then, instead, we have $(k \equiv m) \equiv \sim((k \equiv m) \equiv \mathcal{P})$, which is logically equivalent to $\sim \mathcal{P}$ (being a special case of $q \equiv \sim(q \equiv \mathcal{P})$).)

Next, let us look again at Problem 2.4: What statement could an inhabitant write from which one could deduce that the inhabitant is a female knight? A little thought and ingenuity reveals that the sentence "I am a male knave would work, but isn't there a *systematic* method of solving problems of this type? Yes, there is: If you wanted to find out whether the inhabitant was or was not a female knight, then, using the Nelson Goodman principle, you would ask "$(k{\equiv}m){\equiv}({\sim}m{\wedge}k)$?" (${\sim}m{\wedge}k$ is the statement that the inhabitant is a female knight.) An affirmative answer would indicate that the inhabitant is a female knight; so if the inhabitant wrote "$(k{\equiv}m){\equiv}({\sim}m{\wedge}k)$," you would know that the inhabitant was a female knight. But surely this convoluted expression can be simplified! Just make a truth table for the expression, and you will see that it comes out true only in the single case when m is true and k is false, and thus the expression $(k{\equiv}m){\equiv}({\sim}m{\wedge}k)$ is logically equivalent to the simple expression $m{\wedge}{\sim}k$, which is that the inhabitant is a male knave.

The next two problems of Chapter 2 can be solved by the same systematic method.

Let us now consider some problems in which there is more than one inhabitant involved. For two inhabitants A_1 and A_2, we let k_1 be the proposition that A_1 is a knight, k_2 that A_2 is a knight, m_1 that A_1 is male, and m_2 that A_2 is male (and similarly if there are more inhabitants A_3, A_4,... involved).

Let us consider the following problems: Two masked inhabitants A_1 and A_2 walk onto the stage, and you are told that they are a married couple. They write the following statements:

A_1: We are both knaves.
A_2: That is *not* true!

Which of the two is the husband?

Before considering a systematic solution, let us reason it out informally: Since A_2 contradicted A_1, and the two are of opposite sexes, they must be of the same knight-knave type (both knights or both knaves). Suppose they are both knaves. Then A_1 told the truth, hence A_1 is a female knave. Suppose they are both knights. Then A_1 lied, and hence A_1 is a female knight, so in either case, A_1 is the female, and so A_2 is the husband.

One could also solve this systematically, but the solution is quite tedious! We know the following true facts:

(1) $(m_1{\equiv}k_1){\equiv}({\sim}k_1{\wedge}{\sim}k_2)$.

(2) $(m_2{\equiv}k_2){\equiv}{\sim}({\sim}k_1{\wedge}{\sim}k_2)$.

(3) $m_1{\equiv}{\sim}m_2$.

Fact (1) comes from A_1's statement; (2) comes from A_2's statement, and (3) comes from the fact that A_1 and A_2 are of opposite sexes. Now, one could make a truth table for the complex conjunction $(1 \wedge 2) \wedge 3$, but it involves 16 rows (since there are four variables, k_1, k_2, m_1, m_2, involved), and this is what makes it so tedious (though a computer could handle it easily). Such a truth table reveals that the conjunction comes out *true* only in two of the sixteen rows, and in both rows m_1 is false and m_2 is true, which means that A_1 is female and A_2 is male.

Problem 8.12. Suppose we are again given that A_1 and A_2 are a married couple, and they write the following statements:

A_1: My spouse is a knight.
A_2: We are both knights.

From this it is possible to tell both which one is the husband, and which knight-knave type each is. What is the solution? (It can be found either with a little ingenuity or systematically by using a truth table.)

Silent Knights and Knaves

Let us now revisit the island of Chapter 3 in which all knights tell the truth and all knaves lie, but instead of answering yes/no questions with words, they flash red or black cards, but we don't know what the colors mean.

In accordance with the general principle of Chapter 6, we will define an inhabitant A to be of Type 1 if he flashes red in response to yes/no questions whose correct answer is *yes*. Thus if red means *yes*, then Type 1 inhabitants are knights, whereas if red means *no*, then Type 1 inhabitants are knaves. Thus an inhabitant A is of Type 1 if and only if the following condition holds: A is a knight if and only if red means *yes*. As usual, we let k be the proposition that A is a knight, and we shall let r be the proposition that red means *yes*. Thus the statement that A is of Type 1 is symbolized $k \equiv r$. To find out whether a given proposition $\mathcal{P}$ is true or not, the question to ask is: "$(k \equiv r) \equiv \mathcal{P}$?" ("You are of Type 1 if and only if $\mathcal{P}$?") If a red card is flashed, then $\mathcal{P}$ is true; if a black card is flashed, $\mathcal{P}$ is false.

Suppose, for example, you want to find out whether the native is a knight. You can then ask "$(k \equiv r) \equiv k$?" However, $(k \equiv r) \equiv k$ is simply equivalent to r! You can see this with a truth table, or more interestingly by noting that $(k \equiv r) \equiv k$ is equivalent to $k \equiv (k \equiv r)$, which we already know is equivalent to r (the Nelson Goodman tautology). Thus, instead of asking "$(k \equiv r) \equiv k$?" you can more simply ask "r?" (Does red mean *yes*?) This was Problem 3.3. Problem 8.2 was to find a question that would determine what red means. Well, using the Nelson Goodman principle for this island, you ask "$(k \equiv r) \equiv r$?" But this simply boils down to k! Thus you ask the simple question: "Are you a knight?" which was the solution we gave, and we recall that knights and knaves both answer *yes* to that question, hence whatever color is flashed must mean *yes*.

Problem 8.13. What question could you ask that would make it impossible for the native to flash either red or black?

We could also look at the remaining chapters of Part I from the viewpoint of propositional logic, but by now the reader can do this on his or her own. We wish at this point to continue with propositional logic in general.

Solutions

8.1. A asserts that *if* he is a knight, *then* there is gold on the island. Let us see if this is right. I will look at this in more than one way.

For the first way, suppose he is a knight. Then what he said is true—it really is the case that if he is a knight then there is gold on the island, and since he is a knight (under our assumption), there is gold on the island. This proves that *if* he is a knight, then there is gold, and that's all it proves. I haven't proved that he *is* a knight, nor have I proved that there is gold; all I have proved so far is the hypothetical statement that *if* he is a knight, then there is gold on the island. But this is exactly what he asserted, so he was right when he said that *if* he is a knight, then there is gold. Since he was right, he must be a knight! Thus he really is a knight, his statement was true, hence it further follows that there is gold on the island. So the answer is that he is a knight, and there is gold on the island.

Some of you will doubtless say at this point: "But you haven't considered the case when the native is a knave!" My answer is important: We don't *need* to, because I have already proved that he is a knight!

Nevertheless, I wish to consider the case that he is a knave, because it leads to a second way of solving the problem: Suppose he is a knave. Then his statement must be false. But how can a hypothetical statement, a statement of the form "*if p then q*," be false? Only if p is true and q is false. Thus the only way the statement "If he is a knight then there is gold" can be false is that he is a knight and there is no gold, but then a knight wouldn't make a false statement! So the assumption that he is a knave leads to the contradiction that he is also a knight; hence he can't be a knave. This affords a second way of proving that he must be a knight.

Discussion. If you should visit the Island of Knights and Knaves and meet a native who says "If I am a knight, then Santa Claus exists," it would really follow that the native is a knight and that Santa Claus exists! In one of my books I wrote the following dialogue—a *parody* on a similar situation:

A: Santa Claus exists, if I am not mistaken.
B: Well, of course Santa Claus exists, *if* you are not mistaken.
A: So I was right.
B: Of course!
A: So I am not mistaken.
B: Right.
A: Hence Santa Claus exists!

This is just a humorous version of the famous Curry paradox. Consider the following sentence:

If this sentence is true then Santa Claus exists.

It is easily seen that if the above sentence is true, then Santa Claus exists, and since the sentence asserts just that, it must be true.

8.2. This is really the same as the last puzzle, because if he were a knave, then it would be true that *either* he is a knave *or* there is gold, but knaves don't make true statements, hence he is a knight. Hence it is true that either he is a knave or there is gold, but he is not a knave, so there is gold.

8.3. Obviously, no knight would claim that he is a knave and there is no gold on the island, hence the native is a knave. Hence his statement is false. Now, if there really were no gold there, then it would be true that he is a knave and there is no gold, but knaves don't make true statements. Hence there is gold there.

Thus the solution is that he is a knave and there is gold on the island.

8.4. His assertion is equivalent to the simpler statement that there is both gold and silver on the island. Nothing significant can be deduced from his having said that; it could be that he's a knight and there is both gold and silver, or it could be that he's a knave and there is not both gold and silver there.

8.5. This is a very different story! If he were a knave, both his statements would be false, but this is not possible, because the only way his second statement could be false is that there is gold but not silver, which would make his first statement true! (For any two propositions p and q, it is impossible for both p and $p \Rightarrow q$ to be false, although the single proposition $p \wedge (p \Rightarrow q)$ could be false!)

Thus the solution is that the native is a knight and there is both gold and silver on the island.

8.6. Suppose that the judge had found out that the defense attorney was a knave. Then he would have known that the prosecutor's statement was true, but from that he could not possibly determine whether the defendant was innocent or guilty. On the other hand, suppose that he had found out that the defense attorney was a knight. Then he would have known that the prosecutor's statement was false, but the only way it could be false is that the defendant was guilty but had no accomplice. Since the judge *did* know, it must be true that the judge found out that the defense attorney was a knight, and hence that the defendant was guilty (but had no accomplice).

Incidentally, what the defense attorney said was the most stupid thing possible, since it led to the conviction of his client!

8.7. If you knew that the native was a knight, you could ask him: "Will you answer *no* to this question?" (It is impossible for him to correctly answer either *yes* or *no*.) Since, however, you don't know whether he is a knight or a knave, you use a Nelson Goodman device and instead ask him: "Are you a knight if and only if you will answer *no* to this question?"

Let k be the proposition that he is a knight, and let n be the proposition that he will answer *no* to the question. You are asking him whether the equivalence $k \equiv n$ holds. Suppose he is a knight and answers *yes*. Then k holds but n doesn't hold, hence $k \equiv n$ doesn't hold. Yet the knight affirmed that it did hold, contrary to the fact that knights don't answer questions wrongly. Now suppose that he is a knight and answers *no*. Then k and n are both true, hence $k \equiv n$ does hold, yet the knight falsely denies it! Now suppose he is a knave and answers *yes*. Then k and n are both false, hence $k \equiv n$ does hold, so the knave *correctly* affirmed that it holds, which is not possible for a knave. Finally, suppose he is a knave and answers *no*. Then k is false but n is true, hence $k \equiv n$ doesn't hold, so the knight correctly *denied* that $k \equiv n$ holds, which a knave cannot do.

This proves that neither a knight nor a knave can answer either *yes* or *no* to this question.

Actually, we have done more work than we needed to (part of which was a repetition of our former argument proving the Nelson Goodman principle). If we use the Nelson Goodman principle (which we have already established), we have the following swift proof: We know that, for any proposition $\mathcal{P}$, if we ask whether $k \equiv \mathcal{P}$ holds, then the native answers *yes* if and only if $\mathcal{P}$ does hold. Well, we take n for $\mathcal{P}$, so if you ask whether $k \equiv n$ holds, then he answers *yes* if and only if n is true—in other words, he answers *yes* if and only if he answers *no*, which means that he either answers both *yes* and *no*, or neither. Since he doesn't answer both, he answers neither.

8.8. Now, suppose you ask the same question to someone who is neither a knight nor a knave. The curious thing then is that he cannot answer falsely; he has the option of saying *yes* or *no*, but in either case his answer will be correct.

Reason. You are asking "$k \equiv n$?" Since he is neither a knight nor a knave, k is automatically false. Suppose he answers *yes*. Then n is also false, hence $k \equiv n$ does hold, so *yes* was a correct answer! On the other hand, if he answers *no*, then n is true, and since k is false, $k \equiv n$ doesn't hold, so *no* was the right answer! Thus *yes* and *no* are both correct answers.

Another way of looking at it is this: Since k is false, $k \equiv n$ is equivalent to $\sim n$, and asking "$\sim n$?" is tantamount to asking "Will you answer *yes* to this question?" If you ask this of anyone, both *yes* and *no* are correct answers.

8.9. The defendant is innocent (and both are knights).

8.10. The defendant is guilty (and both are knights).

8.11. The defendant is guilty (and A_1 is a knight and A_2 is a knave).

8.12. A_1 is a female knight and A_2 is a male knave.

8.13. We recall Problem 8.7, in which we showed that a native of the knight-knave island of Chapter 1 cannot answer the question "Are you a knight if and only if you will answer *no* to this question?" Well, for the present island, the analogous question to ask is "Are you of Type 1 if and only if you will flash a black card in reply to this question?"

Let t_1 be the proposition that he is of Type 1 and b be the proposition that he flashes a black card. You are now asking him whether $t_1 \equiv b$ holds. Well, suppose he is of Type 1 and flashes a red card. Then

t_1 is true, b is false, hence $t_1 \equiv b$ is false, so the correct answer to the question is *no*, but a native of Type 1 cannot flash red to questions whose correct answer is *no*! Can he flash black? Well, if he flashes black, then t_1 and b are both true, hence $t_1 \equiv b$ does hold, so the correct answer is *yes*, but a native of Type 1 cannot flash black to questions whose correct answer is *yes*. Now suppose that he is of Type 2 and flashes red. Then t_1 and b are both false, hence $t_1 \equiv b$ is true, and hence the correct answer is *yes*, but a native of Type 2 cannot then flash red. And if he flashes black, then t_1 is false and b is true, hence $t_1 \equiv b$ is false, hence the correct answer is *no*—but a native of Type 2 cannot then flash black.

Again, we did more work than we had to: By the Nelson Goodman principle for this island, if you ask whether $t_1 \equiv b$ holds, then the native flashes red if and only if b holds—thus he flashes red if and only if he flashes black, and since he cannot flash both, he must flash neither.

- Chapter 9 -

Variable Liars

Note. If you find the problems of this chapter to be too difficult, you might find it helpful to read the next chapter before this one.

Boolean Islands

We shall now visit an interesting cluster of islands in which, on each island, the lying or truth-telling habits can vary from day to day—that is, an inhabitant might lie on some days and tell the truth on other days, but on any given day, he or she lies the entire day or tells the truth the entire day.

Such an island will be called a *Boolean island* (in honor of the 19th Century mathematician George Boole, who discovered its basic principles) if and only if the following three laws hold:

N: For any inhabitant *A* there is an inhabitant who tells the truth on all and only those days on which *A* lies.

C: For any inhabitants *A* and *B* there is an inhabitant C who tells the truth on all and only those days on which *A* and *B* both tell the truth.

D: For any inhabitants *A* and *B* there is an inhabitant C who tells the truth on all and only those days on which either *A* tells the truth or *B* tells the truth (or both). (In other words, C lies on those and only those days on which *A* and *B* both lie.)

My friend Inspector Craig of Scotland Yard, of whom I have written much in some of my earlier puzzle books, was as interested in logic as in crime detection. He heard about this cluster of islands of variable liars, and his curiosity prompted him to make a tour of them. The first one he visited was called Conway's Island, after Captain Conway, who was its leader. Craig found out that conditions **N** and **C** both hold on this island. After finding this out, he thought about the matter and came to the conclusion that condition **D** must also hold, and hence that this island must be a Boolean island. Craig was right—condition **D** does logically follow from conditions **N** and **C**.

Problem 9.1. Why does **D** follow from **N** and **C**?

Note. Unlike all the other chapters, the solutions to the problems of this chapter will not be given until the next chapter, when the reader will already know some more basic facts about propositional logic. I will, however, give you one hint concerning the above problem, which should also be quite helpful in solving the remaining problems in this chapter.

We are given that Conway's Island satisfies condition **N**. Well, for any inhabitant A, let A' be an inhabitant who tells the truth on those and only those days on which A lies. We are also given that this island obeys condition **C**, and so, for any inhabitants A and B, let $A \cap B$ be an inhabitant who tells the truth on all and only those days on which A and B both tell the truth. What can you say about $A' \cap B$—his truth-telling habits, that is? What about $A \cap B'$? What about $(A \cap B')'$? What about $(A' \cap B)'$? What about $A' \cap B'$? What about $(A' \cap B')'$? Can't you find *some* combination of A and B, using the operations $\cap$ and $'$, that tells the truth on those and only those days on which at least one of A, B tells the truth?

The next island visited by Craig was known as Diana's Island, named after its queen. It satisfied conditions **N** and **D**.

Problem 9.2. Is this island necessarily a Boolean island?

On Irving's Island (of variable liars), Craig found out that condition **N** holds, as well as the following condition:

> **I**: For any inhabitants A and B there is an inhabitant C who tells the truth on all and only those days on which either A lies or B tells the truth (or both).

Problem 9.3. Prove that Irving's Island is necessarily a Boolean island.

Problem 9.4. Does every Boolean island necessarily satisfy condition **I** of Irving's Island?

The next island visited by inspector Craig was Edward's Island and satisfied conditions **C**, **D**, **I**, as well as the following condition:

> **E**: For any inhabitants *A* and *B* there is an inhabitant *C* who tells the truth on all and only those days on which *A* and *B* behave alike—i.e., both tell the truth or both lie.

PROBLEM 9.5. Is Edward's Island necessarily a Boolean island?

PROBLEM 9.6. Does a Boolean island necessarily satisfy condition **E**?

An island is said to satisfy *condition* **T** if at least one inhabitant tells the truth on all days, and to satisfy *condition* **F** if at least one inhabitant lies on all days.

PROBLEM 9.7. Which, if any, of the conditions **T**, **F** must necessarily hold on a Boolean island?

PROBLEM 9.8. Suppose an island satisfies conditions **I** and **T**. Is it necessarily a Boolean island?

PROBLEM 9.9. What about an island satisfying conditions **I** and **F**; is it necessarily a Boolean island?

Jacob's Island, visited by inspector Craig, was a very interesting one. Craig found out that the island satisfies the following condition:

> **J**: For any inhabitants *A* and *B*, there is an inhabitant *C* who tells the truth on all and only those days on which *A* and *B* both lie.

After learning about condition **J**, and after some thought, Craig came to a startling realization: From just the single condition **J**, it must follow that the island is a Boolean island!

PROBLEM 9.10. Why does it follow?

Solomon's Island also turned out to be quite interesting. When Craig arrived on it, he had the following conversation with the resident sociologist:

Craig:	Is this island a Boolean island?
Sociologist:	No.
Craig:	Can you tell me *something* about the lying and truth-telling habits of the residents here?
Sociologist:	For any inhabitants *A* and *B*, there is an inhabitant *C* who tells the truth on all and only those days on which either *A* lies or *B* lies (or both).

This interview puzzled inspector Craig; he felt that something was wrong. After a while he realized for sure that something *was* wrong—the sociologist was either lying or mistaken!

Problem 9.11. Prove that Craig was right.

Problem 9.12. Concerning the last problem, is it possible that the sociologist was lying? (Remember that this is an island of *variable liars,* in which in any one day, an inhabitant either lies the entire day or tells the truth the entire day.)

Here is another condition that some of the islands of this cluster obey:

E′: For any inhabitants A and B there is an inhabitant C who tells the truth on all and only those days on which either A tells the truth and B lies, or B tells the truth and A lies.

Problem 9.13. Does a Boolean island necessarily satisfy condition **E**′?

Problem 9.14. Someone once conjectured that if an island satisfies condition **N**, then conditions **E** and **E**′ are equivalent—each implies the other. Was this conjecture correct?

Problem 9.15. Suppose that an island satisfies condition **E**′ as well as condition **I** of Irving's Island. Is such an island necessarily a Boolean island?

I′: For any inhabitants A and B there is an inhabitant C who tells the truth on all and only those days on which A tells the truth and B lies.

Problem 9.16.

(a) Show that if an island obeys condition **N** then it obeys **I** if and only if it obeys **I**′.

(b) Show that every Boolean island satisfies **I**′.

(c) Show that any island obeying conditions **E** and **I**′ must be a Boolean island.

Partial Boolean Islands and Craig's Laws

We have now considered several conditions—**N**, **C**, **D**, **I**, **E**, **T**, **F**, **J**, **S**, **I**′ and **E**′—any of which an island of variable liars may or may not satisfy.

We will call an island of variable liars a *partial* Boolean island if it satisfies some, but not necessarily all, of the above conditions.

Inspector Craig visited several partial Boolean islands in this cluster as well as Boolean ones. When he got home, he became very interested in the possible interrelationships between the abovementioned conditions—which conditions imply which? He discovered several interesting laws, among which are the following:

Craig's Laws:

(1) Any island satisfying condition **I** also satisfies condition **D**. (Craig, however, was not the first to discover this.)

(2) Any island satisfying condition **I′** also satisfies condition **C**. (Craig may well have been the first one to discover this.)

(3) Any island satisfying conditions **E** and **I** also satisfies condition C. (This law was also discovered quite independently by the logician Raymond Smullyan.)

(4) Any island satisfying conditions **E** and **C** also satisfies condition **D**.

(5) Any island satisfying conditions **E** and **D** also satisfies condition **C**.

(6) Any island satisfying conditions **E′** and **D** also satisfies condition **C**.

(7) Any island satisfying conditions **E′** and **C** also satisfies condition **D**.

(8) Any island satisfying **C** and **E′** also satisfies **I′**.

(9) Any island satisfying **D** and **E′** also satisfies **I′**.

(10) Any island satisfying **I** and **I′** must be a Boolean island.

Care to try proving Craig's laws? It will be a good and instructive workout. At any rate, all will fall into place in the next chapter.

- Chapter 10 -

Logical Connectives and Variable Liars

Let me begin by saying that all the problems of Chapter 9 are really problems of this chapter in disguise!

Some Standard Results

Suppose a man from Mars comes down to our planet and starts studying propositional logic. At one point he says: "I understand the meaning of $\sim$ (not) and $\wedge$ (and), but what does the symbol $\vee$, or the word *or*, mean? Can you explain it to me in terms of $\sim$ and $\wedge$, which I already understand?"

What he is asking for is a definition of $\vee$ in terms of $\sim$ and $\wedge$—i.e., he wants a formula using two propositional variables—say, p and q—that uses only the connectives $\sim$ and $\wedge$ and that is equivalent to the formula $p \vee q$.

Problem 10.1. Find such a formula. Also, how does this solve Problem 9.1?

Next, a lady from Venus claims to understand the meaning of $\sim$ and $\vee$, but not of $\wedge$. How would you explain it to her?

Problem 10.2. A man from Saturn understands $\sim$ and $\Rightarrow$, but not $\wedge$ or $\vee$. Can $\wedge$ and $\vee$ be defined from $\sim$ and $\Rightarrow$? How does this relate to Problem 9.3?

Problem 10.3. A certain lady from Saturn understands $\sim$, $\wedge$ and $\vee$, but not $\Rightarrow$. Can $\Rightarrow$ be defined in terms of $\sim$ and $\vee$, or in terms of $\sim$ and $\wedge$? How is this related to Problem 9.4?

Problem 10.4. An inhabitant of Uranus understands $\wedge$, $\vee$, $\Rightarrow$ and $\equiv$, but not $\sim$. Can $\sim$ be defined in terms of $\wedge$, $\vee$, $\Rightarrow$ and $\equiv$? How is this related to Problem 9.5?

Problem 10.5. Obviously $\equiv$ is definable from $\sim$, $\wedge$, $\vee$ and $\Rightarrow$—indeed, from $\wedge$ and $\Rightarrow$ alone: $p{\equiv}q$ is equivalent to $(p{\Rightarrow}q)\wedge(q{\Rightarrow}p)$. Also, $p{\equiv}q$ is equivalent to $(p\wedge q)\vee({\sim}p\vee{\sim}q)$. Doesn't this obviously solve Problem 9.6?

As was discussed in Chapter 7, some systems of propositional logic employ, in addition to propositional variables, propositional *constants* f and t (standing for *truth* and *falsehood*, respectively). In such systems, the *formulas* are those expressions that are built up from propositional variables and t and f, using the logical connectives. (For example, the expressions $p{\Rightarrow}t$, $f{\Rightarrow}(p{\Rightarrow}(q{\Rightarrow}t))$, $f\wedge(p\vee{\sim}t)$ are all formulas.) It is to be understood that, in any interpretation of such formulas, t must always be given the value *truth* and f the value *falsehood*. Other systems of propositional logic take just the propositional constant f and *define* t to be $f{\Rightarrow}f$. We, however, will take both t and f.

Problem 10.6. Is t definable from the connectives $\sim$, $\wedge$, $\vee$ and $\Rightarrow$? What about f? How does this relate to Problem 9.7? (The answers to these questions are pretty darn obvious, aren't they?)

Problem 10.7. Are the connectives $\sim$, $\wedge$ and $\vee$ all definable from $\Rightarrow$ and t? How does this relate to Problem 9.8?

Problem 10.8. Can the connectives $\sim$, $\wedge$ and $\vee$ all be defined from $\Rightarrow$ and f? How does this relate to Problem 9.9?

Joint Denial

There is a logical connective from which *all* logical connectives are definable: $p{\downarrow}q$, read "p and q are both false." This connective is called *joint denial*. It has the following truth table:

p	q	$p{\downarrow}q$
T	T	F
T	F	F
F	T	F
F	F	T

Problem 10.9. Prove that the connectives $\sim$, $\vee$, $\wedge$ and $\Rightarrow$ are all definable from $\downarrow$. How does this relate to Problem 9.10?

THE SHEFFER STROKE

There is another logical connective from which all logical connective are definable - namely $p|q$, read "At least one of p, q is false," or "p is incompatible with q." It has the following truth table:

p	q	$p\|q$
T	T	F
T	F	T
F	T	T
F	F	T

The symbol $|$ in this use is called the *Sheffer stroke*, and was discovered by H. M. Sheffer [12]. It is known that joint denial and the Sheffer stroke are the only binary connectives from which all the other connectives are definable.

Suppose now that we have an unknown binary connective h, and we make a truth table for phq. This can be done in exactly 16 possible ways:

p	q	phq
T	T	a
T	F	b
F	T	c
F	F	d

where each of the letters a, b, c, d is replaced by either T or F. There are two possible choices for a, and with each of these, there are two possible choices for b, so there are four possible choices for a and b. With each of these four, there are two possible choices for c, hence there are eight possible choices for a, b and c. With each of these eight, there are two possible choices for d, and so there are 16 possible choices altogether. Thus there are 16 possible truth tables for phq, so there are exactly 16 different binary connectives h. Some of them have already been named ($\wedge$, $\vee$, $\Rightarrow$, $\downarrow$, $|$), others not. We saw in Chapter 7 how, given any truth table, one can find a formula whose truth table it is—a formula using the connectives $\sim$, $\wedge$ and $\vee$. But these connectives are all definable from $\downarrow$ alone, or from the Sheffer stroke alone. Thus all 16 binary connectives are definable from $\downarrow$, or from the stroke.

PROBLEM 10.10. Show that $\sim$, $\wedge$, $\vee$ and $\Rightarrow$ are all definable from the Sheffer stroke. How does this solve Problem 9.11?.

PROBLEM 10.11. Same as Problem 9.12.

Some Other Connectives

The Connective $\not\equiv$

We read $p\not\equiv q$ as "p is *not* equivalent to q," or "One of the propositions p, q is true and the other false," or "p or q is true, but not both." Thus $p\not\equiv q$ is equivalent to $\sim(p\equiv q)$, also equivalent to $(p\wedge\sim q)\vee(\sim p\wedge q)$, also equivalent to $(p\vee q)\wedge\sim(p\wedge q)$.

Thus $p\not\equiv q$ is really *exclusive disjunction.*

Problem 10.12. Same as Problem 9.13.

Problem 10.13. Same as Problem 9.14.

Problem 10.14. Can all the connectives be defined from $\not\equiv$ and $\Rightarrow$? How is this related to Problem 9.15?

The Connective $\not\Rightarrow$

We read $p\not\Rightarrow q$ as "p does *not* imply q," or "p is true and q is false." $p\not\Rightarrow q$ is thus equivalent to $\sim(p\Rightarrow q)$ and also to $p\wedge\sim q$.

Problem 10.15. Show that all connectives are definable from $\not\equiv$ and $\not\Rightarrow$. How does this solve part (c) of Problem 9.16? (Parts (a) and (b) are already quite obvious.)

Further Results

I have long been interested in various interrelationships between logical connectives; which connectives are definable from which? These questions are all related to partial Boolean islands and Craig's laws. Here are some results along these lines.

Problem 10.16. We already know that $\vee$ is definable from $\sim$ and $\Rightarrow$ ($p\vee q$ is equivalent to $\sim p\Rightarrow q$). Curiously enough, $\vee$ is definable from $\Rightarrow$ alone! (This fact is fairly well known.) How is $\vee$ definable from $\Rightarrow$? (The solution is not at all obvious.) How does this prove Craig's law (1) of Chapter 9?

Problem 10.17. It is also true that $\wedge$ is definable from $\not\Rightarrow$. How? And how does this prove Craig's law (2)?

Let us recall Problem 8.1, in which a native of the knight-knave island said "If I'm a knight, then there is gold on this island," from which we concluded that the native must be a knight and there must be gold on the island. This led me to the discovery that $\wedge$ is definable from $\Rightarrow$ and $\equiv$.

Problem 10.18. How can $\wedge$ be defined in terms of $\Rightarrow$ and $\equiv$? Also, how is this related to the above knight-knave problem? Also, how does this result prove Craig's law (3) of Chapter 9?

Problem 10.19. Surprisingly enough, $\vee$ is definable from $\equiv$ and $\wedge$. How? (The solution is quite tricky!) This also proves Craig's law (4).

Problem 10.20. Also, $\wedge$ is definable from $\equiv$ and $\vee$. How? This also proves Craig's law (5).

Problem 10.21. Prove that $\wedge$ is definable from $\not\equiv$ and $\vee$. This proves Craig's law (6).

Problem 10.22. Also, $\vee$ is definable from $\not\equiv$ and $\wedge$. How? This proves Craig's law (7).

Problem 10.23. Prove that $\not\Rightarrow$ is definable from $\wedge$ and $\not\equiv$, and show how it proves Craig's law (8).

Problem 10.24. Prove that $\not\Rightarrow$ is definable from $\not\equiv$ and $\vee$, and show how it proves Craig's law (9).

Problem 10.25. Prove that all logical connectives are definable from $\Rightarrow$ and $\not\Rightarrow$, and relate this to Craig's law (10).

By now, the reader has probably guessed why in the last chapter I chose the letters **N**, **C**, **D**, **I**, **E**, **J**, **S**, **E**$'$, **I**$'$, the way I did, for the various conditions of the islands. Well, I chose **N** for *negation*, **C** for *conjunction*, **D** for *disjunction*, **I** for *implication*, **E** for *equivalence*, **J** for *joint denial*, **S** for the Sheffer *stroke*, **E**$'$ for in*equivalence* ($\not\equiv$), and **I**$'$ for the opposite of *implication* ($\not\Rightarrow$).

Bonus Problem. If an island of variable liars satisfies conditions **N** and **E**, is it necessarily a Boolean island? Equivalently, are all connectives definable from $\sim$ and $\equiv$?

Solutions

10.1. Concerning the problem of the Martian's request to define $\vee$ in terms of $\sim$ and $\wedge$, to say that *p or q* is true (in the sense that *at least one of them* is true) is equivalent to saying that they are not both false. The formula $\sim p \wedge \sim q$ says that p and q are both false (p is false and q is false), and so $\sim(\sim p \wedge \sim q)$ says that it is *not* the case that they are both false—in other words, it says that at least one of

p, q is true. Thus the formula $\sim(\sim p \wedge \sim q)$ is logically equivalent to $p \vee q$ (the two formulas have the same truth table—the same last column, that is), so $p \vee q$ can thusly be defined from $\sim$ and $\wedge$.

Now let us consider the corresponding problem of Conway's Island. We are given conditions **N** and **C** and are to decide condition **D**. In this and similar problems, we will employ Craig's terminology and notation (which I believe he got from reading the works of George Boole): We will say that an inhabitant is *opposed* to inhabitant A if he lies on those and only those days on which A tells the truth. If there is such an inhabitant, we let A' be such a one (it makes no difference which one). For any inhabitants A and B, we will say that an inhabitant is *conjunctively* related to A and B if he tells the truth on all and only those days on which A *and* B both tell the truth, and if there is such an inhabitant, we let $A \cap B$ be any one such. We will say that an inhabitant is *disjunctively* related to A and B if he tells the truth on all and only those days on which A tells the truth *or* B tells the truth (or both), and we let $A \cup B$ be such an inhabitant, if there is one.

Now consider two inhabitants A and B of Conway's Island. By condition **N**, there is an A' opposite to A and a B' opposite to B, and then by condition **C** there is an $A' \cap B'$ conjunctively related to A and B, and then there is an $(A' \cap B')'$ who is opposite to $A' \cap B'$. Then $(A' \cap B')'$ must be *disjunctively* related to A and B, because on any day on which either A or B is truthful, A' or B' lies, hence $A' \cap B'$ lies, hence $(A' \cap B')'$ is truthful. Conversely, on any day on which $(A' \cap B')'$ is truthful, $A' \cap B'$ lies, hence either A' or B' lies, hence either A or B is truthful. Thus on any day, $(A' \cap B')'$ is truthful if and only if either A or B (or both) is truthful. Thus $(A' \cap B')'$ is disjunctively related to A and B, so we can take $A \cup B$ to be $(A' \cap B')'$ (just as $p \vee q$ is equivalent to $\sim(\sim p \wedge \sim q)$).

10.2. To see that $\wedge$ is definable from $\sim$ and $\vee$, we note that to say that p *and* q are both true is to say that it is not the case that either one is false, which is symbolically expressed by $\sim(\sim p \vee \sim q)$. Thus $p \wedge q$ is equivalent to $\sim(\sim p \vee \sim q)$.

Similarly, on Diana's Island, on which conditions **N** and **D** hold, given any inhabitants A and B, the inhabitant $(A' \cup B')'$ is conjunctively related to A and B, and thus can be takes as $A \cap B$. Thus condition **C** also holds, so Diana's Island is indeed a Boolean island.

10.3. $p \wedge q$ is equivalent to $\sim(p \Rightarrow \sim q)$, so $\wedge$ is definable from $\sim$ and $\Rightarrow$. Once we have $\sim$ and $\wedge$, we can get $\vee$ (as we have seen), but, more directly, $p \vee q$ is equivalent to $\sim p \Rightarrow q$ (not to be confused with $\sim(p \Rightarrow q)$. As to Irving's Island, for any inhabitants A and B, we are given that there is an inhabitant who is truthful on just those days on which either A lies or B tells the truth. We let $A \rightsquigarrow B$ be such an inhabitant. Then $(A \rightsquigarrow B')'$ is an inhabitant who is truthful on just those days on which A and B are both truthful. Also, $A' \rightsquigarrow B$ is truthful on just those days on which either A or B is truthful. Thus conditions **C** and **D** both hold on Irving's Island (as well as condition **N**, which is given), so Irving's Island is indeed a Boolean island.

10.4. $p \Rightarrow q$ is equivalent to $\sim p \vee q$. Also, $p \Rightarrow q$ is equivalent to $\sim(p \wedge \sim q)$. Thus, concerning Problem 9.4, on a Boolean island, $A' \cup B$ is the inhabitant who tells the truth on just those days on which A lies or B tells the truth (and $(A \cap B')'$ is also such an inhabitant).

10.5. We will first solve Problem 9.5: We are given that Edward's Island satisfies conditions **C**, **D**, **I**, and **E** (**E** for *equivalence*), and the question is whether this island is necessarily a Boolean island. Now, it *might* be a Boolean island, but it is *not* necessarily one. It could be, for all we know, that all the inhabitants are always truthful on all days, in which case conditions **C**, **D**, **I** and **E** all hold, but not condition **N**! Thus Edward's Island is not necessarily a Boolean island.

From this it follows that $\sim$ is *not* definable from $\wedge$, $\vee$, $\Rightarrow$ and $\equiv$, for, if it were, then condition **N** would logically follow from conditions **C**, **D**, **I**, and **E**, which we have seen is not the case.

10.6. Of course it does: $(A \cap B) \cup (A' \cap B')$ tells the truth on just those days on which A and B behave alike. And so does $(A \rightsquigarrow B) \cap (B \rightsquigarrow A)$.

10.7. Obviously t is equivalent to any tautology, such as $(p \vee \sim p)$, and f is equivalent to any contradiction, such as $(p \wedge \sim p)$. On a Boolean island, conditions **T** and **F** must both hold, since $A \cup A'$ is an example of an inhabitant who tells the truth on all days, and $A \cap A'$ is an example of one who lies on all days.

10.8. Let us first do Problem 9.8, in which we were asked whether an island satisfying conditions **I** and **T** is necessarily a Boolean island. The answer is *no*, because any island on which the inhabitants are truthful on all days satisfies conditions **I** and **T**, but not **N**. This also proves that $\sim$ is not definable from $\Rightarrow$ and t.

10.9. This is a very different story: To begin with, $\sim$ is definable from $\Rightarrow$ and f, since $\sim p$ is equivalent to $p \Rightarrow f$. Once we have $\sim$ and $\Rightarrow$, we can get all the other connectives, as we have seen. This also shows that any island satisfying conditions **I** and **F** must be a Boolean island, because if we let f be any inhabitant who lies on all days, then, for any inhabitant A, the inhabitant $A \Rightarrow f$ tells the truth on just those days on which A lies, hence condition **N** is satisfied (as well as the given condition **I**).

10.10. $\sim p$ is equivalent to $p \downarrow p$, so $\sim$ is definable from $\downarrow$. Next, $p \vee q$ is equivalent to $\sim(p \downarrow q)$, or in terms of $\downarrow$ alone, $p \vee q$ is equivalent to $(p \downarrow q) \downarrow (p \downarrow q)$. Thus $\vee$ is definable from $\downarrow$. Once we have $\sim$ and $\vee$, we can get all the other connectives, as we now know.

NOTE. The applications of the above to Boolean islands should be obvious. From now on, the applications of the solutions of the remaining problems to the problems of Chapter 9 will be left to the reader.

10.11. $\sim p$ is equivalent to $p|p$, and $p \wedge q$ is equivalent to $\sim(p|q)$ (which in turn is equivalent to $(p|q)|(p|q)$). Once we have $\sim$ and $\wedge$, we can get everything.

10.12. If the sociologist had been lying, then both his answers would have to be false, which is not possible, since condition **S** does imply that the island is Boolean (because $\sim$ and $\wedge$ are definable from the Sheffer stroke). Therefore the sociologist was simply mistaken.

10.13. Of course it does, since $p \not\equiv q$ is equivalent to $(p \wedge \sim q) \vee (\sim p \wedge q)$.

10.14. Of course the conjecture is correct, since $p \equiv q$ is equivalent to $\sim(p \not\equiv q)$ and $p \not\equiv q$ is equivalent to $\sim(p \equiv q)$.

10.15. The answer is *yes*: $\sim$ is definable from $\not\equiv$ and $\Rightarrow$, because $\sim p$ is equivalent to $p \not\equiv (p \Rightarrow p)$. Once we have $\sim$ and $\Rightarrow$, we can get all the other connectives.

Alternatively, the Sheffer stroke is immediately definable from $\not\equiv$ and $\Rightarrow$, since $p|q$ is equivalent to $p \not\equiv (p \Rightarrow q)$.

10.16. $\sim p$ is equivalent to $p \equiv (p \not\Rightarrow p)$. Also $p|q$ is equivalent to $p \equiv (p \not\Rightarrow q)$. Also $p \downarrow q$ is equivalent to $q \equiv (p \not\Rightarrow q)$.

10.17. $p \vee q$ is equivalent to $(p \Rightarrow q) \Rightarrow q$.

10.18. $p \wedge q$ is equivalent to $p \not\Rightarrow (p \not\Rightarrow q)$.

10.19. Let us first re-consider the knight-knave problem, only now from the viewpoint of the translation method of Chapter 8. The native has asserted that *if* he is a knight, *then* there is gold on the island. Let k be the proposition that he is a knight and g be the proposition that there is gold on the island. He has asserted $k \Rightarrow g$, so the reality of the situation is that he is a knight if and only if his assertion is true—symbolically, $k \equiv (k \Rightarrow g)$, from which we inferred both k and g. Conversely, it is easily seen that if k and g are both true, then so is $k \equiv (k \Rightarrow g)$. Thus $k \equiv (k \Rightarrow g)$ is *equivalent* to $k \wedge g$. Indeed, for *any* propositions p and q, the proposition $p \equiv (p \Rightarrow q)$ is equivalent to $p \wedge q$. Thus $\wedge$ is definable from $\equiv$ and $\Rightarrow$.

10.20. This is indeed tricky! $p \wedge q$ is equivalent to $(p \vee q) \equiv (p \equiv q)$. (I don't recall how I ever discovered this weird fact!)

10.21. Equally curious: $p \vee q$ is equivalent to $(p \wedge q) \equiv (p \equiv q)$.

10.22. $p \wedge q$ is equivalent to $(p \vee q) \not\equiv (p \not\equiv q)$. (This means that conjunction is definable from the inclusive or and the exclusive or.)

10.23. $p \vee q$ is equivalent to $(p \wedge q) \not\equiv (p \not\equiv q)$.

10.24. $p \not\Rightarrow q$ is equivalent to $p \not\equiv (p \wedge q)$.

10.25. $p \not\Rightarrow q$ is equivalent to $q \not\equiv (p \vee q)$.

10.26. The Sheffer stroke is definable from $\Rightarrow$ and $\not\Rightarrow$: $p|q$ is equivalent to $p \Rightarrow (p \not\Rightarrow q)$. Thus all connectives are definable from $\Rightarrow$ and $\not\Rightarrow$.

BONUS PROBLEM. The answer is *no*; not all connectives are definable from $\sim$ and $\equiv$—or, equivalently, an island of variable liars satisfying conditions **N** and **E** is not necessarily a Boolean island; in fact, it can fail to satisfy condition **C** (conjunction).

Consider the following island: It has just eight inhabitants: A, B, C, T, A', B', C', T'.

- A tells the truth on Monday and Tuesday, but no other days;
- B tells the truth on Monday and Wednesday, but no other days;
- C tells the truth on Tuesday and Wednesday, but no other days;
- T tells the truth on all days.

A', B', C', and T' are respective opposites of A, B, C, and T (thus, e.g., A' lies on Monday and Tuesday, but no other days; T' always lies).

It is immediate that this island satisfies condition **N** (for if X is any one of the inhabitants A, B, C, or T, then X' tells the truth on just those days on which X lies). We now show that this island satisfies condition **E**. Well, let us say that an inhabitant Z is **E**-*related* to X and Y if Z tells the truth on all and only those days on which X and Y behave alike (both tell the truth or both lie). We are to show that for any inhabitants X, Y there is some Z which is **E**-related to X and Y.

Case 1: X and Y Are the Same. Obviously X is **E**-related to X and X.

Case 2: One of X, Y Is T. Obviously X is **E**-related to X and T.

Case 3: One of X, Y Is T'. There is no day on which X and T' both tell the truth (since T' never tells the truth), so T' tells the truth on just those days on which X and T' both tell the truth (which is no days at all). Thus T' is **E**-related to X and T'.

Case 4: X Is Distinct from Y and Neither One Is Either T or T'. We leave it to the reader to verify that if X, Y, Z are three distinct inhabitants, each of whom is one of A, B, or C, then

(1) Z' is **E**-related to X and Y.

(2) Z' is **E**-related to X' and Y'.

(3) Z is **E**-related to X and Y'.

(4) Z is **E**-related to X' and Y.

This takes care of all possible cases, so condition **E** does hold. Thus **N** and **E** both hold, but condition **C** fails, because the only day on which A and B both tell the truth is Monday, and none of the inhabitants tells the truth only on Monday.

- CHAPTER 11 -

THE TABLEAU METHOD

We will shortly be studying *first-order logic,* which, as we shall see, is an enormous advance over propositional logic and which, in fact, is an adequate logical framework for the entire body of mathematics! For this vital subject, the method of truth tables will not at all suffice, so we now turn to the method known as *tableau,* which in this chapter will be treated on the elementary propositional level, and which will be extended to first-order logic in a later chapter.

We are now dealing with *formulas,* as defined in Chapter 7. We are defining an *interpretation* of a formula to be a specification as to which of its propositional variables are to be interpreted as *true* and which as *false.* For a formula with only one propositional variable p, there are two and only two interpretations: (1) p true; (2) p false. For a formula with two propositional variables—say, p and q—there are four interpretations: (1) p true, q true; (2) p true, q false; (3) p false, q true; (4) p false, q false. With three propositional variables, there are eight interpretations, and, in general, for n variables there are 2^n possible interpretations. A truth table for a formula X is nothing more than a systematic device for determining under which interpretations X is true and under which X is false—each line in the truth table corresponds to one of the interpretations. We recall that a formula is called a *tautology* if it is true under all interpretations, and it is the tautologies that are of particular interest.

THE METHOD OF TABLEAUX

We begin by noting that under any interpretation the following eight facts hold (for any formulas X, Y):

(1) • If $\sim X$ is true, then X is false.
- If $\sim X$ is false, then X is true.

(2) • If a conjunction $X \wedge Y$ is true, then X, Y are both true.
- If a conjunction $X \wedge Y$ is false, then either X is false or Y is false.

(3) • If a disjunction $X \vee Y$ is true, then either X is true or Y is true.
- If a disjunction $X \vee Y$ is false, then X, Y are both false.

(4) • If $X \Rightarrow Y$ is true, then either X is false or Y is true.
- If $X \Rightarrow Y$ is false, then X is true and Y is false.

These eight facts provide the basis for the tableau method.

SIGNED FORMULAS

At this stage it will prove useful to introduce the symbols T and F to our object language and define a *signed formula* as an expression TX or FX, where X is an unsigned formula. (Informally, we read "TX" as "X is true" and "FX" as "X is false.")

DEFINITION. Under any interpretation, a signed formula TX is called *true* if X is true and *false* if X is false. And a signed formula FX is called *true* if X is false and *false* if X is true.

Thus the truth value of TX is the same as that of X; the truth value of FX is the same as that of $\sim X$.

By the *conjugate* of a signed formula we mean the result of changing "T" to "F" or "F" to "T" (thus the conjugate of TX is FX; the conjugate of FX is TX).

ILLUSTRATION OF THE METHOD OF TABLEAUX

Before we state the eight rules for the construction of tableaux, we shall illustrate the construction with an example.

Suppose we wish to prove the formula $[p \vee (q \wedge r)] \Rightarrow [(p \vee q) \wedge (p \vee r)]$. The following tableau does this; the explanation is given immediately following the tableau.

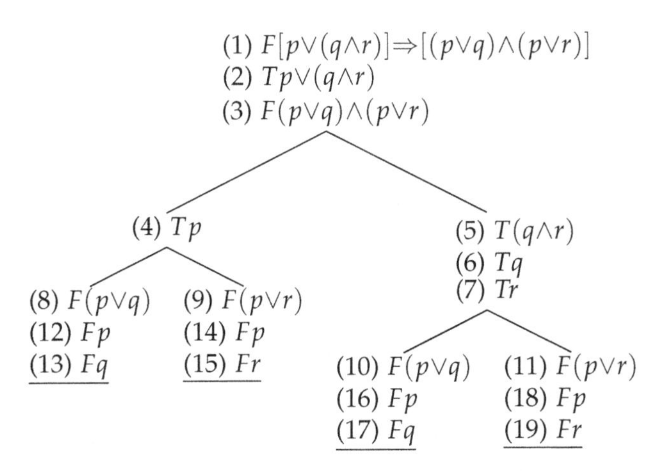

EXPLANATION. The tableau was constructed as follows. We want to derive a contradiction from the assumptions that the formula

$$[p\vee(q\wedge r)]\Rightarrow[(p\vee q)\wedge(p\vee r)]$$

is false. So our first line consists of this formula preceded by the letter F. Now, a formula of the form $X\Rightarrow Y$ can be false only if X is true and Y is false. Thus (in the language of tableaux) TX and FY are *direct* consequences of the (signed) formula $F(X\Rightarrow Y)$. So we write the lines (2) and (3) as *direct* consequences of line (1). Now let us look at line (2); it is of the form $T(X\vee Y)$, where $X=p$, $Y=(q\wedge r)$. We cannot draw any *direct* conclusion about the truth value of X nor about the truth value of Y; all we can infer is that *either* TX *or* TY. So the tableau *branches* into two columns; one for each possibility. Thus line (2) *branches* into lines (4) and (5). Line (5), viz. $T(q\wedge r)$, immediately yields Tq and Tr as direct consequences; we thus have lines (6) and (7). Now look at (3). It is of the form $F(X\wedge Y)$. This means that *either* FX *or* FY. We also know that either (4) or (5) holds. So for *each* of the possibilities (4), (5) we have the two possibilities FX, FY. There are hence now four possibilities. So each of the branches (4), (5) branches again into the possibilities FX, FY. More specifically, (4) branches to (8), (9), and (5) branches to (10), (11) (which are respectively the same as (8), (9)). Lines (12), (13) are direct consequences of (8); (14), (15) are direct consequences of (9); (16), (17) of (10); and (18), (19) of (11).

We now look at the leftmost branch and we see that (12) is a direct contradiction of (4) (i.e., it is the conjugate of (4)), so we put a bar after (13) to signify that this branch leads to a contradiction. Similarly, (14) contradicts (4), so we "close" the branch leading to (15) by putting a bar after (15). The next branch is closed by virtue of (17) and (6). Finally, the rightmost branch is closed by virtue of (19) and (7). Thus all branches lead to a contradiction, so line (1) is untenable. Thus

$$[p\vee(q\wedge r)]\Rightarrow[(p\vee q)\wedge(p\vee r)]$$

can never be false in any interpretation, so it is a tautology.

Remark. The numbers to the left of the lines are only for the purpose of identification in the above explanation; we do not need them for the actual construction.

Remark. We could have closed some of our branches a bit earlier; lines (13), (15) are superfluous. In subsequent examples we shall close a branch as soon as a contradiction appears (a contradiction is of the form of two formulas FX, TX).

Rules for the Construction of Tableaux

We now state all the rules in schematic form; explanations immediately follow. For each logical connective, there are two rules: one for a formula preceded by T, the other for a formula preceded by F:

$$(1)\quad \frac{T{\sim}X}{FX} \qquad \frac{F{\sim}X}{TX}$$

$$(2)\quad \frac{T(X\wedge Y)}{\begin{array}{c}TX\\ TY\end{array}} \qquad \begin{array}{c}F(X\wedge Y)\\ \wedge\\ FX \quad FY\end{array}$$

$$(3)\quad \begin{array}{c}T(X\vee Y)\\ \wedge\\ TX \quad TY\end{array} \qquad \frac{F(X\vee Y)}{\begin{array}{c}FX\\ FY\end{array}}$$

$$(4)\quad \begin{array}{c}T(X\Rightarrow Y)\\ \wedge\\ FX \quad TY\end{array} \qquad \frac{F(X\Rightarrow Y)}{\begin{array}{c}TX\\ FY\end{array}}$$

Some Explanations. Rule (1) means that from $T{\sim}X$ we can directly infer FX (in the sense that we can subjoin FX to any branch passing through $T{\sim}X$) and that from $F{\sim}X$ we can directly infer TX. Rule (2) means that $T(X\wedge Y)$ directly yields both TX and TY, whereas $F(X\wedge Y)$ *branches* into FX and FY. Rules (3) and (4) can now be understood analogously.

Signed formulas, other than signed variables, are of two types: (A) those that have *direct* consequences (viz. $F{\sim}T$, $T{\sim}X$, $T(X\wedge Y)$, $F(X\vee Y)$, $F(X\Rightarrow Y)$) and (B) those that *branch* (viz. $F(X\wedge Y)$, $T(X\vee Y)$, $T(X\Rightarrow Y)$).

It is practically desirable, in constructing a tableau, that when a line of type (A) appears on the tableau, we simultaneously subjoin its consequences to *all* branches that pass through the line. Then that line need never be used again. And in using a line of type (B), we divide *all* branches that pass through the line into sub-branches, and the line need never be used again. For example, in Tableau 1 below, we use (1) to get (2) and (3), and (1) is never used again. From (2) we get (4) and (5), and (2) is never used again. Line (3) yields (8), (9), (10), (11), and (3) is never used again, etc.

If we construct a tableau in the above manner, it is not difficult to see that after a finite number of steps we must reach a point where every line has been used (except, of course, for signed variables, which are never used at all to create new lines). At this point our tableau is *complete* (in a precise sense that we will subsequently define).

One way to complete a tableau is to work systematically downward, i.e., never to use a line until all lines above it (on the same branch) have been used. Instead of this procedure, however, it turns out to be more efficient to give priority to lines of type (A)—i.e., to use up all such lines at hand before using those of type (B). In this way, one will avoid repeating the same formula on different branches; rather, it will have only one occurrence *above* all those branch points.

As an example of both procedures, let us prove the formula

$$[p \Rightarrow (q \Rightarrow r)] \Rightarrow [(p \Rightarrow q) \Rightarrow (p \Rightarrow r)].$$

Tableau 1 works systematically downward; Tableau 2 uses the second suggestion. For the convenience of the reader, we put to the right of each line the number of the line from which it was inferred.

It is apparent that Tableau 2 is quicker to construct than Tableau 1, involving 13 rather than 23 lines.

As another practical suggestion, one might put a checkmark to the right of a line as soon as it has been used. This will subsequently aid the eye in hunting upward for lines that have not yet been used. (The checkmark may be later erased, if the reader so desires.)

The method of tableaux can also be used to show that a given formula is a logical consequence of a given finite set of formulas. Suppose we wish to show that $X \Rightarrow Z$ is a logical consequence of the two formulas $X \Rightarrow Y$, $Y \Rightarrow Z$. We could, of course, simply show that $[(X \Rightarrow Y) \wedge (Y \Rightarrow Z)] \Rightarrow (X \Rightarrow Z)$ is a tautology by constructing a truth table for it. Alternatively, we can construct a tableau starting with the three signed formulas

$$T(X \Rightarrow Y),$$
$$T(Y \Rightarrow Z),$$
$$F(X \Rightarrow Z),$$

and show that all branches close.

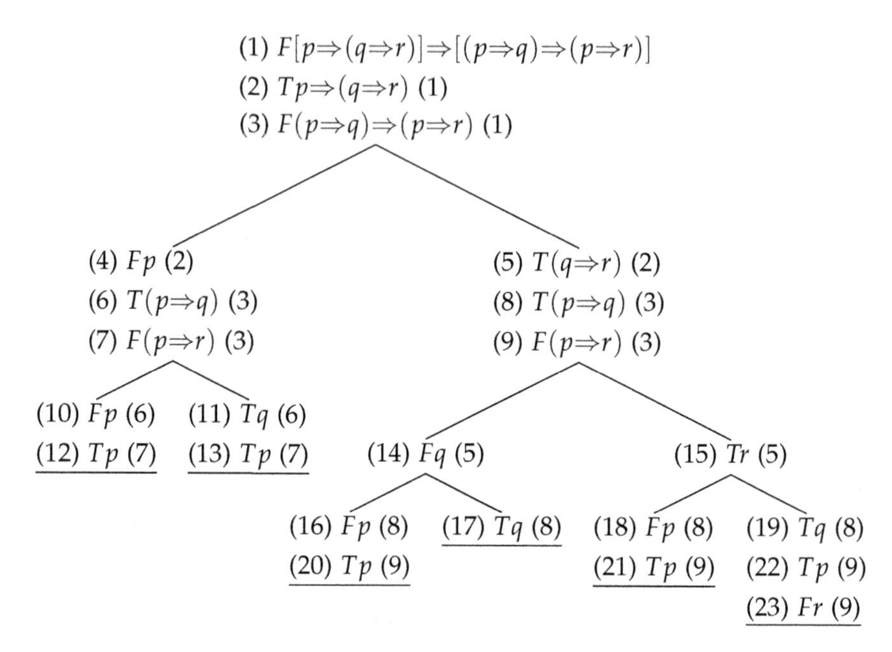

Tableau 1.

(1) $F[p\Rightarrow(q\Rightarrow r)]\Rightarrow[(p\Rightarrow q)\Rightarrow(p\Rightarrow r)]$
(2) $Tp\Rightarrow(q\Rightarrow r)$ (1)
(3) $F(p\Rightarrow q)\Rightarrow(p\Rightarrow r)$ (1)
(4) $T(p\Rightarrow q)$ (3)
(5) $F(p\Rightarrow r)$ (3)
(6) Tp (5)
(7) Fr (5)

(8) Fp (2) | (9) $T(q\Rightarrow r)$ (2)

(10) Fp (4) | (11) Tq (4)

(12) Fq (9) | (13) Tr (9)

Tableau 2.

In general, to show that Y is logically implied by $X_1, \ldots, X_n$, we can construct either a closed tableau starting with $F[(X_1 \wedge \ldots \wedge X_n) \Rightarrow Y]$ or one starting with

$$\begin{array}{c} TX_1 \\ \vdots \\ TX_n \\ FY. \end{array}$$

Tableaux Using Unsigned Formulas

Our use of the letters T and F, though perhaps heuristically useful, is theoretically quite dispensable—simply delete every T and substitute $\sim$ for F. (In which case, incidentally, the first half of Rule (1) becomes superfluous.) The rules then become

(1) $$\frac{\sim\sim X}{X}$$

(2) $$\frac{X \wedge Y}{\begin{array}{c} X \\ Y \end{array}} \qquad \begin{array}{c} \sim(X \wedge Y) \\ \sim X \quad \sim Y \end{array}$$

(3) $$\begin{array}{c} X \vee Y \\ X \quad Y \end{array} \qquad \frac{\sim(X \vee Y)}{\begin{array}{c} \sim X \\ \sim Y \end{array}}$$

(4) $$\begin{array}{c} X \Rightarrow Y \\ \sim X \quad Y \end{array} \qquad \frac{\sim(X \Rightarrow Y)}{\begin{array}{c} X \\ \sim Y \end{array}}$$

In working with tableaux that use unsigned formulas, "closing" a branch means terminating the branch as soon as two formulas appear, one of which is the *negation* of the other. A tableau is called *closed* if every branch is closed. A branch is called *open* if it is not closed.

By a tableau *for* a formula X, we mean a tableau that starts with X. If we wish to prove a formula X to be a tautology, we construct a tableau not for the formula X but for its negation $\sim X$.

EXERCISE 11.1. By the tableau method, prove the following tautologies:

(a) $q \Rightarrow (p \Rightarrow q)$.

(b) $((p \Rightarrow q) \wedge (q \Rightarrow r)) \Rightarrow (p \Rightarrow r)$.

(c) $((p \Rightarrow r) \wedge (q \Rightarrow r)) \Rightarrow ((p \vee q) \Rightarrow r)$.

(d) $((p \Rightarrow q) \wedge (p \Rightarrow r)) \Rightarrow (p \Rightarrow (q \wedge r))$.

(e) $\sim(p \wedge q) \Rightarrow (\sim p \vee \sim q)$.

(f) $\sim(p \vee q) \Rightarrow (\sim p \wedge \sim q)$.

(g) $(p \wedge (q \vee r)) \Rightarrow ((p \wedge q) \vee (p \wedge r))$.

(h) $(p \vee (q \wedge r)) \Rightarrow ((p \vee q) \wedge (p \vee r))$.

(i) $((p \Rightarrow q) \wedge (\sim p \Rightarrow q)) \Rightarrow q$.

(j) $(((p \wedge q) \Rightarrow r) \wedge (p \Rightarrow (q \vee r))) \Rightarrow (p \Rightarrow r)$.

A Unifying Notation

In this book, we are taking $\sim$, $\wedge$, $\vee$ and $\Rightarrow$ as independent logical connectives—the so-called starters—even though we could have defined some of them in terms of the others. Some treatments of propositional logic take only $\sim$ and $\Rightarrow$ as starters and define the others in terms of them; other treatments start with only $\sim$ and $\vee$; others with $\sim$ and $\wedge$; others with $\sim$, $\wedge$ and $\vee$; still other with all four of $\sim$, $\wedge$, $\vee$, $\Rightarrow$—which is what we are doing. The advantage of using many starters is that proofs *within* the system tend to be shorter and more natural; but proofs *about* the system (the *metatheory*, which we study later on) tend to be much longer and involve analyzing many different cases. On the other hand, those systems that use few starters have advantages in the quickened metatheory, but proofs *within* the system tend to be long and unnatural. Well, in [17] was presented a unifying scheme that enables one, so to speak, to have one's cake and eat it too—that is, it combines the advantages of many starters with the advantages of few. For example, in our present setup, there are eight types of signed formulas (other than signed variables)—two for each of the four connectives $\sim$, $\wedge$, $\vee$, $\Rightarrow$—and hence eight tableau rules. Well, our unified scheme allows us to collapse these eight rules into only two! Moreover, in the next section of this chapter, we will be raising some vital questions about the adequacy of the tableau method (questions about the metatheory) where answers would require an analysis of eight separate cases, were it not for the unifying scheme, which makes it possible to reduce the number of cases to two. Does this whet your appetite? Well, here is the scheme (which will play a vital role throughout this volume).

We use the Greek letter α to stand for any signed formula of one of the five forms $TX \wedge Y$, $FX \vee Y$, $FX \Rightarrow Y$, $T\sim X$, $F\sim X$. For every such formula α, we define the two signed formulas α_1 and α_2 as follows:

if $\alpha = T(X \wedge Y)$, then $\alpha_1 = TX$ and $\alpha_2 = TY$;
if $\alpha = F(X \vee Y)$, then $\alpha_1 = FX$ and $\alpha_2 = FY$;
if $\alpha = F(X \Rightarrow Y)$, then $\alpha_1 = TX$ and $\alpha_2 = FY$;
if $\alpha = T{\sim}X$, then α_1 and α_2 are both FX;
if $\alpha = F{\sim}X$, then α_1 and α_2 are both TX.

For perspicuity, we summarize these definitions in the following table:

α	α_1	α_2
$T(X \wedge Y)$	TX	TY
$F(X \vee Y)$	FX	FY
$F(X \Rightarrow Y)$	TX	FY
$T{\sim}X$	FX	FX
$F{\sim}X$	TX	TX

We refer to α_1 and α_2 as the *components* of α. We note that under any interpretation, α is true if and only if α_1 and α_2 are *both* true. Accordingly, we refer to an α-formula as one of *conjunctive* type, or of *type A*.

We let β be any signed formula of one of the five forms $FX \wedge Y$, $TX \vee Y$, $TX \Rightarrow Y$, $T{\sim}X$, $F{\sim}X$, and we define its components β_1 and β_2 as in the following table:

β	β_1	β_2
$T(X \vee Y)$	TX	TY
$F(X \wedge Y)$	FX	FY
$T(X \Rightarrow Y)$	FX	TY
$T{\sim}X$	FX	FX
$F{\sim}X$	TX	TX

We note that, under any interpretation, β is true if and only if either β_1 is true or β_2 is true, and we accordingly refer to a β-formula as one of *disjunctive* type, or of *type B*.

Some special remarks are in order for signed formulas of the form $T{\sim}X$ or $F{\sim}X$: These are the only types of formulas that we have classified as of both type α and type β, because, in both cases, its two components are identical, and therefore to say that both components are true is equivalent to saying that at least one of them is true— so we can regard them either as an α or as a β.[1]

Using our unifying α, β-notation, we note the pleasant fact that our eight tableau rules can now be succinctly lumped into the following two:

RULE A. $\dfrac{\alpha}{\begin{array}{c}\alpha_1\\ \alpha_2\end{array}}$ RULE B. $\dfrac{\beta}{\beta_1 \quad \beta_2}$

[1]In [17], we took them only as α, but the present scheme works better for this volume.

REMARK. Of course, in constructing a tableau, we should treat $T{\sim}X$ and $F{\sim}X$ as α's, since it is pointless to extend a branch to two identical branches! More generally, given any β-type formula whose two components are identical (such as $TX \vee X$), one should simply adjoin a single copy of the component rather than branch to two identical copies—in our case, adjoin TX rather than branch to TX and TX!

Some Properties of Conjugation

We recall that by the conjugate of TX we mean FX, and by the conjugate of FX we mean TX. We shall use the notation $\overline{X}$ to mean the *conjugate* of X (thus $\overline{TX} = FX$, and $\overline{FX} = TX$). Conjugation obeys the following nice symmetric laws:

J_0. (a) $\overline{X}$ is distinct from X;
(b) $\overline{\overline{X}} = X$.

J_1. The conjugate of any α is some β, and the conjugate of any β is an α.

J_2. If β is the conjugate of α, then β_1 is the conjugate of α_1 and β_2 is the conjugate of α_2.

Laws J_0 and J_1 are obvious. As to J_2, this must be verified case by case. For example, consider the case where $\alpha{=}(FX{\Rightarrow}Y)$ and $\beta{=}(TX{\Rightarrow}Y)$. Then $\alpha_1{=}TX$ and $\beta_1{=}FX$, which is the conjugate of α_1. Also $\alpha_2{=}FY$ and $\beta_2{=}TY$, and so β_2 is the conjugate of α_2.

EXERCISE 11.2. Verify law J_2 for the remaining cases.

(a) $\alpha{=}(TX{\wedge}Y)$, $\beta{=}(FX{\wedge}Y)$.

(b) $\alpha{=}(TX{\vee}Y)$, $\beta{=}(TX{\vee}Y)$.

We can also use a unifying α-β notation for *unsigned* formulas—just delete T and replace F by $\sim$. The tables then become:

α	α_1	α_2
$X{\wedge}Y$	X	Y
${\sim}(X{\vee}Y)$	${\sim}X$	${\sim}Y$
${\sim}(X{\Rightarrow}Y)$	X	${\sim}Y$
${\sim}{\sim}X$	X	X

β	β_1	β_2
${\sim}(X{\wedge}Y)$	${\sim}X$	${\sim}Y$
$X{\vee}Y$	X	Y
$X{\Rightarrow}Y$	${\sim}X$	Y
${\sim}{\sim}X$	X	X

Now, what about *conjugation* for unsigned formulas? How should we define it? Well, we will take $X{\wedge}Y$ and ${\sim}(X{\wedge}Y)$ to be conjugates of each other, and similarly with $X{\vee}Y$ and ${\sim}(X{\vee}Y)$, and similarly with $X{\Rightarrow}Y$

and $\sim(X \Rightarrow Y)$; but now the delicate case is that of formulas of the form $\sim\sim X$, or $\sim\sim\sim X$, or any other formula involving multiple negations. We want our definition of conjugation to be such that law J_0 (b) holds ($\overline{\overline{X}} = X$). Well, the following scheme works: Let Y be any formula that is not itself a negation of any formula, and let X be either Y or Y preceded by one or more negations (i.e., $X{=}Y$, or $\sim Y$, or $\sim\sim Y$, or $\sim\sim\sim Y$, etc.). If the number of negations is even, then take $\overline{X}$ to be the result of adding one more negation, but if the number is odd, then take away one negation. So, for example, if Y is not itself a negation, then $\overline{Y}{=}\sim Y$, $\overline{\sim Y}{=}Y$, $\overline{\sim\sim Y}{=}\sim\sim\sim Y$, $\overline{\sim\sim\sim Y}{=}\sim\sim Y$, etc. In this way, $\overline{\overline{X}}$ is always X. Indeed, laws J_0, J_1 and J_2 all hold under this definition of conjugation for unsigned formulas.

Discussion. There are other logical connectives that we could have taken as starters that would fit nicely into our unified α-β scheme—for example, $\downarrow$ (joint denial) and $|$ (the Sheffer stroke). (Recall that $p{\downarrow}q$ means that p and q are both false, and $p|q$ means that p and q are not both true.)

Problem 11.1. Suppose we had taken these two as independent connectives. What should be the tableau rules for $TX{\downarrow}Y$, $FX{\downarrow}Y$, $TX|Y$, $FX|Y$? Also, which of them are α's and which are β's?

The Biconditional

We have not given tableau rules for the biconditional $\equiv$ (if and only if), because we could simply take $X{\equiv}Y$ as an abbreviation of

$$(X \wedge Y) \vee (\sim X \wedge \sim Y)$$

(or alternatively of the equivalent formula $(X \Rightarrow Y) \wedge (Y \Rightarrow X)$). If we alternatively take $\equiv$ as an independent connective, then the obviously appropriate tableau rules are the following:

Rule A. $TX{\equiv}Y$ branches into: TX, TY | FX, FY

Rule B. $FX{\equiv}Y$ branches into: TX, FY | FX, TY

(To say that $X{\equiv}Y$ is true is to say that X and Y are both true or both false. To say that $X{\equiv}Y$ is false is to say that one of them is true and the other false, thus either TX and FY or FX and TY.)

Let us note that if we took $X{\equiv}Y$ as an abbreviation of $(X \wedge Y) \vee (\sim X \wedge \sim Y)$, our tableau for the T-case would have gone as follows:

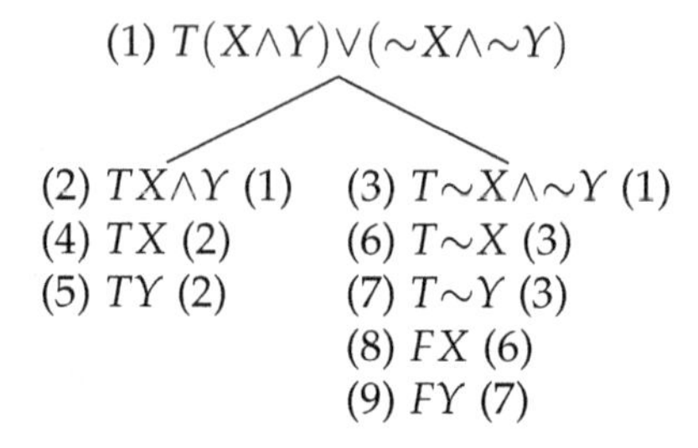

It really leads to the same thing, since on the left branch we have TX and TY, and on the right branch FX and FY.

If a tableau is constructed for $F(X \wedge Y) \vee (\sim X \wedge \sim Y)$, it is easily seen that one branch will contain TX and FY and the other FX and TY.

Suppose we instead took $X \equiv Y$ as an abbreviation of $(X \Rightarrow Y) \wedge (Y \Rightarrow X)$. Let us see what the tableau for $T(X \Rightarrow Y) \wedge (Y \Rightarrow X)$ looks like.

(1) $T(X \Rightarrow Y) \wedge (Y \Rightarrow X)$
(2) $TX \Rightarrow Y$
(3) $TY \Rightarrow X$

(4) FX (2) | (5) TY (2)

FY(3) | TX (3) | FY (3) | TX (3)

We now have four branches, but only two *open* ones, and again one of them contains FX and FY and the other, TX and TY.

(The reader should now try to make a tableau for $F(X \Rightarrow Y) \wedge (Y \Rightarrow X)$.)

We thus see that it makes little difference whether we take $\equiv$ independent of the other connectives or not, but if we do, and use the rule given above, it eliminates unnecessary work.

We remark that very few (if any) treatments of propositional logic take $\equiv$ as a starter, and we also remark that $\equiv$ does not fit at all into our unifying α-β scheme.

EXERCISE 11.3. By tableau, prove the following:

(a) $p \equiv (q \equiv r)) \equiv ((p \equiv q) \equiv r)$.

(b) $((p \Rightarrow q) \wedge (q \Rightarrow p)) \equiv ((p \wedge q) \vee (\sim p \wedge \sim q))$.

(c) $(p \equiv (p \Rightarrow q)) \equiv (p \wedge q)$.

(d) $(q \equiv (p \Rightarrow q)) \equiv (p \vee q)$.

(e) $(p \Rightarrow q) \equiv (p \equiv (p \wedge q))$.

Some Vital Questions: Prolegomena to Metatheory

A proof procedure for propositional logic is called *correct* if it proves only tautologies and *complete* if it can prove all tautologies. The truth-table method is obviously both correct and complete. As to the tableau method, it is easy to show that it is correct, but the key question is whether the method is complete. That is, if X is a tautology, can we be sure that we can get a tableau starting with FX to close? Before the reader answers this question too hastily, we wish to point out that if we delete some of the rules for the construction of tableaux, it will still be true that a *closed* tableau for FX *always* indicates that X is a tautology, but if we delete too many of the rules, then we may not have enough power to derive a closed tableau for FX whenever X is a tautology. As examples, if we delete the first half of the conjunction rule (that from $TX \wedge Y$ we can directly infer both TX and TY) then it would be impossible to prove the tautology $(p \wedge q) \Rightarrow p$, though it would still be possible to prove $p \Rightarrow (q \Rightarrow (p \wedge q))$. On the other hand, if we delete the second half of the rule but retain the first half, then we could prove $(p \wedge q) \Rightarrow p$ but we couldn't prove $p \Rightarrow (q \Rightarrow (p \wedge q))$. So the question is whether we really have enough rules to prove *all* tautologies! The answer to that question is not so obvious, and will be given in a later chapter. (Meanwhile the more ambitious reader might like to try his or her hand at it: Can you either prove that the tableau method *is* complete, or find a tautology that is *not* provable by the method?)

A Related Question

Given two formulas X and Y, suppose that X and $X \Rightarrow Y$ are both provable by the tableau method. Does it necessarily follow that Y is provable by this method? Of course, if X and $X \Rightarrow Y$ are both tautologies, so is Y (because Y is true under all interpretations in which X and $X \Rightarrow Y$ are both true, so if X and $X \Rightarrow Y$ are both true under all interpretations, so is Y). Therefore, since the tableau method is correct, if X and $X \Rightarrow Y$ are both provable by tableau, then both are tautologies, hence Y is also a tautology—but this is as far as we can go, since we have not yet answered the question of whether all tautologies are provable by tableau! If it turns out that the method is complete, it would indeed then follow that Y is provable (if X and $X \Rightarrow Y$ both are), but what if the method is *not* complete?

Synthetic Tableaux

Suppose we add to our tableau rules the following:

This means that at any stage of the construction of a tableau, we can pull an arbitrary formula out of the air, and then let any (or all) of the branches split to TY and FY (or we can even start a tableau with the two branches TY and FY).

Let us call this rule the *splitting rule*, and let us call a tableau that uses this rule a *synthetic tableau* (and we will call *analytic* those tableaux that do not use the splitting rule). This rule is certainly sound, since, in any interpretation, one (and only one) of the signed formulas TY and FY is true, and so any formula provable by a synthetic tableau is also a tautology. Now an interesting question arises: Can one prove any more formulas with the splitting rule than without it? Does the addition of the splitting rule increase the class of provable formulas? Is it possible that without the splitting rule, the tableau method is incomplete, but adding this rule makes the system complete? All these questions will be answered later on.

One thing I will tell you now: We previously raised the question of whether if X and $X \Rightarrow Y$ are both provable by the tableau method, Y is necessarily provable, and the answer was not immediate. For synthetic tableaux, however, the answer *is* immediate and affirmative.

Problem 11.2. Why is it true that for synthetic tableaux, if X and $X \Rightarrow Y$ are both provable, so is Y? More constructively, given a closed tableau $\mathfrak{T}_1$ for FX and a closed tableau $\mathfrak{T}_2$ for $FX \Rightarrow Y$, how would you combine them (using the splitting rule) to obtain a closed tableau for FY?

Another Related Question

In mathematics, it is customary in proving a theorem X to use in the proof a previously proved theorem Y. Suppose we incorporate this into our tableau system by adding the rule that at any stage of the construction of a tableau, one may, on any branch, introduce TY, where Y is any formula previously proved. Does this increase the class of provable formulas? Well, without the splitting rule the answer is not obvious, and will be given in a later chapter. But with the splitting rule, the answer is easily seen to be *no*.

Problem 11.3. Why is this?

Solutions

11.1. The tableau rules for $\downarrow$ and $|$ are the following:

$$\frac{TX{\downarrow}Y}{\begin{array}{c}FX\\FY\end{array}} \qquad \frac{FX|Y}{\begin{array}{c}TX\\TY\end{array}} \qquad \begin{array}{c}T(X|Y)\\ \diagup\;\diagdown \\ FX \quad FY\end{array} \qquad \begin{array}{c}F(X{\downarrow}Y)\\ \diagup\;\diagdown \\ TX \quad TY\end{array}$$

Thus $TX{\downarrow}Y$ and $FX|Y$ are the α's, and the others are the β's.

11.2. We are given a closed tableau $\mathfrak{T}_1$ for FX and a closed tableau $\mathfrak{T}_2$ for $FX{\Rightarrow}Y$, and we construct a closed synthetic tableau for FY as follows:

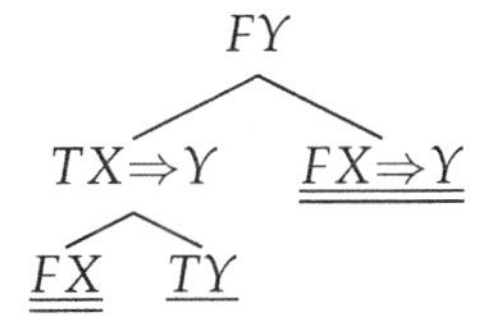

Explanation. We start the tableau with FY. We then immediately split to $TX{\Rightarrow}Y$ (left branch) and $FX{\Rightarrow}Y$ (right branch). The double bar under $FX{\Rightarrow}Y$ indicates that one *can* get a closure under it (by using tableau $\mathfrak{T}_2$). Next, the left branch $TX{\Rightarrow}Y$ itself branches to FX and TY. The TY clashes with FY and that sub-branch closes. The FX extends to the tableau $\mathfrak{T}_1$, which is closed.

11.3. We are given a proof of a formula Y, a closed tableau $\mathfrak{T}$ whose origin is FY. We wish to prove a formula X. At any stage of a tableau starting with FX, in place of arbitrarily adding TY as a branch, if we use the splitting rule, then we can alternatively split the branch thus:

$$\begin{array}{c}\diagup\;\diagdown \\ TY \quad \underline{\underline{FY}}\end{array}$$

But we can then close FY by copying the tableau $\mathfrak{T}$. So, in effect, we have added TY.

- Chapter 12 -

All and Some

Propositional logic, which we have just studied, is only the *beginning* of the logic we need for mathematics and science. The real meat comes in the field known as *first-order logic,* which is one of the main subjects of this book. It uses the logical connectives of propositional logic, with the addition of the notions of *all* and *some* ("some" in the sense of *at least one*), which will be symbolically treated in the next chapter. In this chapter, we treat these two notions on an informal and intuitive level.

To begin with, suppose a member of a certain club says: "All Frenchmen in this club wear hats." Now, suppose it turns out that there are no Frenchmen in the club, then how would you regard the statement—*true, false,* or *neither*? I guess many would say *false,* many would say *neither* (in other words, *inapplicable,* or *meaningless*) and perhaps some would say *true.* Well, it will shock many of you to hear that in logic, mathematics and computer science, the statement is regarded as *true*! The statement means nothing more nor less than that there are no Frenchmen in the club who don't wear hats. Thus if there are no Frenchmen at all in the club, then there are certainly no Frenchmen in the club who don't wear hats, and so the statement is true.

In general, the statement "All A's are B's" is regarded as automatically true ("*vacuously true*" is the technical term) if there are no A's. So, for example, shocking as it may seem, the following statement is true:

All unicorns have five legs.

The only way the above sentence can be falsified is to exhibit at least one unicorn who doesn't have five legs, which is not possible, since there are no unicorns. It is also true that all unicorns have six legs! *Anything* one says about *all* unicorns is vacuously true. On the other hand, the statement that all *horses* have six legs is easily shown to be false; just exhibit a horse with only four legs!

I realize that all this takes some getting used to, but, in the next chapter, we will see the practical use of this way of looking at it. For now, we will consider some knight-knave problems involving the notions of *all* and *some*.

Problems

Abercrombie once visited a whole cluster of knight-knave islands of the simple type of Chapter 1, in which all knights tell the truth and all knaves lie. He was interested in the proportions of knights to knaves on the islands and, also, whether there was any correlation between lying and smoking.

Problem 12.1. On the first island he visited, all the inhabitants said the same thing: "All of us here are of the same type."

What can be deduced about the inhabitants of that island?

Problem 12.2. On the next island, all the inhabitants said: "Some of us are knights and some are knaves."

What is the composition of that island?

Problem 12.3. On the next island, Abercrombie one day interviewed all the inhabitants but one, who was asleep. They all said: "All of us are knaves." The next day, Abercrombie met the inhabitant who was asleep the day before, and asked him: "Is it true that all the inhabitants of this island are knaves?" The inhabitant then answered (*yes* or *no*). What answer did he give?

Problem 12.4. On the next island, Abercrombie was particularly interested in the smoking habits of the natives. They all said the same thing: "All knights on this island smoke."

What can be deduced about the knight-knave distributions and the smoking habits of the natives?

Problem 12.5. On the next island, each one said: "Some knaves on this island smoke."

What can be deduced from this?

Problem 12.6. On the next island all were of the same type, and each one said: "If I smoke, then all the inhabitants of this island smoke."

What can be deduced?

Problem 12.7. On the next island again all were of the same type, and each one said: "If any inhabitant of this island smokes, then I smoke."

What can be deduced?

Problem 12.8. On the next island, too, all were of the same type, and each one said: "Some of us smoke, but I don't."

What follows?

Problem 12.9. Suppose that on the same island, instead, each inhabitant made the following two statements: "Some of us smoke." "I don't smoke."

What would you conclude?

Problem 12.10. The next island visited by Abercrombie was inhabited by two tribes—Tribe A and Tribe B. All the members of Tribe A said: "All the inhabitants of this island are knights." "All of us smoke."

Each member of Tribe B said: "Some of the inhabitants of this island are knaves." "No one on this island smokes."

What can be deduced?

Problem 12.11. On this island, as well as the next two visited by Abercrombie, there are male and female knights and knaves. The female knights lie and the female knaves tell the truth, whereas the males act as before (male knights are truthful and male knaves are not). All the inhabitants of this island (male and female) said the same thing: "All inhabitants of this island are knights."

What can be deduced?

Problem 12.12. On the next island, all the men said: "All the inhabitants of this island are knaves." Then the women were asked whether it was true that all the inhabitants were knaves. They all gave the same answer (*yes* or *no*).

What answer did they give?

Problem 12.13. On the next island, all the men said: "All the inhabitants are knights and they all smoke." The women all said: "Some of the inhabitants are knaves. All the inhabitants smoke."

What can be deduced?

Problem 12.14. On the next island visited by Abercrombie, he met six natives, named Arthur, Bernard, Charles, David, Edward, and Frank, who made the following statements:

Arthur:	Everyone here smokes cigarettes.
Bernard:	Everyone here smokes cigars.
Charles:	Everyone here smokes either cigarettes or cigars or both.
David:	Arthur and Bernard are not both knaves.
Edward:	If Charles is a knight, so is David.
Frank:	If David is a knight, so is Charles.

Is is possible to determine of any one of these that he is a knight, and if so, which one or ones?

Problem 12.15. It has been related that one day a god came down from the sky and classified each inhabitant of the earth as either *special* or *non-special*. As it turns out, for each person x, x was special if and only if it was the case that either everyone was special or no one was special. Which of the following three statements logically follows from this?

(1) No one is special.

(2) Some are special and some are not.

(3) Everyone is special.

Problem 12.16. According to *another* version of the above story, it turned out that for each person x, x was special if and only if some of the people were special and some were not. If this version is correct, then which of the above statements (1), (2), (3) logically follow?

Problem 12.17. On a certain planet, each inhabitant was classified as either *good* or *evil*. A statistician from our planet visited that planet and came to the correct conclusion that for each inhabitant x, x was good if and only if it was the case that all the good inhabitants had green hair. Which of the following three statements logically follow?

(1) All of them are good.

(2) None of them are good.

(3) Some of them are good and some are not.

Also, which of the following three statements follow?

(4) All of them have green hair.

(5) None of them have green hair.

(6) Some of them have green hair and some do not.

PROBLEM 12.18. On another planet, again each inhabitant is classified as either *good* or *evil*. It turns out that for each inhabitant x, x is good if and only if there is at least one evil inhabitant who has green hair. Which of (1)–(6) above logically follows?

PROBLEM 12.19. On a certain island there is a barber named Jim who shaves all those inhabitants who don't shave themselves. Does Jim shave himself or doesn't he?

PROBLEM 12.20. On another island there is a barber named Bill who shaves *only* those who don't shave themselves. (In other words, he never shaves an inhabitant who shaves himself.) Does Bill shave himself or doesn't he?

PROBLEM 12.21. What would you say about an island on which there is a barber who shaves *all those* and *only those* inhabitants who don't shave themselves? (In other words, if an inhabitant shaves himself, the barber won't shave him, but if the inhabitant doesn't shave himself, then the barber shaves him.) Does this barber shave himself or doesn't he? What would you say about such a barber?

NOTE. Be sure to read the important discussion following the solution!

PROBLEM 12.22 (VALID AND SOUND SYLLOGISMS). A syllogism is an argument consisting of a major premise, a minor premise and a conclusion. A syllogism is called *valid* if the conclusion really does follow from the premises, regardless of whether the premises themselves are true. A syllogism is called *sound* if it is valid and if the premises themselves are true. For example, the following syllogism is sound:

All men are mortal. (major premise)
Socrates is a man. (minor premise)
∴ Socrates is mortal. (conclusion)

(The symbol "∴" abbreviates "therefore.")

The following syllogism, though obviously not sound, *is* valid!

All men have green hair.
Socrates is a man.
∴ Socrates has green hair.

This syllogism is unsound, because the major premise ("all men have green hair") is simply not true. But the syllogism is valid, because if it were really the case that all men had green hair, then Socrates, being a man, would have to have green hair. Both the above syllogisms are special cases of the following general syllogism:

All A's are B's.
x is an A.
$\therefore x$ is a B.

Is the following syllogism valid?

Everyone loves my baby.
My baby loves only me.
∴ I am my own baby.

In the delightfully humorous book *The Devil's Dictionary* by Ambrose Bierce, he gives this example of a syllogism in the following definition of *logic*:

> **Logic**, n. The art of thinking and reasoning in strict accordance with the limitations and incapacities of the human misunderstanding. The basis of logic is the syllogism, consisting of a major and minor premise and a conclusion—thus:
>
> *Major Premise*: Sixty men can do a piece of work sixty times as quickly as one man.
> *Minor Premise*: One man can dig a posthole in sixty seconds; therefore
> *Conclusion*: Sixty men can dig a posthole in one second.

Solutions

12.1. Since they all said the same thing, they really are all of the same type, so what they said was true. Thus they are all knights.

12.2. Since they all said the same thing, then it is not possible that some are knights and some are knaves, hence they all lied. Thus they are all knaves.

12.3. All the inhabitants interviewed on the first day said the same thing, so they are all of the same type. They are obviously not knights (no knight would say that all the inhabitants (which includes himself) are knaves), and so they are all knaves. Therefore their statements were all false, and so the sleeping native cannot also be a knave. Since he is a knight, he obviously answered *no*.

12.4. Again, all the inhabitants are of the same type (since they all said the same thing). Suppose they are all knaves. Then their statements are false: It is false that all knights on the island smoke. But the only way it can be false is if there is at least one knight on the island who doesn't smoke, but that is not possible by our assumption that all of them are knaves! So our assumption that they are all knaves leads to a contradiction, and hence they are all knights. It then further follows that all the inhabitants are knights and all of them smoke.

12.5. Again, all the natives must be of the same type. If they were knights, they certainly wouldn't say that some knaves on the island smoke (since this would imply that some of them are knaves). Thus they are all knaves, and since their statements are therefore false, it follows that none of them smoke.

12.6. We are given that they are all of the same type. Now, consider any one of the natives. He says that if he smokes, then all of them smoke. The only way that it could be false is if he smokes but not all of them smoke. But since they *all* said that, the only way the statements could be false is if each one of the inhabitants smokes, yet not all of them smoke, which is obviously absurd. Therefore the statements cannot be false, and so the inhabitants are all knights. Since their statements are all true, there are two possibilities: (1) None of them smoke (in which case their statements are all true, since a false proposition implies any proposition); (2) All of them smoke. And so all the inhabitants are knights, and all that can be deduced about their smoking is that either none of them smoke or all of them smoke, but there is no way to tell which.

12.7. Again, all the inhabitants are of the same type. Each inhabitant claims that if any inhabitant smokes, then he does; and if that were false, then some inhabitant smokes, but the given inhabitant doesn't. Since they *all* say that, it follows that if the statements were false, we would have the contradiction that some inhabitant smokes, but each one doesn't. Thus the natives are all knights, and again either all of them smoke or none of them smoke, and again there is no way to tell which.

12.8. Again, they are all of the same type. They couldn't be knights, because if their statements were true, then some of them smoke, yet each one doesn't, which is absurd. Hence they are all knaves. Since their statements are false, it follows that for each inhabitant x, either it is false that some inhabitants smoke, or it is false that x doesn't smoke—in other words, either none of them smoke or x smokes. It could be that none of them smoke. If that alternative doesn't hold, then for each inhabitant x, x smokes, which means that all of them smoke. So the solution is that all of them are knaves, and either all of them smoke or none of them smoke, and there is no way to tell which.

12.9. What you *should* conclude is that the situation is impossible!

Reason. If the situation occurred, then for all the same reasons as in the last problem, all the inhabitants would have to be knaves. But this time, for each inhabitant x, *both* of his statements are false, which means that nobody smokes, yet x smokes, which is absurd!

(This is another interesting case where a knave can assert the conjunction of two statements, but cannot assert them separately.)

12.10. All members of Tribe A are of the same type, and all members of Tribe B are of the same type. Since the members of Tribe B have contradicted the members of Tribe A, it cannot be that the members of both tribes are knights, so the members of Tribe A made false statements, and are therefore knaves. It then further follows that the members of Tribe B are knights, because they correctly said that some of the inhabitants of the island are knaves (and indeed, the members of Tribe A *are* knaves). Then their second statements were also true, hence no one on this island smokes. Thus Tribe A consists of knaves, Tribe B consists of knights, and no one on the island smokes.

12.11. Since all the inhabitants said that all the inhabitants are knights, it is impossible that they all are, because if they were, the female knights wouldn't have made the true statement that they are. Therefore all inhabitants of this island lie, and thus the males are all knaves and the females are all knights.

12.12. Obviously all the men are knaves (male knights would never say that all the inhabitants are knaves), and their statements are therefore false, which implies that at least one of the women is a knight. And since all the women said the same thing, it follows that all the women are knights. Hence they all falsely answered *yes*.

12.13. If the men were knights, then all the inhabitants would be knights and all of them would smoke. But then the female knights would never have made the true statement that they all smoke. Hence the men are all knaves. It further follows that the women are all knaves (because they truthfully said that some of the inhabitants are knaves). Since the women are truthful, all the inhabitants do smoke (as the women said). So all the inhabitants of the island are knaves and they all smoke.

12.14. If either everyone there smokes cigarettes or everyone there smokes cigars, then certainly everyone there smokes either cigarettes or cigars. Hence if either Arthur or Bernard is a knight, then so is

Charles. Thus if Arthur and Bernard are not both knaves, then Charles is a knight. Hence if David is a knight, so is Charles (because if David is a knight, then Arthur and Bernard are not both knaves). Thus Frank's statement is true, so Frank is definitely a knight.

Edward's type cannot be determined: It could be that either everyone smokes cigarettes or everyone smokes cigars, in which case it would follow that Edward is a knight (why?), or it could be that everyone smokes either cigarettes or cigars, yet some don't smoke cigarettes and some don't smoke cigars, in which case it would follow that Edward is a knave (why?). Frank is the only one whose type can be determined.

12.15. Let p be the proposition that either everyone is special or no one is special. Also, for each person x, let us abbreviate the statement that x is special by Sx. (In general, in symbolic logic, given any property P and any individual object x, the proposition that x has the property P is abbreviated Px.) Recall that two propositions are called *equivalent* if they are either both true or both false. Well, we are given that for each person x, the proposition Sx is equivalent to p (x is special if and only if p is true—i.e., if and only if either all or none are special). Then for any two people x and y, the propositions Sx and Sy must be equivalent to each other, since both of them are equivalent to p. This means that for any two people, either they are both special or neither one is special, from which it follows that either *all* the people are special, or none of them are special—in other words, the proposition p is true! Then, since for each person x, the proposition Sx is equivalent to p, it follows that for each person x, the proposition Sx is true—in other words, everyone is special!

12.16. Again, for each person x, let Sx be the proposition that x is special. Now let q be the proposition that some people are special and some are not (which is the *opposite* of the proposition that either all or none are special). In the present version, Sx is equivalent to q, for each person x, and so, as in the solution to the last problem, either all are special or none are special (because for every two persons x and y, the propositions Sx and Sy are equivalent, each being equivalent to q). Thus q is false, and since for any person x, the proposition Sx is equivalent to q, Sx must be false. So, according to this version, no one is special.

12.17. For each inhabitant x, let Gx be the proposition that x is good. Let p be the proposition that all the good inhabitants have green hair.

For each inhabitant x, we are given that Gx is equivalent to p, so by the same reasoning as in the last two problems, either all the inhabitants are good, or none of them are good. Suppose that none of them are good. Then for each inhabitant x, the proposition Gx is false, and since Gx is equivalent to p, it follows that p must be false. But the only way that p can be false—the only way that it can be false that all the good inhabitants have green hair—is that there is at least one good inhabitant who doesn't have green hair, which of course implies that at least one inhabitant is good, which is impossible under our assumption that none of the inhabitants are good. Thus the supposition that none of the inhabitants are good leads to a contradiction, and hence must be false. Therefore it is not the case that none of the inhabitants are good, yet we have seen that either all or none are good, hence it must be that all are good. Furthermore, since Gx is true for each inhabitant x, and Gx is equivalent to p, p must be true, which means that every good inhabitant has green hair, and since each inhabitant is good, the conclusion is that all the inhabitants are good and all have green hair.

12.18. Again, let Gx be the proposition that x is good. Let q be the proposition that there is at least one evil (not good) inhabitant who has green hair. We are now given that Gx is equivalent to q, for each inhabitant x. Again, it follows that either all the inhabitants are good or none of them are. Suppose that all of them are good. Then for any inhabitant x, the proposition Gx holds, and since Gx is equivalent to q, q must be true—it must be true that at least one evil inhabitant has green hair—and hence there must be at least one evil inhabitant, contrary to the supposition that all the inhabitants are good. Hence the supposition must be false, so none of the inhabitants are good. Furthermore, since Gx is false for each inhabitant x, and Gx is equivalent to q, q must be false. Thus it is not true that at least one evil inhabitant has green hair, yet all the inhabitants are evil, hence none of them have green hair. The conclusion is that all the inhabitants are evil and none of them have green hair.

Remark. As will be shown in the next chapter, the last four problems above are really Problems 10.1, 10.2, 10.4, and 10.5 over again, but in a new dress.

12.19. If Jim didn't shave himself, he would be one of those who don't shave themselves, but Jim must shave all such inhabitants, hence he would have to shave himself, which is a contradiction. Therefore he must shave himself. (Stated otherwise, if he didn't shave himself, he would fail to shave someone who doesn't shave himself, which goes against the given condition.)

12.20. We are given that Bill never shaves anyone who shaves himself. Therefore if Bill shaved himself, he *would* be shaving someone who shaved himself, which is contrary to what is given. Therefore Bill cannot shave himself.

12.21. What you *should* say is that the situation is impossible; there *cannot* be such a barber!

REASON. If there were such a barber, then since he shaves all those who don't shave themselves, he must shave himself (Problem 12.19), but on the other hand, since he shaves *only* those who don't shave themselves, he cannot shave himself (Problem 12.20). Thus there can be a barber who shaves *all* those who don't shave themselves, and there can be *another* barber who shaves *only* those who don't shave themselves, but there cannot be a barber who does both.

DISCUSSION. This "Barber Paradox" is a popularization by Bertrand Russell of a famous paradox, also by Russell, that had grave significance for the entire foundation of mathematics. Toward the end of the 19th Century, there appeared a monumental work by Gottlob Frege on set theory that attempted to derive all mathematics from a few basic principles of logic. The main axiom of Frege's system was that, given any property, there existed the set of all things having that property. This seems intuitively evident enough, but Russell showed that it conceals a logical contradiction! Here is Russell's paradox:

Call a set *extraordinary* if it is a member of itself, and *ordinary* if it is not a member of itself. For example, a set of chairs is not itself a chair, thus it is not a member of itself, so is *ordinary*. On the other hand, the set of all entities—if there were such a set—would itself be an entity, hence would be one of its own members, and so would be an *extraordinary* set. Whether there really exist extraordinary sets is open to question, but there certainly exist ordinary ones—indeed, all sets commonly encountered are ordinary. Well, by Frege's principle there exists the set—call it $\mathcal{O}$—of all ordinary sets. Thus $\mathcal{O}$ contains all and only those sets that do not contain themselves (just like the barber who shaves all and only those who do not shave themselves). Is the set $\mathcal{O}$ ordinary or extraordinary? Suppose $\mathcal{O}$ is ordinary. Then, since *all* ordinary sets are members of $\mathcal{O}$, $\mathcal{O}$ must be a member of $\mathcal{O}$, which makes $\mathcal{O}$ extraordinary! This is clearly a contradiction. On the other hand, suppose $\mathcal{O}$ is extraordinary. Thus $\mathcal{O}$ is a member of itself, but since only ordinary sets are members of

$\mathcal{O}$, $\mathcal{O}$ must be one of the ordinary sets, which is again a contradiction! Thus there cannot exist a set $\mathcal{O}$ containing all and only those sets that don't contain themselves (just as there cannot be a barber who shaves all and only those people who don't shave themselves).

Frege's principle is thus seen to lead to a contradiction, hence has to be modified. This has subsequently been done in several ways, which constitutes the field known as *set theory*.

12.22. Funny as it may seem at first, the argument *is* valid! Since *everyone* loves my baby, it follows that my baby, being a person, loves my baby. Thus my baby loves my baby, but also loves *only* me. It then follows that my baby is the same person as me!

Of course the argument, though valid, cannot be sound; the premises can't both be true, since they lead to the absurd conclusion that I am my own baby.

- Chapter 13 -

Beginning First-Order Logic

Now we shall approach first-order logic more systematically.

Introducing $\forall$ and $\exists$

In first-order logic we use letters x, y, z, with or without subscripts, to stand for arbitrary objects of the domain under discussion. What the domain is depends on the application in question. For example, if we are doing algebra, the letters x, y, z stand for arbitrary *numbers*. If we are doing geometry, the letters x, y, z might stand for *points* in a plane. If we are doing sociology, the objects in question might be *people*. First-order logic is extremely general and is thus applicable to a wide variety of disciplines.

Given any property P and any object x, the proposition that x has the property P is neatly symbolized Px. Now, suppose we wish to say that *every* object x has property P; how do we render this symbolically? Well, here is where we introduce the symbol $\forall$—called the *universal quantifier*—which stands for "all," or "every." Thus $\forall x$ is read "for all x" or "for every x," and so the English sentence "Everything has the property P" is symbolically rendered $\forall x Px$ (read "For every x, Px").

Although in ordinary English, the word "some" *tends* to have a plural connotation, in logic it means only "at least one"; it does not mean "two or more," only "one or more." This is important to remember! Now, in first-order logic, the phrase "there exists at least one" is symbolized by the single symbol $\exists$, called the *existential quantifier*.Then the proposition that at least one object x has the property P is symbolized $\exists x Px$. So first-order logic uses the logical connectives $\sim$, $\wedge$, $\vee$, $\Rightarrow$ of propositional logic and the quantifiers $\forall$ and $\exists$.

Suppose we now use the letters x, y, z to stand for unspecified *people*. Let Gx stand for "x is *good*." Then $\forall x Gx$ says "Everyone is good" and $\exists x Gx$ says "Some people are good" or "At least one person is good" or "There exists a good person." Now, how do we symbolically say that no one is good? One way to do this is $\sim\exists x Gx$ (it is *not* the case that there exists a good person). An alternative way is $\forall x(\sim Gx)$ (for every person x, x is not good).

Let us now abbreviate "x goes to heaven" by Hx. How do we symbolize "All good people go to heaven?" Well, this can be re-stated: "For every person x, if x is good, then x goes to heaven," and hence is symbolized $\forall x(Gx \Rightarrow Hx)$ (for all x, x is good implies x goes to heaven). What about "*Only* good people go to heaven"? One way to symbolize this is $\forall x(\sim Gx \Rightarrow \sim Hx)$ (for all people x, if x is not good, then x doesn't go to heaven). An equivalent rendition is $\forall x(Hx \Rightarrow Gx)$ (for any person x, if x goes to heaven, then x must be good). What about "Some good people go to heaven"? Remember now, this means only that *at least one* good person goes to heaven, in other words, there exists a good person x who goes to heaven, or, equivalently, there exists a person x who is both good and goes to heaven, so the symbolic rendition is simply $\exists x(Gx \wedge Hx)$. What about "No good person goes to heaven"? This is simply $\sim\exists x(Gx \wedge Hx)$. What about "No good person fails to go to heaven"? This is then $\sim\exists x(Gx \wedge \sim Hx)$ (there is no person who is good and fails to go to heaven). This is only a roundabout way of saying, however, that all good people go to heaven, so $\sim\exists x(Gx \wedge \sim Hx)$ is equivalent to $\forall x(Gx \Rightarrow Hx)$.

Now let's consider the old saying "God helps those who help themselves." Actually, there is a good deal of ambiguity here; does it mean that God helps *all* those who help themselves, or that God helps *only* those who help themselves, or does it mean that God helps *all* those and *only* those who help themselves? Well, let us abbreviate "x helps y" by xHy, and let "g" abbreviate "God." Then "God helps *all* those who help themselves" would be symbolized $\forall x(xHx \Rightarrow gHx)$. What about "God helps *only* those who help themselves"? One rendition is $\forall x(gHx \Rightarrow xHx)$ (for all x, God helps x implies that x helps x). Another is $\forall x(\sim xHx \Rightarrow \sim gHx)$ (for all x, if x doesn't help x, then God doesn't help x). Another is $\sim\exists x(gHx \wedge \sim xHx)$ (there is no person x such that God helps x and x doesn't help x).

Let us consider some more translations.

Problem 13.1. Give symbolic renditions of the following (using xHy for x helps y, and g for God):

(a) God helps only those who help God.

(b) God helps all and only those who help themselves.

Problem 13.2. Let h stand for Holmes (Sherlock Holmes) and m for Moriarty. Let us abbreviate "x can trap y" by xTy. Give symbolic renditions of the following:

(a) Holmes can trap anyone who can trap Moriarty.

(b) Holmes can trap anyone whom Moriarty can trap.

(c) Holmes can trap anyone who can be trapped by Moriarty.

(d) If anyone can trap Moriarty, then Holmes can.

(e) If everyone can trap Moriarty, then Holmes can.

(f) Anyone who can trap Holmes can trap Moriarty.

(g) No one can trap Holmes unless he can trap Moriarty.

(h) Everyone can trap someone who cannot trap Moriarty.

(i) Anyone who can trap Holmes can trap anyone whom Holmes can trap.

Problem 13.3. Let us abbreviate "x knows y" by xKy. Express the following symbolically:

(a) Everyone knows someone.

(b) Someone knows everyone.

(c) Someone is known by everyone.

(d) Everyone knows someone who doesn't know him.

(e) There is someone who knows everyone who knows him.

It is interesting to note that the words "anyone" and "anybody" sometimes mean *everyone*, and sometimes mean *someone*. For example, the sentence "Anybody can do it" means that *everybody* can do it, but in the sentence "If anybody can do it, then John can do it" (or "John, if anybody, can do it"), the word "anybody" means *somebody*.

PROBLEM 13.4. Let Dx abbreviate "x can do it" and j abbreviate *John*. Symbolically express the sentence "John, if anybody, can do it."

PROBLEM 13.5. Let Dx abbreviate "x can do it," let j abbreviate *John* and let $x = y$ abbreviate "x is identical with y." How would you express the proposition that John is the *only* one who can do it?

Here are some examples from arithmetic. The letters x, y, z will now stand for arbitrary natural numbers instead of people. (The natural numbers are $0, 1, 2, 3, 4, \ldots$, etc.—that is, 0 together with the positive whole numbers.) The symbol $<$ stands for "is less than," so for any numbers x and y, $x<y$ is read: "x is less than y."

PROBLEM 13.6. Using the symbol $<$ and logical connectives and quantifiers, express symbolically the following statements about natural numbers (i.e., where by "number" is meant "natural number"):

(a) x is greater than y.

(b) For every number there is a greater number.

(c) For every number there is a lesser number.

(d) Every number is greater than some number.

Is statement (c) true or false?

INTERDEPENDENCE OF $\forall$ AND $\exists$

It is possible to define $\exists$ from $\forall$ and the logical connectives, and vice versa, as the following two problems indicate.

PROBLEM 13.7. Let Gx stand for the proposition that x is good. The statement $\forall xGx$ says that everyone is good, and $\exists xGx$ says that at least one person is good. Now, suppose you are living in a country where, for some odd reason, it is illegal to use the symbol $\forall$, but you are allowed to use $\exists$. You wish to express the proposition that *everyone* is good. How can you do that using only the quantifier $\exists$ (as well as any of the logical connectives $\sim$, $\wedge$, $\vee$, $\Rightarrow$)?

PROBLEM 13.8. In another country you are allowed to use the symbol $\forall$, but not the symbol $\exists$. How would you then express the proposition that at least one person x is good?

Formulas, Validity, Satisfiability

Formulas

As the reader already knows, in first-order logic we use the logical connectives $\sim$, $\wedge$, $\vee$, $\Rightarrow$ and the quantifiers $\forall$ and $\exists$; and we use small letters x, y, z, w, with or without subscripts, as variables standing for arbitrary objects of the domain under discussion. These variables are called *individual* variables, and the domain under discussion is called the *range* of the variables. For example, if we are doing number theory, we say that the individual variables *range* over the natural numbers.

We also use capital letters to stand for *properties* of individuals; these letters are called 1-place *predicates*. We also have 2-place predicates standing for (2-place) *relations* between individuals, and 3-place predicates for 3-place relations. (Examples of 3-place relations among individuals x, y, z might be, in arithmetic, $x+y=z$, or, in geometry, "point x is *between* points y and z," or if x, y, z stand for *people*, we might have a 3-place relation such as "x *introduced* y to z.") More generally, for each positive integer n, we have letters called n-place predicates standing for n-place relations—though, for our purposes, we will not need to go beyond 3-place predicates.

By an *atomic formula* we mean either an expression Px, where x is an individual variable and P is a 1-place predicate, or an expression Rxy (also written xRy) where R is a 2-place predicate (and x and y are individual variables), or, in general, any n-place predicate followed by n occurrences of individual variables. Starting with the atomic formulas, one builds the entire class of *first-order formulas* by the following rules: Given formulas F and G already constructed, the expressions $\sim F$, $(F \wedge G)$, $(F \vee G)$, $(F \Rightarrow G)$ are again formulas (as in propositional logic), and $\forall x G$, $\exists x G$, where x is any individual variable, are also formulas. Thus we have the following rules:

(1) Any atomic formula is a formula.

(2) For any formulas F and G, the expression $\sim F$, $(F \wedge G)$, $(F \vee G)$, $(F \Rightarrow G)$ are again formulas.

(3) For any formula F and any individual variable x, the expressions $\forall x F$ and $\exists x F$ are again formulas.

The formula $\forall x F$ is called the *universal quantification* of F with respect to x, and $\exists x F$ is called the *existential quantification* of F with respect to x.

As in propositional logic, in displaying a formula, we can drop superfluous parentheses if no ambiguity results.

Free and Bound Occurrences

Before defining these important notions, we look at some examples:

In the arithmetic of the natural numbers, consider the equation

$$x = 5y. \tag{13.1}$$

This equation, as it now stands, is neither true nor false, but becomes so when we assign values to the variables x and y. (For example, if we take x to be 10 and y to be 2, we have a truth, but if we take x to be 12 and y to be 7, we have an obvious falsehood.) The important thing to notice now is that the truth or falsity of (13.1) depends both on a choice of value for x and a choice for y. This reflects the purely formal fact that x and y occur *freely* in (13.1).

Now consider

$$\exists y(x = 5y). \tag{13.2}$$

The truth or falsity of (13.2) depends on x, but not on any choice for y. Indeed, we could restate (13.2) in a form in which the variable y does not even occur—namely, "x is divisible by 5." And this reflects the fact that x has a free occurrence in (13.2) but y does not—y is *bound* in (13.2).

Suppose x has an occurrence in a formula F. As soon as one puts $\forall x$ or $\exists x$ in front of F, all free occurrences of x become bound—that is, all occurrences of x in $\forall xF$ are bound, and likewise with $\exists xF$. Here are the precise rules determining freedom and bondage:

(1) In an atomic formula, all occurrences of variables are free.

(2) The free occurrences of a variable in $\sim F$ are the same as those in F. The free occurrences of a variable in $F \wedge G$ are those of F and those of G. Likewise with $\vee$ or $\Rightarrow$ instead of $\wedge$.

(3) All occurrences of a variable x in $\forall xF$ are *bound* (not free), but for any variable y distinct from x, the free occurrences of y in $\forall xF$ are those in F itself. Similarly with $\exists$ in place of $\forall$.

A formula is called *closed* if no variable is free in it; otherwise it is called *open*.

Interpretations

An *interpretation* of a formula is given by first specifying the domain of individuals, then assigning to each predicate a property of, or a relation among, the individuals, and, lastly, if there are free variables in the formula, assigning an individual to each of the free variables. Once an interpretation is given, the formula then becomes either true or false under the interpretation.

Let us consider some examples: Take the formula $\exists xRxy$. In this formula, x is bound and y is free. For one interpretation, let the domain of individuals be the set of all people who have ever lived in this world. Define Rxy to be the relation "x is the parent of y." If we take for y the person Abraham Lincoln, then the formula is certainly true under that interpretation (Lincoln certainly had a parent), but if we take y to be Adam or Eve, the sentence is then probably false.

Suppose we alternatively define Rxy to mean that x is *married* to y. Then, $\exists xRxy$ simply says than that y is married, which is true for some y's and false for others. Thus $\exists y\exists xRxy$ is true under this interpretation, but $\forall y\exists xRxy$ is false (it is false that everyone is married).

Let us now consider another interpretation of this same formula—an interpretation in the domain of natural numbers. A number x is said to *properly* divide y if x divides y but $x \neq y$ and $x \neq 1$. Now let us interpret Rxy to mean that x properly divides y. Then $\exists xRxy$ is true for some y's and false for others (for example, it is true for $y=12$ but false for $y=7$). A number is called *prime* if no number properly divides it and *composite* otherwise. Under our present interpretation, $\exists xRxy$ simply says that y is composite (and thus the formula $\sim\exists xRxy$ says that y is prime).

Validity and Satisfiability

A formula is called *valid in a given domain* if it is true under *every* interpretation in that domain, and is called *logically valid*—or just *valid*, for short—if it is valid in all domains that contain at least one element. Thus a valid formula is one that is true under all possible interpretations in all possible domains, except perhaps a domain that contains no elements.

A word about sets that contain no elements—so-called empty sets. This notion may seem strange at first, but a good example of an empty set is the set of all people in a theater after everyone has left. Another example is the set of all even prime numbers greater than 2. There simply are no such numbers, because any even number greater than 2 is properly divisible by 2, so the set of all even primes greater than 2 is empty.

Thus a formula is called *valid* if it is valid in all *non-empty* domains. Now, some curious things happen with interpretations in an *empty* domain! We recall that if a certain club contains no Frenchmen, then the statement that *all* Frenchmen in the club wear hats is true. *Anything* one says about *all* Frenchmen in the club is true if there are no Frenchmen in the club. In general, whatever one says about *all* members of an empty domain is true—the formula $\forall xPx$ is true for an empty domain, regardless of what the property P is. The formula $\forall xPx$ is thus *valid in an empty domain*—though it is of course not valid in any non-empty domain, and hence is not valid. Thus $\forall xPx$ is an example of an invalid formula that is nevertheless valid in an empty domain.

A formula is said to be *satisfiable in a given domain* if it is true under at least one interpretation in the domain, and it is called *satisfiable*, period, if it is satisfiable in at least one *non-empty* domain. Let us note that to say that a formula F is not valid is equivalent to saying that its negation $\sim F$ is satisfiable, and that to say that F is valid is to say that $\sim F$ is not satisfiable. Also F is satisfiable if and only if $\sim F$ is not valid, and F is unsatisfiable if and only if $\sim F$ is valid.

We have seen that the formula $\forall xPx$ is valid in an empty domain but not valid in any non-empty domain. Is there a formula that is valid in all non-empty domains but not valid in an empty domain? Yes, there is. Better yet, there is a formula that is valid in all non-empty domains but is not only not valid in an empty domain, but not even satisfiable in an empty domain.

Problem 13.9. Find such a formula.

If the reader is in the mood for trying a really difficult one, here it is:

Problem 13.10. Find a formula that is not satisfiable in any *finite* non-empty domain, but is satisfiable in some infinite domains.

Tautologies

A formula X of first-order logic is said to be an *instance* of a formula Y of propositional logic if X is obtainable from Y by substituting formulas of first-order logic for the propositional variables in Y. For example, $\exists xQx \vee \sim\forall yPy$ is an instance of $p \vee \sim q$ (it is obtainable by substituting $\exists xQx$ for p and $\forall yPy$ for q). Now, X is called a *tautology* if it is an instance of a tautology of propositional logic. For example, $\forall xQx \vee \sim\forall xQx$ is a tautology, because it is an instance of the propositional tautology $p \vee \sim p$. Even if one did not know what the symbol $\forall$ meant, one would know that *whatever* it meant, $\forall xQx \vee \sim\forall xQx$ must be true, because for *any* proposition, either it or its negation must be true. Thus the truth of $\forall xQx \vee \sim\forall xQx$ is obtainable just from propositional logic; it can be shown by a truth table, taking $\forall xQx$ as a unit. Another tautology is $(\forall xPx \wedge \forall xQx) \Rightarrow \forall xPx$—it is of the form (an instance of) $(p \wedge q) \Rightarrow p$. Now, the formula $(\forall xPx \wedge \forall xQx) \Rightarrow \forall x(Px \wedge Qx)$, though valid, is *not* a tautology! It is valid, because if *every* x has property P and *every* x has property Q, then of course every x has both properties P and Q. But it is not an instance of any tautology of propositional logic. Also, to realize the validity of the formula, one must know what the symbol $\forall$ means—for example, if one reinterpreted $\forall$ to mean "there exists," instead of "for all," the formula wouldn't always be true (if some element has the property P and some element has the property Q, it doesn't necessarily follow that some single element has both properties P and Q). All tautologies are of course valid, but they constitute only a fragment of the class of valid formulas! It is the valid formulas that we are now interested in, and we seek a general proof procedure for valid formulas. Such a procedure is the method of *tableaux*, to which we now turn.

Tableaux for First-Order Logic

Tableaux for formulas of first-order logic use the eight rules of propositional logic together with four new rules for the quantifiers, which we will shortly present. First for some examples.

Suppose we wish to prove the formula $\exists xPx \Rightarrow \sim\forall x \sim(Px \vee Qx)$. As with propositional logic, we start the tableaux with F followed by the formula we wish to prove.

$$(1)\quad F\exists xPx \Rightarrow \sim\forall x \sim(Px \vee Qx)$$

Next, we use rules from propositional logic to extend the tableaux thus. (Remember, we put to the right of each line the number of the line from which it was inferred.)

$$\begin{array}{lll} (2) & T\exists xPx & (1) \\ (3) & F\sim\forall x\sim(Px\vee Qx) & (1) \\ (4) & T\forall x\sim(Px\vee Qx) & (3) \end{array}$$

There is no rule from propositional logic that is now applicable, so we start working on the quantifiers. By (2), there is at least one x such that Px; let a be the name of any such x. So we add

$$\begin{array}{lll} (5) & TPa & (2) \end{array}$$

Next we look at line (4), which says that whatever x we take, it is not the case that Px or Qx; in particular, it is not the case that Pa or Qa, so we add:

$$\begin{array}{lll} (6) & T\sim(Pa\vee Qa) & (4) \end{array}$$

At this point, (5) and (6) constitute a clear inconsistency in propositional logic, and we can stop right here, or alternatively close the tableau using only tableau rules of propositional logic thus:

$$\begin{array}{lll} (7) & FPa\vee Qa & (6) \\ (8) & \underline{FPa} & (7) \end{array}$$

(Line 8 clashes with line 5.)

In the future, to save unnecessary work, whenever we obtain a branch that contains an inconsistency in propositional logic, we can put a bar under it and treat it as closed—since we know it *can be* closed (in fact, closed using only tableau rules of propositional logic).

Let us consider another example: We wish to prove the formula

$$(\forall x Px \wedge \exists x(Px \Rightarrow Qx)) \Rightarrow \exists x Qx.$$

Here is a tableau that does this (explanations follow):

(1) $F(\forall x Px \wedge \exists x(Px \Rightarrow Qx)) \Rightarrow \exists x Qx$	
(2) $T\forall x Px \wedge \exists x(Px \Rightarrow Qx)$	(1)
(3) $F\exists x Qx$	(1)
(4) $T\forall x Px$	(2)
(5) $T\exists x(Px \Rightarrow Qx)$	(2)
(6) $TPa \Rightarrow Qa$	(5)
(7) TPa	(4)
(8) FQa	(3)

(9) $\underline{FPa}$ (6) (10) $\underline{TQa}$ (6)

EXPLANATIONS. Lines (2), (3), (4), (5) were obtained only by tableau rules of propositional logic. Now, line (5) says that $Px \Rightarrow Qx$ holds for at least one x, and so we let a be such an x, and thus get line (6). Line (4) says that Px holds for every x, and so in particular, Pa holds, which gives us line (7). Line (3) says that it is false that there is any x such that Qx, so, in particular, Qa is false, which gives us line (8). Then, of course, line (6) branches to (9) and (10), and the tableau then closes.

TABLEAU RULES FOR FIRST-ORDER LOGIC

It is technically convenient to have a group of letters called *parameters*, which we use to substitute for free occurrences of variables in formulas. We shall use the letters a, b, c, with or without subscripts, as parameters. For any formula G and any variable x and any parameter a, by $G_x(a)$ we shall mean the result of substituting a for every *free* occurrence of x in G. Thus:

(1) If G is an *atomic* formula, then $G_x(a)$ is the result of substituting a for every occurrence of x in G. (Of course, if x does not occur in G, then $G_x(a)$ is G itself.)

(2) For any formula, the formula $(\sim G)_x(a)$ is $\sim G_x(a)$—that is, to substitute a for all free occurrences of x in $\sim G$, we first substitute a for all free occurrences of x in G, thus obtaining $G_x(a)$, and then take its negation $\sim G_x(a)$.

(3) For any formulas G and H, the formula $(G \wedge H)_x(a)$ is $G_x(a) \wedge H_x(a)$. (That is, to substitute a for all free occurrences of x in $G \wedge H$, we first substitute a for all free occurrences of x in G, obtaining $G_x(a)$, then do the same thing with H, obtaining $H_x(a)$, and then take the conjunction $G_x(a) \wedge H_x(a)$.) Similarly,

$$(G \vee H)_x(a) = G_x(a) \vee H_x(a),$$
$$(G \Rightarrow H)_x(a) = G_x(a) \Rightarrow H_x(a),$$
$$(G \equiv H)_x(a) = G_x(a) \equiv H_x(a).$$

(4) (a) For any variable y distinct from x, the formula $(\forall y G)_x(a)$ is $\forall y G_x(a)$. (That is, the result of substituting a for all free occurrences of x in $\forall y G$ consists of the universal quantification $\forall y$ followed by the result $G_x(a)$ of substituting a for all free occurrences of x in G.)

(b) But for the variable x itself, $(\forall x G)_x(a)$ is simply $\forall x G$, since x has no free occurrence in $\forall x G$.

(5) Similarly, $(\exists y G)_x(a) = \exists y G_x(a)$ if $y \neq x$, but $(\exists x G)_x(a)$ is simply $\exists x G$.

A common, and useful, notational convention is to let $\phi(x)$ be a formula in which the variable x is the one to be acted on; then take $\phi(a)$ to be the result of substituting a for all free occurrences of x. (Thus if G is the formula $\phi(x)$, then $\phi(a)$ is $G_x(a)$.)

THE TABLEAU RULES FOR THE QUANTIFIERS

We have four quantifier rules—one for each of $T\forall x\phi(x)$, $F\forall x\phi(x)$, $T\exists x\phi(x)$, $F\exists x\phi(x)$. None of the rules are branching.

RULE $T\forall$. From $T\forall x\phi(x)$, we may directly infer $T\phi(a)$, where a is any parameter.

RULE $T\exists$. This is more delicate! From $T\exists x\phi(x)$ we may directly infer $T\phi(a)$, *provided a is a parameter that has not been used before*!

Here is the reason for the proviso: Suppose in the course of a proof we show that there is some x having property P; we then can say: "Let a be such an x." Now, suppose we later show that there is some x having property Q; we cannot legitimately say: "Let a be such an x," because we have already committed the symbol a to being the name of some x having property P, and we do not know that there is some x having *both* property P *and* property Q! So we must take a new symbol b, and say: "Let b be an x having property Q."

RULE $F\forall$. This is similar to Rule $T\exists$: From $F\forall x\phi(x)$ we may directly infer $F\phi(a)$, again *provided that a is new*!

Here's why: $F\forall x\phi(x)$ says that it is false that $\phi(x)$ holds for every x, which is equivalent to saying that there is at least one x such that $\phi(x)$ is false; we let a be such an x, and so write $F\phi(a)$. Again, a must be a new parameter for the same reasons that we explained in Rule $T\exists$.

(We might remark that instead of Rule $F\forall$, we could just as well have taken as a rule the following: "From $F\forall x\phi(x)$ we can directly infer $T\exists x{\sim}\phi(x)$." But then it would take three steps to get $F\phi(a)$ from $F\forall x\phi(x)$—namely $F\forall x\phi(x)$, $T\exists x{\sim}\phi(x)$, $T{\sim}\phi(a)$ (by Rule $T\exists$), and then $F\phi(a)$.)

RULE $F\exists$. From $F\exists x\phi(x)$ we may infer $F\phi(a)$ for *any* parameter a (no restriction necessary).

You see, $F\exists x\phi(x)$ says that it is false that there is *any* x such that $\phi(x)$—in other words, $\phi(x)$ is false for *every* x, and so in particular, $\phi(a)$ is false, and thus $F\phi(a)$. (Again, this rule could be replaced by the rule "From $F\exists x\phi(x)$ we can infer $T\forall x{\sim}\phi(x)$.")

Let us now review the four rules in schematic form:

RULE $T\forall$. $$\frac{T\forall x\phi(x)}{T\phi(a)}$$ (a is any parameter)

RULE $F\exists$. $$\frac{F\exists x\phi(x)}{F\phi(a)}$$ (a is any parameter)

RULE $T\exists$. $$\frac{T\exists x\phi(x)}{T\phi(a)}$$ (a must be new)

RULE $F\forall$. $$\frac{F\forall x\phi(x)}{F\phi(a)}$$ (a must be new)

Rules $T\forall$ and $F\exists$ are collectively called *universal* rules ($F\exists x\phi(x)$ asserts the *universal* fact that for *every* element a, it is false that $\phi(a)$). The rules $T\exists$ and $F\forall$ are called *existential* rules ($F\forall x\phi(x)$ asserts the *existential* fact that there exists at least one element a such that $\phi(a)$ is false.

For *unsigned* formulas, the quantificational tableau rules are

RULE $\forall$. $$\frac{\forall x\phi(x)}{\phi(a)}$$ (a is any parameter)

RULE $\sim\exists$. $$\frac{\sim\exists\phi(x)}{\sim\phi(a)}$$ (a is any parameter)

RULE $\exists$. $$\frac{\exists x\phi(x)}{\phi(a)}$$ (a must be new)

RULE $\sim\forall$. $$\frac{\sim\forall\phi(x)}{\sim\phi(a)}$$ (a must be new)

UNIFIED NOTATION

We recall the unifying α, β notation for propositional logic. It will save us considerable circumlocution to extend this now to first-order logic.

We continue to use α and β the same way as we did in propositional logic, except that "formula" is now construed as a *closed* formula of first-order logic. We now add two more categories, γ and δ, as follows.

γ shall be any formula of the universal type—$T\forall x\phi(x)$ or $F\exists x\phi(x)$—and, for any parameter a, by $\gamma(a)$ we shall mean $T\phi(a)$, $F\phi(a)$, respectively. We refer to any such formula $\gamma(a)$ as a *component* of γ. δ shall be any formula of the *existential* type—either $T\exists x\phi(x)$ or $F\forall x\phi(x)$—and by $\delta(a)$ we shall respectively mean $T\phi(a)$, $F\phi(a)$. We refer to any such formula $\delta(a)$ as a *component* of δ. Our universal rules $T\forall$ and $F\exists$ are now both subsumed under Rule C below, and our existential rules are both subsumed under Rule D below:

RULE C. $\dfrac{\gamma}{\gamma(a)}$ (a is any parameter)

RULE D. $\dfrac{\delta}{\delta(a)}$ (provided a is new)

And, to review the propositional rules,

RULE A. $\dfrac{\alpha}{\begin{array}{c}\alpha_1\\ \alpha_2\end{array}}$

RULE B. $\begin{array}{c}\beta\\ \beta_1 \quad \beta_2\end{array}$

Thus, using our unifying notation, our 12 tableau rules for first-order logic are collapsed into four. This completes the list of tableau rules for first-order logic. We note that Rule B is the only branching rule; all the quantificational rules for first-order logic are direct.

For *unsigned* formulas, we let γ be any formula of the form $\forall x\phi(x)$ or $\sim\exists x\phi(x)$, and by $\gamma(a)$ we respectively mean $\phi(a)$, $\sim\phi(a)$. We let δ be any formula of the form $\exists x\phi(x)$ or $\sim\forall x\phi(x)$, and by $\delta(a)$ we respectively mean $\phi(a)$, $\sim\phi(a)$.

Let us now try another tableau—let us prove the formula $\forall x\forall y(Px\Rightarrow Py)\Rightarrow(\forall xPx\vee\forall x\sim Px)$:

(1) $F\forall x\forall y(Px\Rightarrow Py)\Rightarrow(\forall xPx\vee\forall x\sim Px)$
(2) $T\forall x\forall y(Px\Rightarrow Py)$ (1)
(3) $F\forall xPx\vee\forall x\sim Px$ (1)
(4) $F\forall xPx$ (3)
(5) $F\forall x\sim Px$ (3)
(6) FPa (4)

(7) $F{\sim}Pb$ (5)
(8) TPb (7)
(9) $T\forall y(Pb \Rightarrow Py)$ (2)
(10) $TPb \Rightarrow Pa$ (9)

(11) $\underline{FPb}$ (10) (12) $\underline{TPa}$ (10)

Discussion. In line (7), in accordance with Rule D, I couldn't use the parameter a a second time, so I had to take a new parameter b. Now, in line (9), how did I know that I had best use the parameter b rather than a, or some other parameter? I knew because I informally went through a proof in my mind before I made the tableau, and then made the tableau accordingly. This brings us to some important points.

Tableaux for propositional logic are purely routine things; it makes no essential difference in which order one uses the rules; if one order leads to a closure, then any other order will, also. But with first-order tableaux, the situation is entirely different: For one thing, in constructing a tableau in propositional logic, if one does not repeat any formula, then the tableau must terminate in a finite number of steps; in a tableau for first-order logic, the process can go on indefinitely, because when we use a universal formula γ, we can infer $\gamma(a)$, $\gamma(b)$,... with no limit to the number of parameters we can use. I might remark that an existential formula δ need be used only once; one is *allowed* to use it more than once, but there is never any advantage to doing so; with a γ, we have no idea in advance how many parameters have to be used to effect a closure. Moreover, if one doesn't do things in the right order, the tableau might run on forever without ever closing, even though the tableau could be closed by proceeding differently. Isn't there some systematic procedure that, if followed, will guarantee closure, if closure is possible? Yes, there is, and we will consider one in a later chapter. Following such a procedure is purely mechanical, and it is easy to program a computer to do so. A tableau constructed with intelligence and creative ingenuity, however, often closes much more quickly than one constructed by following the mechanical procedure! (In general, mechanical procedures are unimaginative and incapable of seizing clever strategies.)

In using the universal rule (Rule C), it is best to use parameters that are already on the tree, rather than new ones. Which parameters to use is where ingenuity comes in!

When proving a biconditional $X \equiv Y$, it produces less clutter to make one tableau starting with TX and FY and another starting with FX and TY.

EXERCISE 13.1. By tableaux, prove the following formulas:

(a) $\forall y(\forall xPx \Rightarrow Py)$.

(b) $\forall x(Px \Rightarrow \exists xPx)$.

(c) $\sim\exists yPy \Rightarrow \forall y(\exists xPx \Rightarrow Py)$.

(d) $\exists xPx \Rightarrow \exists yPy$.

(e) $(\forall xPx \wedge \forall xQx) \equiv \forall x(Px \wedge Qx)$.

(f) $(\forall xPx \vee \forall xQx) \Rightarrow \forall x(Px \vee Qx)$.
The converse, $\forall x(Px \vee Qx) \Rightarrow (\forall xPx \vee \forall xQx)$, is not valid. Why?

(g) $\exists x(Px \vee Qx) \equiv (\exists xPx \vee \exists xQx)$.

(h) $\exists x(Px \wedge Qx) \Rightarrow (\exists xPx \wedge \exists xQx)$.
The converse is not valid. Why?

EXERCISE 13.2. In the next group, C is any formula in which x does not occur free (and hence for any parameter a, the formula $C_x(a)$ is C itself).

(a) $\forall x(Px \vee C) \equiv (\forall xPx \vee C)$.

(b) $\exists x(Px \wedge C) \equiv (\exists xPx \wedge C)$.

(c) $\exists xC \equiv C$.

(d) $\forall xC \equiv C$.

(e) $\exists x(C \Rightarrow Px) \equiv (C \Rightarrow \exists xPx)$.

(f) $\exists x(Px \Rightarrow C) \equiv (\forall xPx \Rightarrow C)$.

(g) $\forall x(C \Rightarrow Px) \equiv (C \Rightarrow \forall xPx)$.

(h) $\forall x(Px \Rightarrow C) \equiv (\exists xPx \Rightarrow C)$.

(i) $\forall x(Px \equiv C) \Rightarrow (\forall xPx \vee \forall x \sim Px)$.

THE DRINKING FORMULA

Here is a formula whose validity comes as a surprise to most people. In presenting this formula to a graduate logic class, I prefaced it with the following joke:

> A man at a bar suddenly slammed down his fist and said: "Gimme a drink, and give everyone in the house a drink, because when I drink, *everyone* drinks!" So drinks were happily passed around the house. A few minutes later, the man said: "Gimme another drink, and give everyone else a drink, because when I take another drink, *everyone* takes another drink." So, second drinks were happily passed around the house. Then the man slammed some money on the counter and said: "And when I pay, *everybody* pays!"

This concludes the joke. The interesting thing is that there really must be a person x such that if x drinks, then *everybody* drinks! [$\exists x(Dx \Rightarrow \forall y Dy)$, Dy meaning that y drinks.] Before we prove this by a tableau, consider the following informal argument: Either everybody drinks or not everybody drinks. Suppose everybody drinks: $\forall y Dy$. Since $\forall y Dy$ is true, it is implied by any proposition; hence for *any* x, $Dx \Rightarrow \forall y Dy$ holds, so of course there is some x such that $Dx \Rightarrow \forall y Dy$ (in fact any x will do). Now consider the case when not everyone drinks. Then there is at least one person—call him Jim—who doesn't drink. Let j abbreviate "Jim." Thus Dj is false, and since a false proposition implies every proposition, it follows that $Dj \Rightarrow \forall y Dy$ is true! And so there does exist an x—namely j—such that $Dx \Rightarrow \forall y Dy$.

Now let's prove the formula $\exists x(Dx \Rightarrow \forall y Dy)$ by a tableau.

(1) $F\exists x(Dx \Rightarrow \forall y Dy)$	
(2) $FDa \Rightarrow \forall y Dy$	(1)
(3) TDa	(2)
(4) $F\forall y Dy$	(2)
(5) FDb	(4) [We cannot write FDa.]
(6) $FDb \Rightarrow \forall y Dy$	(1)
(7) $\underline{TDb}$	(6)

(My graduate students have dubbed this formula the *drinking* formula!)

A more natural proof of the above formula is possible if we use a *synthetic* tableau, as we will now see.

SYNTHETIC TABLEAUX

As with propositional logic, we define a *synthetic tableau* as one using the splitting rule (at any stage we can take any formula X and let any branch split to the two formulas TX, FX).

With a synthetic tableau, we can obtain a proof of the drinking formula that comes closer to the informal argument previously given. Recall that, in that informal argument, we split the argument into two cases— $\forall x Dx$ is true, $\forall x Dx$ is false. We carry that idea into the tableau, which begins thus:

$$(1)\ F\exists x(Dx \Rightarrow \forall y Dy)$$

$$(2)\ T\forall y Dy \qquad (3)\ F\forall y Dy$$

We have thus split right after (1). Now we work on the right branch and continue the tableau thus:

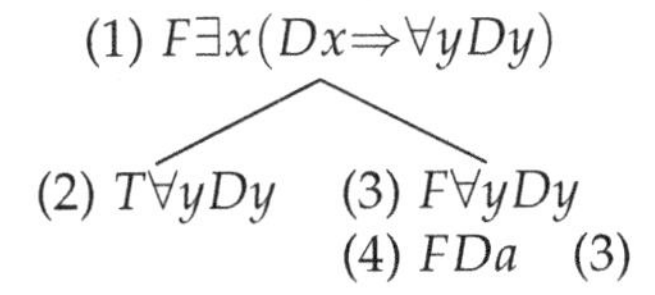

We are free to use the parameter a again in both (2) and (3), so we finish the tableau thus:

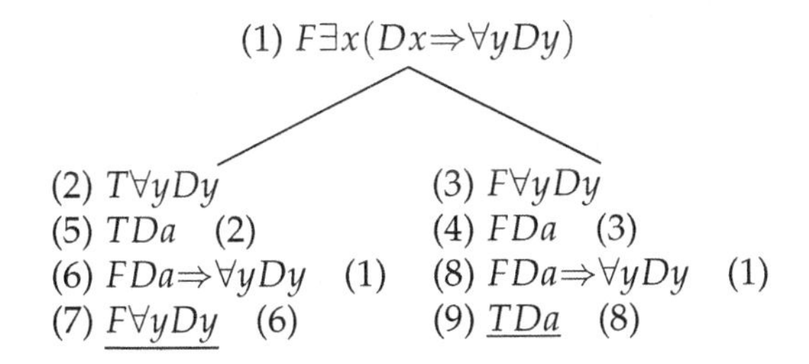

We might note that in the analytic tableau proof of the formula, we had to use two parameters, whereas in the synthetic tableau proof we need only one.

PROBLEM 13.11. Is it necessarily true that there exists a person x such that if anybody drinks, then x drinks?

PROBLEM 13.12 (ANOTHER CURIOUS ONE). Let us define a *lover* as anyone who loves at least one person. Now, suppose we are given the following two facts:

(1) Everyone loves a lover.

(2) John loves Mary.

Does it logically follow from (1) and (2) that Iago loves Othello?

LOGICAL CONSEQUENCE

We say that a formula Y is a *logical consequence* of a formula X if Y is true in all interpretations (in a non-empty domain) in which X is true, or, what is the same thing, if $X \Rightarrow Y$ is logically valid. To prove that Y is a logical consequence of X, it suffices to construct a closed tableau starting with TX, FY.

We say that Z is a logical consequence of X and Y if Z is true in all interpretations (in a non-empty domain) in which X and Y are both true or, equivalently, if $(X \wedge Y) \Rightarrow Z$ is logically valid—and a closed tableau starting with TX, TY and FZ will establish this, and similarly with four or more formulas.

PROBLEM 13.13. Consider the following argument:

Everyone has a mother.
Anybody's mother is that person's parent.
∴ Everyone has a parent.

Symbolize the above argument, letting Mxy mean that x is the mother of y, and Pxy mean that x is the parent of y.

EXERCISE 13.3. After symbolizing the above argument, show by a tableau that the conclusion is really a logical consequence of the two premises.

KNIGHTS AND KNAVES REVISITED

Let us now look at some of the problems of Chapter 12 from the viewpoint of first-order logic.

ALL OR NOTHING. We are on an island where each inhabitant is either a knight or knave. For any inhabitant x, we let Kx be the proposition that x is a knight. Then $\sim Kx$ says that x is a knave. Whenever an inhabitant x asserts a proposition $\mathcal{P}$, we know that if x is a knight then $\mathcal{P}$ is true, and if x is a knave then $\mathcal{P}$ is false—in other words, x is a knight if and only if $\mathcal{P}$ is true. Thus we translate "x asserts $\mathcal{P}$" as $Kx \equiv \mathcal{P}$. Now, in Problem 12.1, each inhabitant asserted that all the inhabitants were of the same type, all knights or all knaves. Thus each inhabitant x asserted $\forall x Kx \vee \forall x \sim Kx$, so for each x we have $Kx \equiv (\forall x Kx \vee \forall x \sim Kx)$. Since this holds for each x, we have $\forall x(Kx \equiv (\forall x Kx \vee \forall x \sim Kx))$. We saw in the solution that $\forall x Kx$ must hold (all the inhabitants are knights). The essence of this problem is that the following formula is logically valid, and can be proved by a tableau:

$$\forall x(Kx \equiv (\forall x Kx \vee \forall x \sim Kx)) \Rightarrow \forall x Kx.$$

EXERCISE 13.4. By a tableau, prove the above formula.

Let us now compare this problem with Problem 12.15, which was that certain inhabitants of earth were classified as *special* or *non-special* and that for each inhabitant x, x is special if and only if either everyone is special or no one is special. Letting Sx mean that x is special, we thus have $\forall x(Sx \equiv (\forall xSx \vee \forall x{\sim}Sx))$, and the solution concludes that $\forall xSx$ (everyone is special), so the logical content of the problem is that the formula $\forall x(Sx \equiv (\forall xSx \vee \forall x{\sim}Sx)) \Rightarrow \forall xSx$ is logically valid. But this is the same as the formula a few lines back, except that "S" has replaced "K"! Thus, as stated in the solutions in Chapter 12, the two problems are really the same, and the reader can now see why.

SMOKING KNAVES. In Problem 12.2, each inhabitant said: "Some of us are knights and some of us are knaves." Thus each x asserted $\exists xKx \wedge \exists x{\sim}Kx$, so the reality is $\forall x(Kx \equiv (\exists xKx \wedge \exists x{\sim}Kx))$. The conclusion was $\forall x{\sim}Kx$ (all are knaves), therefore $\forall x{\sim}Kx$ is a logical consequence of $\forall x(Kx \equiv (\exists xKx \wedge \exists x{\sim}Kx))$.

EXERCISE 13.5. Prove this by a tableau.

It should now be obvious to the reader that Problem 12.16 is really the same thing.

SMOKING KNIGHTS. Let us now look at Problem 12.4. Each inhabitant asserted that all the knights smoke. Thus $\forall x(Kx \equiv \forall x(Kx \Rightarrow Sx))$ (where Sx means that x smokes). The conclusion was that all are knights and all smoke ($\forall x(Kx \wedge Sx)$).

EXERCISE 13.6. Construct a closed tableau starting with $T\forall x(Kx \equiv \forall x\ (Kx \Rightarrow Sx))$ and $F\forall x(Kx \wedge\ Sx)$.

NONSMOKING KNAVES. The logical content of Problem 12.5 is that the statement $\forall x(Kx \wedge {\sim}Sx)$ is a logical consequence of $\forall x(Kx \equiv \exists x(Sx \wedge {\sim}Kx))$.

EXERCISE 13.7. Prove this with a tableau.

We close this chapter leaving two vital questions up in the air:

(1) Can more formulas be proved with synthetic tableaux than with analytic tableaux?

(2) Can all valid formulas be proved with synthetic tableaux? Can they all be proved with analytic tableaux?

The answers to these questions constitute *major* discoveries in the field of mathematical logic! We will see the answers in a later chapter, but we must first turn to the subject of infinity and mathematical induction. We will take temporary leave of the study of first-order logic and return to it later, after we have laid the adequate groundwork.

Solutions

13.1. (a) $\forall x(gHx \Rightarrow xHg)$.
Alternatively, $\forall x(\sim xHg \Rightarrow \sim gHx)$.

(b) $\forall x(gHx \equiv xHx)$.

13.2. (a) $\forall x(xTm \Rightarrow hTx)$.

(b) $\forall x(mTx \Rightarrow hTx)$.

(c) Same as (b).

(d) $\exists x(xTM) \Rightarrow hTm$.

(e) $\forall x(xTm) \Rightarrow hTm$.

(f) $\forall x(xTh \Rightarrow xTm)$.

(g) Same as (f)!

(h) $\forall x \exists y(xTy \wedge \sim yTm)$.

(i) $\forall x(xTh \Rightarrow \forall y(hTy \Rightarrow xTy))$.
Alternatively, $\forall x \forall y((xTh \wedge hTy) \Rightarrow xTy)$. (This is a bit simpler.)

13.3. (a) $\forall x \exists y xKy$.

(b) $\exists x \forall y xKy$.

(c) $\exists x \forall y yKx$.

(d) $\forall x \exists y(xKy \wedge \sim yKx)$.

(e) $\exists x \forall y(yKx \Rightarrow xKy)$.

13.4. $\exists x Dx \Rightarrow Dj$.
Alternatively, $\forall x(Dx \Rightarrow Dj)$.

13.5. This is a bit tricky! A solution is $Dj \wedge \forall x(Dx \Rightarrow x=j)$.
An alternative solution is $\forall x(Dx \equiv x=j)$.

13.6. (a) $y<x$ (this says that y is less than x, which is equivalent to saying that x is greater than y).

(b) $\forall x \exists y(x<y)$.

(c) $\forall x \exists y(y<x)$.

(d) Same as (c).

Statement (c) is false, of course; no natural number is less than zero. But false statements can be symbolized just as well as true ones.

13.7. To say that everyone is good is tantamount to saying that there doesn't exist any person who is not good—it is *not* the case that there exists a person x who is not good, which is symbolized $\sim\exists x \sim Gx$.

We thus see that $\forall$ is definable from $\exists$ and $\sim$: For any property P, the expression $\forall x Px$ is equivalent to $\sim\exists x \sim Px$ (there doesn't exist an x that fails to have property P).

13.8. In the other direction, if we start with $\forall$ and $\sim$, we can get $\exists$ thus: Given any property P, to say that there exists an x having property P is tantamount to saying that it is *not* the case that *every x fails* to have property P—$\sim\forall x \sim Px$. In particular, to say that at least one person is good ($\exists x Gx$) is tantamount to saying that it is not the case that everyone is not good ($\sim\forall x \sim Gx$).

13.9. $\forall x Px \Rightarrow \exists x Px$ is one such formula. In a non-empty domain, if every element x in it has property P, then of course there is at least one x in it having property P, so the formula is certainly valid. But for an empty domain, the formula $\forall x Px$ is true for any choice of P, as we have seen, but $\exists x Px$ is certainly false, since there doesn't exist any x at all in an empty domain. Thus for any interpretation of P in an empty domain, $\forall x Px$ is true and $\exists x Px$ is false, so the entire formula $\forall x Px \Rightarrow \exists x Px$ must be false.

13.10. Let F be the conjunction of the following three formulas:

F_1. $\forall x \exists y Rxy$.

F_2. $\sim\exists x Rxx$.

F_3. $\forall x \forall y \forall z ((Rxy \wedge Ryz) \Rightarrow Rxz)$.

That is, F is the formula $(F_1 \wedge F_2) \wedge F_3$. Well, F is satisfiable in the domain of the natural numbers: Take Rxy to mean that x is *less than* y. Then

(1) Of course, for any number x, there is a number y such that x is less than y.

(2) No number x is less than itself.

(3) For any numbers x, y, and z, if x is less than y and y is less than z, then surely x is less than z.

Thus F is satisfiable in the infinite domain of natural numbers. Next we show that if A is any non-empty domain in which F can be satisfied, then A must contain infinitely many elements. Well, let R be any relation in A for which F is true. Since A is non-empty, it contains at least one element, which we will denote a_1. By F_1, the element a_1 must bear the relation R to some element a_2. Could a_2 be the same element as a_1? No; by F_2, it is false that Ra_1a_1. Thus a_2 is distinct from a_1. Next, Ra_2a_3 holds for some element a_3 (by F_1), and a_3 must be distinct from a_2 (again by F_2). Also, since Ra_1a_2 and Ra_2a_3, it follows that Ra_1a_3 (by F_3), and since Ra_1a_1 cannot hold, it follows that a_3 cannot be a_1. Thus a_3 is distinct from a_1, also. Now we have three distinct elements a_1, a_2 and a_3. Next, Ra_3a_4 holds for some element a_4, which by similar reasoning must be different from a_1, a_2 and a_3. Then Ra_4a_5 holds for some new element a_5, and so forth. Thus we get an infinite sequence $a_1, a_2, \ldots, a_n, \ldots$ of elements of the domain.

13.11. Yes, it is necessarily true. Here, of course, "anybody" means *somebody*, and we are to show that there is a person x such that if somebody drinks, then x drinks. Well, either it is true or false that somebody drinks. Suppose it is true. Then let x be any one of the people who drink. Thus Dx is true, hence of course $\exists y Dy \Rightarrow Dx$ is true (since a true proposition is implied by any proposition). On the other hand, if $\exists y Dy$ is false, then for *any* person x, it is true that $\exists y Dy \Rightarrow Dx$ (since a false proposition implies any proposition).

The formula $\exists x(\exists y Dy \Rightarrow Dx)$ is thus valid, and, as an exercise, the reader should prove it with a tableau.

13.12. Yes, it does logically follow! Since John loves Mary, John is a lover. Hence everyone loves John. Hence everyone is a lover. Hence, everyone loves everyone! In particular, Iago loves Othello.

Let Lxy mean "x loves y." Then "x is a lover" is symbolically expressed as $\exists y Lxy$. We have seen that if everyone loves a lover and if there exists so much as one lover, then everybody loves everybody! Well, "everyone loves a lover" is symbolized $\forall x(\exists y Lxy \Rightarrow \forall z Lzx)$. Also, "there exists a lover" is symbolized $\exists x \exists y Lxy$. Of course, "everybody loves everybody" is symbolized $\forall x \forall y Lxy$. Thus the formula $\forall x \forall y Lxy$ is a logical consequence of $\forall x(\exists y Lxy \Rightarrow \forall z Lzx)$ and $\exists x \exists y Lxy$. This can be proved by a tableau, and it is a good exercise for the reader.

13.13. Symbolically, the argument is this:

$$\begin{array}{l} \forall x \exists y Myx \\ \underline{\forall x \forall y (Myx \Rightarrow Pyx)} \\ \therefore\ \forall x \exists y Pyx \end{array}$$

- Part III -

Infinity

- Chapter 14 -

The Nature of Infinity

We shall start with a little problem...

Problem 14.1. In a strange universe, there is a certain club called the *correspondence club* of which the following facts are known:

(1) Every member of the club has written a letter to at least one other member.

(2) No member has ever written a letter to himself.

(3) No member has ever received a letter from more than one member.

(4) There is one member who has never received any letters at all.

The number of members of this club is kept as a strict secret. According to one rumor, there are more than 500 members. According to another rumor, this is not so. Is there any way of determining which of the two rumors is correct?

Finite and Infinite Sets

Perhaps nothing has stirred the imagination of mankind more than the notion of *infinity*. To understand this notion, we must first understand the notion of a *1-1* (*one-to-one*) *correspondence*.

Suppose we look into a theater and see that everyone is seated, no one is standing, all the seats are taken, and no one is sitting on anyone's lap. Then, without having to count either the people or the seats, we know that their numbers must be the same, because the set of people is in a *one-to-one correspondence* with the set of seats, each person corresponding to the seat on which he or she is sitting.

Again, a flock of seven sheep is related to a grove of seven trees in a way in which it is not related to a grove of four trees, because it can be put into a one-to-one correspondence with the grove of seven trees—by, say, tethering each of the sheep to one and only one tree. In general, for any two sets A and B, we say that A can be put into a 1-1 correspondence with B if the elements of A can be paired with the elements of B in such a manner that each element of A and each element of B belong to exactly one of the pairs. If A can be put into a 1-1 correspondence with B, then we will also say that A is *similar* to B—in symbols, $A \cong B$—or that A can be *matched* with B.

When I say that 5 is the number of fingers in my left hand, all that is meant is that this set of fingers can be put into a 1-1 correspondence with the positive integers from 1 to 5, say by pairing my thumb with 1, the next finger with 2, the next with 3, the next with 4, and the pinky with 5. In general, for any positive integer (whole number) n, to say that a set A has n elements is to say that A can be put into a 1-1 correspondence with the set of integers from 1 to n. The process of making such a 1-1 correspondence has a popular name: *counting*. Yes, that's exactly what counting is.

Of course, we also say that the number of elements of a set is *zero* if the set is empty (has no members at all).

We have now defined what it means for a set A to have (exactly) n elements, where n is any *natural number* $(0, 1, 2, 3, 4, \ldots)$ and we now define a set A to be *finite* if there exists a natural number n such that A has n elements, and *infinite* if there is no such n.

Suppose we let $E(A, n)$ mean that the set A has n elements, where n is some natural number. Then the sentence "A is finite" is symbolized $\exists n E(A, n)$, and "A is infinite" is symbolized $\sim\exists n E(A, n)$.

REMARKS. It is intuitively pretty obvious that for any finite set A, if we count the elements of A in one order, and then count them again in a different order, the two numbers arrived at will be the same. This fact can be rigorously proved mathematically, but the proof uses a principle known as *mathematical induction*, which is a subject of the next chapter.

SUBSETS

A set A is said to be a *subset* of a set B—in symbols, $A \subseteq B$—if every element of A is also an element of B. (For example, the set of men in this world is a subset of the set of humans in this world, since every man is human.) If A is a subset of B, but not the whole of B (i.e., if it lacks at least one element of B), then A is said to be a *proper* subset of B. (For example, the set of all men is a *proper* subset of the set of all humans, since every man is human, but not every human is a man (fortunately!).)

The symbol for membership in a set is "$\in$." Thus for any element x and set A, the expression "$x \in A$" is read "x is an element (or a *member*) of A," or "x belongs to A," or, again, "A contains x." The statement "x *doesn't* belong to A" is symbolized $x \notin A$. Thus $x \notin A$ is synonymous with $\sim x \in A$.

Now, to say that A is a subset of B is to say that every element of A is also an element of B—in other words, that for every element x, if x belongs to A, then x belongs to B—and this can be symbolically written $\forall x(x \in A \Rightarrow x \in B)$. The statement "$A$ is a *proper* subset of B" is symbolically rendered $\forall x(x \in A \Rightarrow x \in B) \wedge \exists x(x \in B \wedge x \notin A)$.

A set A is said to be *identical* with a set B—in symbols, $A = B$—if A and B contain exactly the same elements ($\forall x(x \in A \equiv x \in B)$), or, equivalently, if $A \subseteq B$ and $B \subseteq A$.

We recall that a set is called *empty* if it contains no elements at all. (A good example is the set of all people in the theater after everyone has left.)

PROBLEM 14.2. Can there be more than one empty set?

PROBLEM 14.3.

(a) Is it true that the empty set is necessarily a subset of *every* set?

(b) Suppose A is a subset of every set. Is A necessarily empty?

For any finite number of elements $a_1, \ldots, a_n$, by $\{a_1, \ldots, a_n\}$ is meant the set whose elements are exactly $a_1, \ldots, a_n$. Thus, for example, $\{x, y\}$ is the set whose elements are x and y. Also, for any single element x, by $\{x\}$ is meant the set whose only element is x. Such a set $\{x\}$ is called a *unit* set, or a *singleton*. One should not confuse $\{x\}$ with x itself! For example, x itself might be a set containing three elements, whereas $\{x\}$ has only one element—the set x.

We have seen that there is only one empty set, and this set is usually symbolized $\emptyset$. An alternative symbol, which I like because it is particularly suggestive, is $\{\}$.

For any set A, the set of *all* subsets of A is called the *power set* of A, and symbolized $\mathfrak{P}(A)$. If A is a finite set with n elements, then the number of subsets of A (including the empty set) is 2^n.

EXAMPLES. Suppose a mother has 5 children, a, b, c, d and e. She decides to take some of them on a picnic—maybe all, maybe none. In how many ways can this be done? There are only two possibilities for a: she either takes a or leaves a. Thus there are two subsets of $\{a\}$, namely $\{a\}$ and $\{\}$.

Now, with each of the two possibilities for a, there are two possibilities for b, making four possibilities altogether. Thus there are four subsets of $\{a,b\}$: $\{a\}$, $\{\}$, $\{a,b\}$, $\{b\}$. Then with each of these four sets, we either put in c or we don't, thus getting those same four sets without c and another four sets with c added, so the eight subsets of $\{a,b,c\}$ are $\{a\}$, $\{\}$, $\{a,b\}$, $\{b\}$, $\{a,c\}$, $\{c\}$, $\{a,b,c\}$, $\{b,c\}$. And with each of these eight sets, we can either add d or leave d, thus getting 16 subsets of $\{a,b,c,d\}$, and with each of these 16 we can either add e or leave e, thus getting 32 subsets of $\{a,b,c,d,e\}$.

Sizes of Sets

We say that two sets A and B are of *the same size*—in symbols, $A \cong B$—if A can be put into a 1-1 correspondence with B. Now, how should we define what it means for set A to be *smaller* than set B? A natural guess would be that it should mean that A can be put into a 1-1 correspondence with a *proper* subset of B. Well, such a definition would be fine for *finite* sets, but it would raise serious problems for infinite sets, because for infinite sets A and B it can happen that A can be paired with a *proper* subset of B and also B can be paired with a proper subset of A! For example, let E be the set of even positive integers and O the set of odd positive integers. Then, on the one hand, we can match E with a *proper* subset of O by pairing each even number n with the odd number $n+1$, so that 2 would be paired with 3, 4 with 5, 6 with 7, and so forth; and thus E would be in a 1-1 correspondence, not with the whole of O, but with the set of all odd numbers equal to or greater than 3 (1 would be left out). On the other hand, we can match O with a *proper* subset of E by pairing each odd number n with the even number $n+3$, so that 1 is paired with 4, 3 with 6, 5 with 8, and so forth. Thus O is now matched, not with the whole of E, but with E minus the element 2. Now, we certainly wouldn't want to say that E is smaller than O and also that O is smaller than E! (Actually, O is of *the same size* as E, because of the 1-1 correspondence in which each odd number n is paired with the even number $n+1$—1 with 2, 3 with 4, 5 with 6, etc.)

No, the above definition won't work with infinite sets. The proper definition of A being *smaller* than B, or B being *larger* than A, is that A can be put into a 1-1 correspondence with a subset of B, but A *cannot* be put into a 1-1 correspondence with the whole of B! In other words, A can be matched with a subset of B, but *every* 1-1 correspondence from A to a subset of B will leave out at least one element of B. This definition is important to remember!

A curious thing about an infinite set is that it can be put into 1-1 correspondence with a *proper* subset of itself! We will later prove this about infinite sets in general, but, for now, we note that this is obviously true for the set $\mathcal{N}$ of natural numbers—$\mathcal{N}$ can be matched with just the set of positive natural numbers thus:

$$\begin{array}{cccccccc} 0 & 1 & 2 & 3 & \ldots & n & \ldots \\ 1 & 2 & 3 & 4 & \ldots & n+1 & \ldots \end{array}$$

Indeed, $\mathcal{N}$ can be matched with just the set of even natural numbers:

$$\begin{array}{cccccccc} 0 & 1 & 2 & 3 & \ldots & n & \ldots \\ 0 & 2 & 4 & 6 & \ldots & 2n & \ldots \end{array}$$

Perhaps even more startling is the fact, observed by Galileo in 1630, that even the set of perfect squares (the set $\{1, 4, 9, 25, 36, \ldots, n^2, \ldots\}$)—which seems so sparse and gets more and more sparse as we go further and further out—this set can be 1-1 matched with the set of positive integers:

$$\begin{array}{cccccccc} 1 & 2 & 3 & 4 & \ldots & n & \ldots \\ 1 & 4 & 9 & 16 & \ldots & n^2 & \ldots \end{array}$$

What should we say about this? Let P be the set of positive integers and S the set of perfect squares. Isn't there some sense in which P is greater than S? Yes, in the sense that P contains all the elements of S, and more (much more) besides. Nevertheless, P is *numerically* equal to S; the two sets P and S are of the same size!

DENUMERABLE SETS

Are all infinite sets of the same size, or do they come in different sizes? This was a basic question considered by the mathematician Georg Cantor (1845–1918), who is generally regarded as the father of set theory. For twelve years Cantor tried to prove that all infinite sets were of the same size, but in the thirteenth he found a counterexample (which I like to call a "Cantor-example").

A set is called *denumerable,* or *denumerably infinite*, or *enumerable,* if it can be put into 1-1 correspondence with the set of all positive integers. The word "countable" is sometimes used synonymously with "denumerable," but unfortunately the terminology is used a bit differently by different authors, some of whom use "denumerable" to mean either *finite* or denumerably infinite, while others use it to mean only denumerably infinite. Well, to avoid confusion, I will use "denumerable" only for infinite sets, and "countable" for *finite or denumerable.* I will use "enumerable" synonymously with "denumerable."

A 1-1 correspondence between a set A and the set of positive integers is called an *enumeration* of A. An enumeration of a set A can be thought of as an infinite *list* $a_1, a_2, \ldots, a_n, \ldots$, where for each positive integer n, the element a_n is the element of A that is paired with n. The number n is called the *index* of a_n.

PROBLEM 14.4. We have seen that the set of positive integers can be put into a 1-1 correspondence with a proper subset of itself. Is this true for *every* denumerable set?

PROBLEM 14.5. Is an infinite subset of a denumerable set necessarily denumerable?

A set is called *non-denumerable* or *uncountable* if it is infinite but not denumerable. As already mentioned, Cantor first thought that all infinite sets were denumerable. What he did was to consider various sets that on the surface *appeared* to be non-denumerable, but in each case he would find a clever way of enumerating the set. In courses I have given in mathematical logic or set theory, I liked to illustrate these enumerations in the following way. I would tell my class: "Imagine you are all in Hell and I am the Devil. That's not hard to imagine, is it?" (This usually gets a good laugh.) "Now, suppose I tell you that I have written down a positive integer that I have sealed in an envelope. Each day you have one and only one guess as to what the number is. If and when you correctly guess it, you go free; otherwise, you stay here forever. Is there some strategy you can devise to guarantee that you will get out sooner or later?" Of course there is: On the first day you ask whether the number is 1, on the next day whether it is 2, and so forth. Then if the number I have written down is n, you go free on the nth day.

My second test is only a wee bit harder: This time I tell you that I have written down either one of the positive integers $1, 2, 3, \ldots, n, \ldots$, or one of the negative integers $-1, -2, -3, \ldots, -n, \ldots$. Again, you have only one guess per day. Is there now a strategy that will guarantee that you will sooner or later go free? One person foolishly suggested that he would first go through all the positive integers and then through the negative ones. Well, if I wrote a positive integer, this would work fine, but if I wrote a negative integer, he would never get to it. No, the obvious thing is to enumerate the positive and negative integers in the order $1, -1, 2, -2, \ldots, n, -n, \ldots$. So the set of positive and negative integers, which at first blush seems to be twice the size of the positives alone, is really the same size after all.

My next test is definitely harder: This time I tell you that I have written down *two* positive integers, a and b, maybe the same or maybe different. Each time you have one and only one guess as to what the two integers a and b are. You are not allowed to guess one of them on one day and the other on another day; you must correctly guess *both* of them on the same day. Is there now some strategy that will definitely work? The situation seems hopeless, because there are infinitely many possibilities for a, and for each one there are infinitely many possibilities for b. Nevertheless, there is a usable strategy.

Problem 14.6. What strategy will work?

Next, I again write down a number a and then a number b to the right of it. This time, you must guess not only the two numbers written, but also the order in which they were written. Again you have one and only one guess each day.

Problem 14.7. What strategy will work?

In my next test, I tell you that I have written down a *fraction* $\frac{a}{b}$, where a and b are positive integers. Again, you have one and only one guess a day as to what the fraction is.

Problem 14.8. What strategy will work?

After reading the solution, you now see that the set of positive fractions is denumerable, a fact first discovered by Cantor that took the mathematical world by surprise—nay, by shock!

My next test is harder still: This time I have written down some *finite* set of positive integers. I am not telling you either how many integers I have written nor the highest one. Again you have one and only one guess each day.

Problem 14.9. What strategy will now work?

"And so," I said to my class, after going through the solution, "you now see that the set of all *finite* sets of positive integers is denumerable."

"What about the set of *all* sets of integers, finite ones and infinite ones," asked one of my students. "Is that set denumerable?"

"Ah!" I replied, "*That* set is non-denumerable. That is Cantor's great discovery!"

"No one has yet found a way of enumerating that set?" asked another student.

"No one has, and no one ever will," I replied, "because it is logically impossible to enumerate that set."

"How is that known?" asked another.

"Well, let's look at it this way," I replied. "Imagine a book with denumerably many pages—Page 1, Page 2, ..., Page n, ... On each page is written down a description of a set of positive integers. If *every* set of positive integers is listed in the book, the owner of the book wins a grand prize. But I tell you that winning the prize is impossible because, without even looking at the book, I can describe a set of positive integers that couldn't possibly be described on any page of the book."

"What set is that?" asked a student.

"I'll give you some hints," I replied, "in the form of two problems."

PROBLEM 14.10. In a certain community, the inhabitants have formed various clubs. An inhabitant may belong to more than one club. Each inhabitant has one and only one club named after him. An inhabitant is called *sociable* if he belongs to the club named after him, and *unsociable* otherwise. The set of unsociable people have formed a club of their own: all unsociable inhabitants are members, and no sociable inhabitants are allowed in the club.

According to one rumor, the club of unsociable people is named after someone called Jack Brown, but according to another rumor, it is named after Bill Green. Is the first rumor true or false? What about the second rumor?

PROBLEM 14.11. In a certain universe U, *every* set of inhabitants forms a club (even the empty set). The registrar of this universe would like to name each club after an inhabitant in such a way that each club would be named after one and only one inhabitant and each inhabitant would have one and only one club named after him.

Now, if the universe U had only finitely many inhabitants, then the scheme would be impossible, because there would then be more clubs than people (if there were n people, there would be 2^n clubs). This universe U happens to have *infinitely* many inhabitants, however, so the registrar sees no reason why the scheme should be impossible. For years and years he has tried one scheme after another, but every one so far has failed. Is this failure due to lack of cleverness on his part, or is he attempting to do something inherently impossible?

After seeing the solutions of the above two problems, the reader should now be prepared for the "grand" problem!

PROBLEM 14.12. Going back to the book with denumerably many pages, describe a set of positive integers that cannot be listed anywhere in the book. More generally, show the following: For any set A, the set $\mathfrak{P}(A)$ of all subsets of A cannot be put into a 1-1 correspondence with A, but A

can be put into a 1-1 correspondence with a proper subset of $\mathfrak{P}(A)$—in other words, $\mathfrak{P}(A)$ is larger than A! (Cantor's Theorem!)

DISCUSSION. We have now seen a proof of Cantor's famous theorem that for *any* set A, the power set $\mathfrak{P}(A)$ is larger than A. Starting with the denumerable set $\mathcal{N}$ of the natural numbers (which of course is the same size as the set of positive integers), the set $\mathfrak{P}(\mathcal{N})$ is larger than $\mathcal{N}$, hence non-denumerable. But then $\mathfrak{P}(\mathfrak{P}(\mathcal{N}))$ is still larger than $\mathfrak{P}(\mathcal{N})$, and $\mathfrak{P}(\mathfrak{P}(\mathfrak{P}(\mathcal{N})))$ is larger still, and we can keep going at this rate without cessation, and so we see that there must be *infinitely* many different sizes of infinite sets.

The set $\mathfrak{P}(\mathcal{N}))$ is an interesting one, since it happens to be the same size as the set of points on a straight line, and accordingly the size of $\mathfrak{P}(\mathcal{N}))$ is known as the *continuum*. Now comes an interesting question: Is there a set A midway in size between $\mathcal{N}$ and $\mathfrak{P}(\mathcal{N})$—that is, a set larger than $\mathcal{N}$ but smaller than $\mathfrak{P}(\mathcal{N})$? Or is it that $\mathfrak{P}(\mathcal{N})$ is of the next size larger than that of $\mathcal{N}$? Cantor *conjectured* that there was no size intermediate between those of $\mathcal{N}$ and $\mathfrak{P}(\mathcal{N})$, and this conjecture is known as the *continuum hypothesis*. More generally, Cantor conjectured that for *every* infinite set A, there is no set whose size is intermediate between that of A and that of $\mathfrak{P}(A)$, and this conjecture is known as the *generalized* continuum hypothesis. To this day, no one knows whether it is true or false! Some (including this author) regard this as *the* grand unsolved problem of mathematics! This much, however, is now known: In the late 1930s, Kurt Gödel showed that in the most powerful axiomatic set theory yet known, the continuum hypothesis is not *disprovable*. Then in the early 1960s, Paul Cohen showed that the continuum hypothesis is not *provable* from the axioms of set theory. Thus the best axioms of set theory to date are not sufficient to settle the question. Indeed, there are some (notably, those called *formalists*) who regard this as a sign that the continuum hypothesis is neither true nor false in its own right; it merely depends on what axiom system is used. Then there are others (most mathematicians, I believe) called *realists* or *Platonists*, who believe that, quite independent of any axiom system, the generalized continuum hypothesis is either true or false in its own right, but we simply don't know which. Interestingly enough, Gödel himself, who proved that the continuum hypothesis is not disprovable in axiomatic set theory, nevertheless expressed the belief that when more is known about sets, the continuum hypothesis will be seen to be false.

The Gödel and Cohen proofs, by the way, can be found in my book *Set Theory and the Continuum Problem* [26], written with my colleague Professor Melvin Fitting, which you can certainly read after having gone through this book.

BOOLEAN OPERATIONS

For any sets A and B, by $A \cup B$ is meant the set of all things that are in either A or in B, or both. The set $A \cup B$ is called the *union* of A and B.

By the *intersection* of A and B, symbolized $A \cap B$, is meant the set of all things that are in *both* A and B.

NOTE. The symbols $\cup$ and $\cap$ bear obvious resemblances to the logical symbols $\vee$ (or) and $\wedge$ (and), respectively. This is no coincidence, since an element x belongs to $A \cup B$ if and only if x belongs to A *or* x belongs to B, and x belongs to $A \cap B$ if and only if x belongs to A *and* x belongs to B.

By $A - B$ (the *set difference* of A and B) is meant the set of all elements of A that are *not* in B.

NOTE. For any set A and any object x, the result of adjoining x to A is written $A \cup \{x\}$ (the union of A with the set whose only element is x), not $A \cup x$! Also, the result of removing an element x from A is written $A - \{x\}$, not $A - x$.

PROBLEM 14.13. Is the union of two denumerable sets necessarily denumerable?

PROBLEM 14.14. We have seen that the set of all *finite* sets of positive integers is denumerable. What about the set of all *infinite* sets of positive integers? Is that set denumerable or not?

PROBLEM 14.15. Suppose we are given a denumerable sequence $D_1, D_2, \ldots, D_n, \ldots$ of denumerable sets. Let S be the union of all these sets—that is, the set of all elements x that belong to at least one of those sets. Is S denumerable or non-denumerable?

PROBLEM 14.16. Given a denumerable set D, is the set of all finite sequences of elements of D denumerable or non-denumerable?

DISCUSSION. I should now like to say something about a significant difference between denumerable and non-denumerable sets.

Suppose you are immortal and you get a check saying: "Payable at some bank." Well, if there are only finitely many banks in the world, you can be sure of collecting the money one day. In fact, even if there are denumerably many banks—Bank 1, Bank 2, ..., Bank n, ...—you can be sure of collecting one day, though you have no idea how long it will take.

However, if there are non-denumerably many banks, then there is no strategy whatsoever that guarantees that you will *ever* collect! In fact the chances that you will ever collect are infinitely small.

More on Sizes

Problem 14.17. Prove that every infinite set has a denumerable subset. (*Hint:* First show that if an element is removed from an infinite set, what remains is an infinite set.)

We stated earlier, without proof, that every infinite set can be put into a 1-1 correspondence with one of its *proper* subsets. By virtue of the above problem, this can now be seen.

Problem 14.18. Prove that every infinite set can be put into a 1-1 correspondence with a proper subset of itself.

Note. In the next chapter we will show that no *finite* set can be put into a 1-1 correspondence with a proper subset of itself.

Problem 14.19 (A Love Story). There once was a club M of men with non-denumerably many members and a club W of women that was of the same size as M. Therefore, it was possible to marry all the men in M to all the women in W. Before any wedding could take place, however, some problems arose:

(a) A man x outside the club M suddenly joined the club and also wanted to marry one of the women in W. Can $M \cup \{x\}$ be put into a 1-1 correspondence with W?

(b) Worse still, a *denumerable* set D of men all decided to join the club M, and they all wanted wives in W. Is it possible to find wives in W for *all* the men in $M \cup D$, and if so, how?

(c) As if this wasn't bad enough, denumerably many women suddenly decided to leave the club W. Are there enough women left to provide wives for all the men?

The Schröder-Bernstein Theorem

This is a basic result that we will first illustrate with a story: We are again given a universe U in which there are infinitely many men and infinitely many women (maybe non-denumerably many of each). We are given that each man loves one and only one of the women, but some of the women may be unloved, and no two men love the same woman. We are also given that each woman loves one and only one man (but not necessarily a man who loves her) and no two women love the same man, but, again, some of the men may be unloved. The problem is to prove that the set M of men can be put into a 1-1 correspondence with the set W of women.

Better still, it is possible to marry all the men to all the women in such a way that, given any couple, either the man loves his wife, or the wife loves her husband. (Unfortunately, there is no general scheme that will guarantee both.)

This problem is too difficult for most readers to solve without a hint, so here is a hint. Each person x, man or woman, can be placed into one of three groups according to the following scheme: Take a person y, if there is one, who loves x. Then take a person z, if there is one, who loves y, and keep going as long as possible. Then either the process will terminate in some unloved man, in which case we will say that x belongs to Group I, or it will terminate in some unloved woman, in which case we will classify x as being in Group II, or the process will go on forever, in which case we will say that x is in Group III. Now let M_1, M_2, M_3 be the sets of men in the first, second and third groups respectively, and W_1, W_2, W_3 the sets of women in the first, second and third groups respectively.

Can you finish the proof?

PROBLEM 14.20. Finish the proof.

The mathematical intent of the above problem is that if set A can be put into a 1-1 correspondence with a subset of a set B, and B can be put into a 1-1 correspondence with a subset of A, then (the whole of) A can be put into a 1-1 correspondence with (the whole of) B. This is the Schröder-Bernstein Theorem.

In fact, as the proof has shown, if C_1 is a 1-1 correspondence from A into a subset of B and C_2 is a 1-1 correspondence from B into a subset of A, then there is a 1-1 correspondence C from all of A to all of B such that for any element a in A and any element b in B, if a is paired with b under C, then either a was paired with b under C_1, or b was paired with a under C_2.

A SPECIAL PROBLEM. Here is a cute problem that I fell for, and so did some pretty good mathematicians!

PROBLEM 14.21. Let us assume that you and I are immortal. I have an infinite supply of dollar bills at my disposal, and you, to begin with, have none. Today I give you ten bills and you give me one back. Tomorrow I give you ten more, and of the nineteen you then have, you give me one back. And on all subsequent days, I give you ten each day and you give me one. We do this throughout all eternity. The question is, how many bills will remain with you permanently? An infinite number? Zero? Some positive finite number? (I'm sure the answer will shock many of you!)

A REVIEW

Let us now review and record the most important things of this chapter.

DEFINITIONS.

(1) For any two sets A and B, A is said to be a *subset* of B—in symbols, $A \subseteq B$—if every element of A is also an element of B. If A is a subset of B, but not the whole of B, then A is said to be a *proper* subset of B.

(2) By the *empty set* $\emptyset$ is meant the one and only set that has no elements at all. (The empty set is a subset of every set.)

(3) By the *union* $A \cup B$ of two sets A and B is meant the set of all elements that are either in A or in B (or in both). By the *intersection* $A \cap B$ of A and B is meant the set of all elements that are in both A and B. By the *set difference* $A - B$ is meant the set of all elements of A that are not in B. By the *power set* $\mathfrak{P}(A)$ of a set A is meant the set of all subsets of A.

(4) Two sets A and B are said to be of the *same size*—in symbols, $A \cong B$—if A can be put into a 1-1 correspondence with B. We say that A is *smaller* than B—in symbols, $A \prec B$—or that B is *larger* than A, if A can be put into a 1-1 correspondence with a subset of B, but not with the whole of B (in other words, A is of the same size as some subset of B, but not of the same size as B).

(5) For any positive integer n, a set A is said to have n elements if A can be put into a 1-1 correspondence with the set of positive integers from 1 to n. A set is said have 0 elements if it is empty. A set A is called *finite* if there is a natural number n such that A has n elements, and *infinite* if there is no such natural number n. A set is called *denumerable* if it can be put into a 1-1 correspondence with the set of positive integers (or equivalently, with the set of all natural numbers).

THEOREM 14.1 (CANTOR'S THEOREM). *For any set A, its power set $\mathfrak{P}(A)$ is larger than A.*

THEOREM 14.2. *Every infinite set has a denumerable subset.*

THEOREM 14.3. *The union of any finite or denumerable set of denumerable sets is denumerable.*

THEOREM 14.4. *The set of all* finite *subsets of a denumerable set D is denumerable. The set of all* infinite *subsets of D is non-denumerable.*

THEOREM 14.5. *For any denumerable set D, the set of all* finite *sequences of elements of D is denumerable.*

THEOREM 14.6. *Every infinite set A can be put into a 1-1 correspondence with a* proper *subset of itself (i.e., with a subset of A that is not the whole of A.)*

THEOREM 14.7 (SCHRÖDER-BERNSTEIN). $(A \preceq B \wedge B \preceq A) \Rightarrow A \cong B$. *If A can be put into a 1-1 correspondence with a subset of B, and B can be put into a 1-1 correspondence with a subset of A, then A can be put into a 1-1 correspondence with B.*

SOLUTIONS

14.1. It is the first rumor that must be correct; there are indeed more than 500 members, many more; in fact, there must be *infinitely* many members! Here is why.

By (4) there is a member, call him x_1, who has never received letters from any other member. Now, x_1 has written to at least one member (by (1)), but this member cannot be x_1, since no member has written to x_1, so it must be someone else, whom we will call x_2. Now x_2 has written to someone, who couldn't be x_2 (by (2)), nor x_1, to whom no member has written. So x_2 wrote to a new person x_3. Now, x_3 has written to some x_4, who again couldn't be x_1, nor x_2—who has been written to by x_1 and hence not by any one else—nor x_3. Thus x_4 is a new person. Then x_4 must have written to some person x_5 who again cannot be x_1, x_2, x_3 or x_4, by similar reasoning. Then x_5 must have written to some new person x_6, and so on. In this way we generate an infinite sequence of distinct members, so no *finite* number can suffice.

14.2. Of course not! The only way that a set A can be different from a set B is that one of them contains an element not contained in the other. Well, if E_1 and E_2 are both empty, then they contain exactly the same elements—namely, no elements at all.

Thus there is one and only one empty set. This set is denoted by "$\emptyset$."

14.3. (a) Yes, it is true that the empty set $\emptyset$ is a subset of every set, because for any element x, the statement $x \in \emptyset$ is false (no x belongs to the empty set) and a false proposition implies *every* proposition; hence $x \in \emptyset \Rightarrow x \in A$ is true! Thus $\forall x (x \in \emptyset \Rightarrow x \in A)$ is true, which is the statement that $\emptyset$ is a subset of A. Thus $\emptyset$ is a subset of every set.

If the reader doubts that the empty set $\emptyset$ is a subset of every set A, just try to find an element of $\emptyset$ that is *not* a member of A!

[We went through this before in the special case that if there are no Frenchmen in a given club, then it is true that all Frenchmen in the club wear hats: Let F be the set of Frenchmen in the club and H be the set of people who wear hats; then if F is empty, it is true that $F \subseteq H$ (which says that all Frenchmen in the club wear hats).]

(b) Of course it is the only one! If A is a subset of *every* set, then in particular, A is a subset of the empty set, and the only subset of the empty set is the empty set.

14.4. Of course it is. The elements of a denumerable set A can be enumerated in a denumerable sequence $a_1, a_2, \ldots, a_n, \ldots$. Then we can pair each element a_n with a_{n+1} thus:

$$\begin{array}{cccccc} a_1 & a_2 & a_3 & \cdots & a_n & \cdots \\ \downarrow & \downarrow & \downarrow & & \downarrow & \\ a_2 & a_3 & a_4 & \cdots & a_{n+1} & \cdots \end{array}$$

14.5. Yes, it is. Suppose A is denumerable. Then its elements can be arranged in a denumerable sequence $a_1, a_2, \ldots, a_n, \ldots$. Let B be a subset of A. Either there is a greatest number n such that a_n is in B, or there isn't. If there is, then B has n or fewer elements, hence B is finite. If there isn't, then we let b_1 be the first element of the sequence $a_1, a_2, \ldots, a_n, \ldots$ that is in B, b_2 the next such element, and so forth. We thus obtain an enumeration $b_1, b_2, \ldots, b_n, \ldots$ of B.

14.6. For any numbers x and y we let $\{x, y\}$ be the set whose members are x and y. Now, for each positive integer n, there are only finitely many such pair-sets whose highest member is n—exactly n such pair-sets, in fact, namely $\{1, n\}$, $\{2, n\}$, ..., $\{n, n\}$. So we first go through all pairs whose highest member is 1, then all those whose highest member is 2, and so forth. Thus we enumerate the pairs in the order $\{1,1\}$, $\{1,2\}$, $\{2,2\}$, $\{1,3\}$, $\{2,3\}$, $\{3,3\}$, $\{1,4\}$, $\{2,4\}$, $\{3,4\}$, $\{4,4\}$, $\{1,5\}$,....

14.7. For this variant of the last problem, in which you must guess also the order, you simply mention each pair of *distinct* numbers in *both* orders before proceeding to the next pair—that is, you name the ordered pairs in the order $(1,1)$, $(1,2)$, $(2,1)$, $(1,3)$, $(3,1)$, $(2,3)$, $(3,2)$, $(3,3)$,..., $(1, n)$, $(n, 1)$, $(2, n)$, $(n, 2)$,.... (Note that the preceding solution involved *unordered* pairs of integers, while the present solution involves *ordered* pairs—hence the difference in notation:

braces versus parentheses. Thus, for example, $\{5,7\}$ is the unordered pair—i.e., the set—consisting of the integers 5 and 7; the order in which we write them doesn't matter, and $\{5,7\}$ is the same unordered pair as $\{7,5\}$. In contrast, $(5,7)$ is an *ordered* pair—i.e., a sequence in which 5 comes first—and it is different from the ordered pair $(7,5)$.)

14.8. This is really the same as the last problem, the only difference being that in a fraction $\frac{a}{b}$, the number a is written *above* b, instead of to the left of it. Thus we can enumerate the fractions in the order $\frac{1}{1}, \frac{1}{2}, \frac{2}{1}, \frac{1}{3}, \frac{3}{1}, \ldots, \frac{1}{n}, \frac{n}{1}, \frac{2}{n}, \ldots$.

14.9. There is one and only one set whose highest number is 1, namely $\{1\}$ (the set whose only member is 1). There are two sets whose highest number is 2, namely $\{1,2\}$ and $\{2\}$. There are four sets whose highest number is 3, namely $\{3\}, \{1,3\}, \{2,3\}, \{1,2,3\}$. In general there are 2^{n-1} sets whose highest number is n, because for any n, there are 2^n subsets of the set of positive integers from 1 to n (this includes the empty set!), so any set whose highest number is n consists of n together with some subset of the integers from 1 to $n-1$, and there are 2^{n-1} such subsets.

The important thing is that for each n, there are only *finitely* many sets of positive integers whose highest member is n. And so one should first name the empty set. Then name the set whose highest member is 1, then the sets whose highest member is 2 (the order doesn't matter), then those whose highest member is 3, and so forth.

14.10. Both rumors must be false; the club of unsociable people couldn't be named after *anybody*! REASON. Suppose it were named after somebody—say, "Jack." Then Jack's club (i.e., the club named after Jack) contains all unsociable people but no sociable people. Well, is Jack sociable or not? If he is, then by definition of "sociable" he belongs to Jack's club, but only unsociable people belong to Jack's club, and so we have a contradiction. On the other hand, if Jack is unsociable, then, since *all* unsociable inhabitants belong to Jack's club, Jack must belong to Jack's club, which makes him sociable, so we again have a contradiction. Thus the set of unsociable people cannot be named after Jack, or anyone else!

14.11. If such a scheme were possible, we would run into the same contradiction as in the last problem: The set of all inhabitants who do not belong to the club named after them would form a club (every set of inhabitants does), and this club would be named after someone who could neither be nor not be a member without contradiction.

14.12. Let S_1 be the set listed on page 1, S_2 the set listed on page 2,..., S_n the set listed on page n,.... We wish to define a set S of positive integers that is different from every one of the sets $S_1, S_2, \ldots, S_n, \ldots$. Well, we first consider the number 1—whether or not it should go into our set S. We do this by considering the set S_1 listed on page 1 of the book. Either 1 belongs to the set S_1 or it doesn't. If it doesn't, then we shall include it in our set S; but if 1 *does* belong to S_1, then we exclude it from S. Thus whatever future decisions we make concerning the numbers $2, 3, \ldots, n, \ldots$, we have secured the fact that S is distinct from S_1, because, of the two sets S and S_1, one contains 1 and the other doesn't. Next we consider the number 2. We put 2 into our set S if 2 does *not* belong to the set S_2, and we leave 2 out of S if 2 belongs to S_2. This guarantees that S is different from S_2, since one of them contains 2 and the other doesn't. And so on with every positive integer n. We thus take S to be the set of all positive integers n such that n does *not* belong to the set listed on page n. Thus for every n, the set S_n is different from S, because of the two sets S and S_n, one of them contains n and the other doesn't. To make matters a bit more concrete, suppose, for example, that the first ten sets S_1 to S_{10} are the following:

S_1. Set of all even numbers.

*S_2. Set of all positive whole numbers.

S_3. The empty set.

S_4. Set of all numbers greater than 100.

*S_5. Set of all numbers less than 57.

S_6. Set of all prime numbers.

S_7. Set of all numbers that are not prime.

*S_8. Set of all numbers divisible by 4.

S_9. Set of all numbers divisible by 7.

*S_{10}. Set of all numbers divisible by 5.

I have starred those lines n whose line number n belongs to S_n. For example, I didn't star line 1, because 1 is not an even number. Of course 2 belongs to S_2, hence line 2 is starred. The next starred line is 5, since 5 is less than 57. The set S consists of all the numbers whose corresponding lines are *unstarred*, and so among the first ten numbers, the numbers $1, 3, 4, 6, 7$ and 9 are in S, whereas $2, 5, 8$ and 10 are not. (Incidentally, the "starred numbers"—i.e., those numbers whose corresponding lines are starred—are like the *sociable* inhabitants of Problem 14.8, and the unstarred numbers are like the unsociable inhabitants. Thus the set S of all unstarred numbers, which is not matched with any line n, is like the club of unsociable inhabitants, which cannot be named after anyone.)

What we have shown is that given any denumerable sequence $S_1, S_2, \ldots, S_n, \ldots$ of sets of positive integers, there exists a set S of positive integers (namely, the set of all n such that n doesn't belong to S_n) that is different from each of the sets $S_1, S_2, \ldots, S_n, \ldots$. This means that *no* denumerable set of sets of positive integers contains *every* set of positive integers—in other words, *the set of all sets of positive integers is non-denumerable.*

This is a special case of Cantor's theorem. The general case is that for *any* set A, it is impossible to put A into a 1-1 correspondence with the set $\mathfrak{P}(A)$ of all subsets of A. The proof is hardly different from the special case that we have considered (nor any different from the solution of Problem 14.10) and is as follows: Suppose we have a 1-1 correspondence that matches each element x of A with a subset S_x of A. Let S be the set of all elements x such that x doesn't belong to S_x. This set S is such that for no element x is it possible that $S=S_x$, because one of the two sets S and S_x contains x and the other doesn't. And so there is *no* x in A that is matched with S, so in the 1-1 correspondence between A and some of the subsets of A, the set S is left out. Thus A cannot be put into a 1-1 correspondence with *all* of $\mathfrak{P}(A)$.

Of course A can be put into a 1-1 correspondence with a *subset* of $\mathfrak{P}(A)$—for example, we can pair each element x with the unit set $\{x\}$.

This proves that for *every* set A, the power set $\mathfrak{P}(A)$ is larger than A. This is Cantor's theorem.

14.13. Of course it is! Suppose A and B are both denumerable. Then A can be enumerated in a denumerable sequence $a_1, a_2, \ldots, a_n, \ldots$ and B in a denumerable sequence $b_1, b_2, \ldots, b_n, \ldots$. Then $A \cup B$ can be enumerated in the sequence $a_1, b_1, a_2, b_2, \ldots, a_n, b_n, \ldots$.

14.14. It must be non-denumerable. More generally, if A is *any* non-denumerable set, and if we remove denumerably many elements from A, what remains must be non-denumerable.

REASON. Let D be the set of the denumerable elements removed, and let B be the set of remaining elements. If B were denumerable, then the union $D \cup B$ would be denumerable (by the last problem), but $D \cup B$ is the whole of A, which is not denumerable. Hence B cannot be denumerable, so it must be non-denumerable.

Thus for any non-denumerable set A and any denumerable subset D of A, the set $A-D$ is non-denumerable.

In particular, since the set of *all* sets of positive integers is non-denumerable, it follows that if we remove the denumerably many finite ones, then the remaining ones (the infinite ones) constitute a non-denumerable set.

14.15. Each set D_n is itself denumerable, hence its elements can be arranged in a denumerable sequence, and so we can talk of the mth element of the set D_n. Let d_n^m be the mth element of the set D_n. Thus D_1 is arranged in the sequence $d_1^1, d_1^2, \ldots, d_1^m, \ldots$, D_2 is arranged in the sequence $d_2^1, d_2^2, \ldots, d_2^m, \ldots$, and so on. We already know that the set of ordered pairs (m, n) of positive integers is denumerable (Problem 14.7), and we can therefore arrange the elements d_n^m of S in the same order (first all those for which the highest of m and n is 1, then those for which the highest is 2, and so forth). Thus the sequence will begin d_1^1, d_2^1, d_1^2, d_2^2, d_3^1, d_1^3, d_3^2, d_2^3, $d_3^3, \ldots$, d_n^1, d_1^n, d_n^2, $d_2^n, \ldots$

14.16. We first consider the following question: Given a finite set of n objects, how many sequences of length k of these objects are there? Well, there are n sequences of length 1, n^2 $(n \times n)$ sequences of length 2 (since there are n choices for the first term, and with each of these choices there are n choices for the second term). For sequences of length 3, the answer is n^3 $(n \times n \times n)$, and, in general, for any positive integer k, there are exactly n^k sequences of length k of n objects.

From this it follows that for any set of n objects, the set of all sequences of length n *or less* of those objects is finite (it contains $n^1 + n^2 + \cdots + n^n$ members).

Now, given a denumerable set D, we can enumerate its elements in some sequence $a_1, a_2, \ldots, a_n, \ldots$. For each positive n, let S_n be the set of all sequences of length n or less of the first n objects $a_1, \ldots, a_n$. We have just seen that for each n, the set S_n is finite. Also, every finite sequence of elements of D is in some S_n (any sequence of length k of the elements $a_1, \ldots, a_m$ is in S_n for any n greater than m and k). We can thus enumerate all finite sequences of elements of D by starting with those involving the elements of S_1 (in any order), followed by those involving the elements of S_2 other than those in S_1, followed by those in S_3 not in S_2, and so forth.

14.17. Suppose A is infinite and we remove an element x. If the remaining set $A-\{x\}$ were finite, then it would have n elements, for some natural number n, which would mean that before x was removed, A had $n+1$ elements, hence A would be finite, contrary to the assumption that A is infinite. Therefore if an element is removed from an infinite set, what remains is clearly infinite.

Now suppose A is infinite. Then of course A is non-empty, so we can remove an element a_1. What remains is infinite, hence non-empty, and so we can remove another element a_2. Then we can remove a_3, $a_4, \ldots$, and thus generate a denumerable sequence of elements of A. The set of elements of this sequence is a denumerable subset of A.

14.18. Suppose A is infinite. Then, as we now have just seen, A includes a denumerable subset $D=\{d_1, d_2, \ldots, d_n, \ldots\}$. This set can be put into a 1-1 correspondence with the set $\{d_2, d_3, \ldots, d_n, \ldots\}$, and we can let each element of A other than those in D correspond to itself. In this correspondence, A is in a 1-1 correspondence with its proper subset $A-\{d_1\}$.

14.19. We have just seen that for any infinite set A and any element x of A, the set A can be put into a 1-1 correspondence with $A-\{x\}$. From this it follows that for any infinite set A and any element x outside A, the set A can be put into a 1-1 correspondence with $A \cup \{x\}$, because $A \cup \{x\}$ can be put into a 1-1 correspondence with $A \cup \{x\} - \{x\}$, which is the set A.

This takes care of (a): $M \cup \{x\}$ is the same size as M, hence the same size as W, so there is no problem.

As for (b), for *any* non-denumerable set A and any denumerable set D, the set $A \cup D$ is the same size as A, because A includes a denumerable subset D_1, and we can put D_1 into a 1-1 correspondence with $D_1 \cup D_2$ (Problem 14.13), and we can let each element of $A-D_1$ correspond to itself, and then A is in a 1-1 correspondence with $A \cup D$. Thus $M \cup D$ is the same size as M, hence the same size as W, so again there is no problem.

As for (c), it follows from what we have just done that for any infinite set A and any denumerable subset D of A, the set $A-D$ is of the same size as A, because $A-D$ is of the same size as $(A-D) \cup D$, which is A.

This settles (c): Let D_1 be the denumerable set of men who have joined M and D_2 be the denumerable set of women who have left W. Then $M \cup D_1$ is of the same size as M, which in turn is of the same size as W, which in turn is of the same size as $W - D_2$, so $M \cup D_1$ is of the same size as $W - D_2$, which means $M \cup D_1$ is of the same size as $W - D_2$, and so again there is no real problem.

14.20. It is obvious that every unloved man must be in Group I and every unloved woman must be in Group II. Therefore, every person in Group III is loved. Furthermore, we easily see that

(a) Every man in Group I loves a woman, who must be in Group I, and every woman in Group I is loved by some man, and such a man must be in Group I. Therefore if all the men in Group I marry the women they love, these women are all in Group I, and include *all* the women in Group I.

(b) Similarly, if every woman in Group II marries the man she loves, her husband must be in Group II, and *all* men in Group II will then be married to women in Group II.

(c) Every man in Group III both loves a woman in Group III and is loved by a woman in Group III. Also every woman in Group III both loves a man in Group III and is loved by a man in Group III. Therefore, we have our choice of either marrying all the men of Group III to the women they love, or of marrying all the women of Group III to the men they love. (Which of the two is the better choice is a problem I leave to a psychologist.) In either case, all the men in Group III will be married to women in Group III and vice-versa.

14.21. Yes, the answer will shock many of you! The answer depends on *which* bills you give me back! It could be that infinitely many bills remain with you, or none remain with you, or any intermediate finite number might remain, all depending on which bills you give me back.

Look, suppose, on the one hand, that you give me back, each time, a bill from the stack of ten that I just gave you. Then nine bills out of each stack will remain with you permanently, hence infinitely many bills will remain with you permanently. On the other hand, you could give me back all of the first stack in the first ten days, then all of the second stack in the next ten days, and so forth. Then none would remain with you permanently.

Another way to look at it is this: Imagine all the bills are numbered $1, 2, \ldots, n, \ldots$. Then you could, as one strategy, systematically give me back all the bills in the order $1, 2, \ldots, n, \ldots$, in which case you would keep none; alternatively, you could give me back the even-numbered bills and keep the infinitely many odd-numbered ones.

Or still again, you could permanently keep any finite number you wanted, and give me back all the others in serial order.

This whole problem is only a deceptive version of the question of how many natural numbers remain if we remove infinitely many of them. Here the answer is obviously that it depends on which numbers we remove (maybe all of them, maybe just all the even ones, maybe all numbers greater than 27).

- CHAPTER 15 -

MATHEMATICAL INDUCTION

On one of my visits to the Island of Knights and Knaves, I came across an inhabitant who said: "This is not the first time I have said what I am now saying."

PROBLEM 15.1. Was the inhabitant a knight or a knave?

The solution of this problem uses, though in a hidden way, an important result known as *mathematical induction*, which I will soon explain. A forward version of the problem is the following: A man who was in search of the secret of immortality once met a sage, who was reported to be a specialist in this area. He asked the sage: "Is it really possible to live forever?" "Oh, quite easily," replied the sage, "provided you do two things: (a) From now on, never make a false statement; always tell the truth. (b) Now say: I will repeat this sentence tomorrow.' If you do those two things, then I guarantee that you will live forever!"

PROBLEM 15.2. Was the sage right?

Again, mathematical induction was implicitly used in the solution. What is mathematical induction? It is a principle that deserves to be far better known than it apparently is. It is really not all that difficult to grasp, so I shall first state it in its general abstract form and then illustrate it with several examples. In its general form, it is simply that if a certain property holds for 0, and if it never holds for any number without holding for the next number as well, then it must hold for all natural numbers (i.e., the numbers $0, 1, 2, 3, 4, \ldots$, etc.). For example, suppose I tell you that today it is raining on Mars, and that on Mars it never rains on any day without raining the next day as well. Isn't it obvious that from now on it must rain on all days?

Here is another illustration I like to use: Let us imagine that we are all immortal and that, as in the good old days, the milkman delivers milk to the door and we leave an empty bottle with a note in it telling the milkman what to do. Well, suppose we leave the following note:

Never leave milk on one day without leaving milk the next day as well.

The note could alternatively read: "If ever you leave milk on any day, be sure and leave it the next day as well." This version comes to the same thing. Now, the milkman, if he wanted, could *never* leave milk, and he would not be disobeying the order: Suppose a thousand years went by and he hasn't left any milk yet. If we were to ask him why he disobeyed our order, he could rightly reply, "I have not disobeyed your order. Have I ever left milk on one day without leaving milk the next day? Certainly not! I have never left milk at all!" And he could keep going as long as he liked without leaving milk. But suppose that two thousand years later, he decides on a whim to leave milk one day. Then he is committed for all days after that! But, again, he might never leave milk at all, so (assuming that we *do* want milk), our note above is not adequate. The following note *is* adequate, however—it guarantees permanent delivery:

(1) Never leave milk on any day without leaving milk the next day as well. (2) Leave milk today.

This note certainly does guarantee permanent delivery, and perfectly illustrates the principle of mathematical induction.

An alternative approach that is particularly neat, suggested to me by the computer-scientist Alan Tritter, illustrates what might be called the *recursion*, or *Turing machine* approach: Instead of the two sentences of the above note, the following single-sentence note cleverly does the job:

Leave milk today and read this note again tomorrow.

By mathematical induction it follows that if the milkman obeys and continues to obey this order, then he will both read this note and leave milk on all days from today on.

A property of natural numbers is called *hereditary* if it never holds for any natural number without holding for the next number as well. The principle of mathematical induction, more succinctly stated, is that if a hereditary property holds for 0, then it holds for all (natural) numbers.

In proving by mathematical induction that all natural numbers have a certain property P, the proof is divided into two steps:

(1) The *basis step*, which is the proof that 0 has the property P.

(2) The *induction step*, which is the proof that for all n, if n has the property P, $n+1$ has the property P as well. In this step, the assumption that n has the property, from which we deduce that $n+1$ also does, is called *the hypothesis of the induction* or the *inductive hypothesis*.

The principle of mathematical induction is sometimes stated starting with 1 instead of 0, in which case it says that if a hereditary property holds for 1, then it holds for *all* positive natural numbers. More generally, of course, for any natural number n, if a hereditary property holds for n, then it must hold for *all* numbers from n on.

SOME APPLICATIONS

We will soon see an application to geometry using the Pythagorean theorem, which we recall is that for any right triangle with sides a and b and hypotenuse c, we have the relation $a^2+b^2=c^2$.

When I teach this theorem to beginning students, I draw on the board a right triangle with squares on the sides and on the hypotenuse.

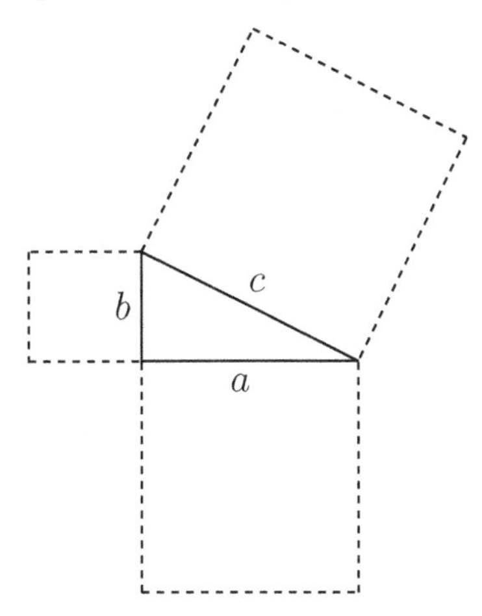

Then I tell the students: "Imagine that those squares are made of valuable gold leaf, and you are offered to take either the one big square or the two small ones. Which would you pick?"

Interestingly enough, about half the class opts for the one big square, and half for the two small ones, and later both camps are equally surprised when shown that it makes no difference.

Now, in geometry, certain constructions are possible with a ruler and a compass. One of the things that can be done is that for any given line segments x and y already constructed, one can construct a right triangle using x and y as sides. Now, suppose we are given a line segment of length one unit. The problem is to show that for *any* positive integer n, one can construct a line segment of length $\sqrt{n}$.

The basic idea behind the proof is that for any n, if a right triangle has sides of lengths 1 and $\sqrt{n}$, then the hypotenuse must have length $\sqrt{n+1}$.

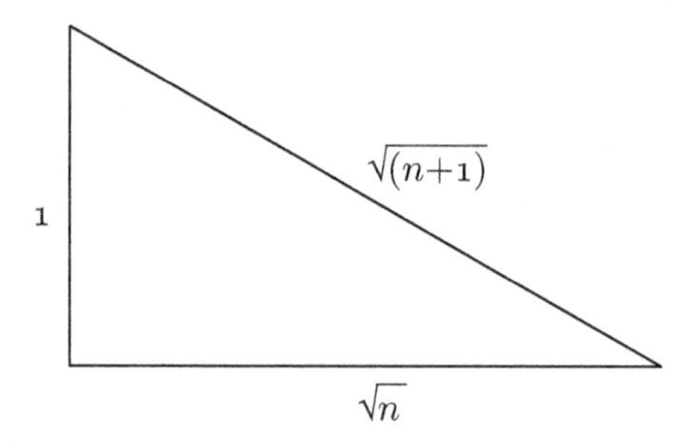

Indeed, let x be the hypotenuse. Then by the Pythagorean theorem, $\sqrt{n}^2+1^2=x^2$, but $\sqrt{n}^2=n$ and $1^2=1$, so $n+1=x^2$, hence $x=\sqrt{n+1}$.

Now, we are given a line segment of length 1 (unit). Then we construct a right triangle whose sides are both length 1, and the hypotenuse will be of length $\sqrt{2}$ ($\sqrt{1^2+1^2}$). Then we can construct a right triangle with sides 1 and $\sqrt{2}$, and the hypotenuse will then have length $\sqrt{3}$. Then we can construct a right triangle with sides 1 and $\sqrt{3}$ which will have a hypotenuse of length $\sqrt{4}$, and so forth. This "and so forth" is but an *informal* mathematical induction. A strict proof using mathematical induction *explicitly* runs as follows: We are to show that, given a line segment of length 1, we can construct, for *every* positive integer n, a line segment of length $\sqrt{n}$. Well, for any n, define $P(n)$ to mean that a segment of length $\sqrt{n}$ can be constructed. We are to show that $P(n)$ holds for all n. By the principle of mathematical induction, it suffices to show two things:

(1) $P(1)$.

(2) For every n, $P(n) \Rightarrow P(n+1)$.

RE (1): BASIS STEP. $P(1)$ means that a segment of length $\sqrt{1}$ can be constructed. Well, $\sqrt{1} = 1$, and we are initially given a segment of length 1, hence P holds for 1.

RE (2): INDUCTION STEP. $P(n) \Rightarrow P(n+1)$ says that if $\sqrt{n}$ can be constructed, so can $\sqrt{n+1}$. We have already seen that this is so by constructing a right triangle with sides 1 and $\sqrt{n}$.

From (1) and (2) it follows by mathematical induction that $P(n)$ holds for *all* positive integers n, i.e., that for every n we can construct a segment of length $\sqrt{n}$.

REMARKS. We have just seen an application of mathematical induction to geometry. It also has a host of applications to algebra. For example, by induction on n one can prove the following laws (which strike me as surprising!):

(1) $1 + \cdots + n = \dfrac{n(n+1)}{2}$,

(2) $1^2 + \cdots + n^2 = \dfrac{n(n+1)(2n+1)}{6}$,

(3) $1^3 + ... + n^3 = \dfrac{[n(n+1)]^2}{4}$.

The reader who likes algebra might like to try these as exercises.

PROBLEM 15.3. Using mathematical induction, show that for any natural number n, if a set A has n elements then its power set $\mathfrak{P}(A)$ has 2^n elements. (We informally showed this earlier.) *Hint:* First show that, for any finite set A and any element x outside A, the number of subsets of $A \cup \{x\}$ is twice the number of subsets of A. (We recall that $A \cup \{x\}$ is the set A with x adjoined.)

PROBLEM 15.4. Suppose P is a property such that

(1) P holds for 0,

(2) for any n, if P holds for all numbers from 0 to n, then P holds for $n+1$.

It seems intuitively obvious that P then holds for all natural numbers, but this can also be proved as a corollary of the mathematical induction principle. How?

NOTE. The proof uses an important and useful trick explained in the solution.

PROBLEM 15.5. Suppose P is a property that obeys the following single condition:

(1) For any n, if P holds for all numbers less than n, then P holds for n.

Does P necessarily hold for all natural numbers?

PROBLEM 15.6. Suppose a property P is such that, for every natural number n, if P fails for n, then P fails for some natural number less than n. Which of the following conclusions are valid?

(a) P fails for all natural numbers.

(b) P holds for all natural numbers.

(c) P holds for some but not all natural numbers.

THE LEAST NUMBER PRINCIPLE

It is intuitively obvious that every non-empty set of natural numbers must contain a least number. This is known as the *least number principle*. Though intuitively obvious in its own right, it can also be established as a corollary of mathematical induction, and, conversely, if we start with the least number principle as an axiom, we can derive from it the principle of mathematical induction.

PROBLEM 15.7. How can the least number principle and the principle of mathematical induction each be derived from the other?

At this point, let us record what we have learned.

THEOREM 15.1 (PRINCIPLE OF MATHEMATICAL INDUCTION). *Suppose a property P satisfies the following two conditions:*

(1) P holds for 0.

(2) For every natural number n, if P holds for n, then P holds for $n+1$.

Then P holds for every natural number.

THEOREM 15.2 (ANOTHER FORM OF INDUCTION). *Suppose a property P satisfies the following two conditions:*

(1) P holds for 0.

(2) For every n, if P holds for all numbers less than or equal to n, then P holds for $n+1$.

Then P holds for every natural number.

THEOREM 15.3 (PRINCIPLE OF COMPLETE INDUCTION). *Suppose P is a property such that for every (natural number) n, if P holds for all numbers less than n, then P holds for n as well. Then P holds for all natural numbers.*

THEOREM 15.4 (PRINCIPLE OF FINITE DESCENT). *Suppose P is such that for all n, if P holds for n, then P holds for some number less than n. Then P holds for no natural number.*

THEOREM 15.5 (LEAST NUMBER PRINCIPLE). *Every non-empty set of natural numbers contains a least member.*

Here is another useful induction principle:

THEOREM 15.6 (PRINCIPLE OF LIMITED INDUCTION). *Suppose k is a positive integer and P is a property satisfying the following two conditions:*

(1) P holds for 1.

(2) For any positive integer less than k, $P(n) \Rightarrow P(n+1)$.

Then P holds for all positive integers from 1 to k.

PROBLEM 15.8. Prove Theorem 15.6.

Here is still another useful induction principle:

THEOREM 15.7. *Suppose a property P of sets satisfies the following two conditions:*

(1) P holds for the empty set.

(2) For any finite set A and any element x outside A, if P holds for A, then P holds for $A \cup \{x\}$.

Then P holds for all finite sets.

PROBLEM 15.9. Prove Theorem 15.7.

In the last chapter we stated, without proof, the following:

THEOREM 15.8. *No finite set can be put into a 1-1 correspondence with any of its* proper *subsets.*

PROBLEM 15.10. Prove Theorem 15.8 (by induction on the number n of elements of the set).

Theorem 15.8 has some important corollaries.

Suppose n and m are *distinct* natural numbers. Then either $m<n$ or $n<m$; hence of the two sets $\{1,\ldots,n\}$, $\{1,\ldots,m\}$, one is a *proper* subset of the other; hence by Theorem 15.8, they cannot be put into a 1-1 correspondence. And so we have the following corollaries.

COROLLARY 15.1. *If $n \neq m$, then $\{1,\ldots,n\}$ cannot be put into a 1-1 correspondence with $\{1,\ldots,m\}$.*

COROLLARY 15.2. *If A is finite, then there is one* and only one *natural number n such that A has n elements.*

This proves the fact (which was actually obvious without proof) that if we count the elements of a finite set, the number we arrive at does not depend on the order in which we count them.

In the last chapter we proved that any infinite set can be put into a 1-1 correspondence with some proper subset of itself, and now we have proved that no finite set can. And so we have

THEOREM 15.9. *A set is infinite* if and only if *it can be put into a 1-1 correspondence with some proper subset of itself.*

A PARADOX

Here is a proof by induction that given any finite set of horses, all of them must be of the same color!

We let $P(n)$ mean that for any set of n horses, all of them are of the same color. We prove by induction on the positive integers that $P(n)$ holds for every positive n.

BASIS STEP. Obviously, given any set of horses with just one horse, all of them are of the same color.

INDUCTION STEP. Suppose n is such that for *every* set of n horses, all of them are of the same color. Now consider any $n+1$ horses $H_1,\ldots,H_n$, H_{n+1}. By the inductive hypothesis, we know $H_1,\ldots,H_n$ are all of the same color. Also all the horses $H_2,\ldots,H_n,H_{n+1}$ are of the same color (because the set $\{H_2,\ldots,H_{n+1}\}$ contains exactly n horses, and for *every* set of n horses, all of them are of the same color.) In particular, H_{n+1} is of the same color as H_n, and hence of the same color as all the horses $H_1,\ldots,H_n$. Thus $H_1,\ldots,H_n,H_{n+1}$ are all of the same color. This completes the induction and hence proves that for any positive n, any n horses are all of the same color.

PROBLEM 15.11. The above proof must obviously contain a fallacy *somewhere*. Where?

SOME MULTIPLE INDUCTIONS. One sometimes proves a result by using an induction *within* an induction, as the following problems will reveal.

PROBLEM 15.12. We are given a denumerable sequence $M_1, M_2, \ldots, M_n, \ldots$ of men and a denumerable sequence $W_1, W_2, \ldots, W_n, \ldots$ of women. We are given the following two facts:

(1) M_1 loves all the women and W_1 is loved by all the men.

(2) For any positive integers x and y, if M_x loves W_y and M_x loves W_{y+1} and M_{x+1} loves W_y, then M_{x+1} loves W_{y+1}.

The problem is to prove that all the men love all the women. *Hint:* Call a woman *desirable* if all the men love her. Show by induction on n that for every n, the woman W_n is desirable. (In going from W_n to W_{n+1}, another induction is involved!)

PROBLEM 15.13. In another universe of infinitely many men $M_1, M_2, \ldots, M_n, \ldots$ and infinitely many women $W_1, W_2, \ldots, W_n, \ldots$, the following facts hold:

(1) All the women love M_1.

(2) For any positive integers m and n, if W_m loves M_n and W_n loves M_m, then W_m loves M_{n+1}.

Prove that all the women love all the men. *Hint:* First show that for any n, if all the women love M_n, then W_n loves all the men.

PROBLEM 15.14. In a certain far-off universe, there is an emperor who has a library of infinitely many books! But there are only denumerably many, and they are numbered Book 1, Book 2,..., Book n,.... Moreover, each book has denumerably many pages: Page 1, Page 2,..., Page n,.... Some of the pages are blank. The following facts hold for his library:

(1) Page 1 of Book 1 is blank.

(2) For each book, if Page k is blank, so is Page $k+1$.

(3) For each n, if all the pages of Book n are blank, then Page 1 of Book $n+1$ is blank.

Suppose you pick a book at random and turn at random to one of the pages. What is the probability that you will find some printing on it?

SOLUTIONS

15.1. If he were a knight, we would get the following contradiction: What he said must have been true, which means that he really *had* made that same statement previously. When he made it then, he was still a knight, hence he must have said it a time before that. Hence also a time before that, hence a time before that,... - and so, unless the guy has lived infinitely far back in the past, he can't be a knight, he must be a knave.

Another way of looking at it is this: Since he said it once, there must have been a first time when he said it, and when he said it then, it was obviously false.

15.2. Of course the sage was right! If I truthfully say: "I will repeat this sentence tomorrow," then tomorrow I will indeed repeat it, and if I do so truthfully, then I will repeat it the day after that, and so forth, through all eternity.

15.3. The subsets of $A \cup \{x\}$ consist of the subsets of A without x adjoined, together with the subsets of A *with* x adjoined, so there are twice as many subsets of $A \cup \{x\}$ as there are of A.

Now for the inductive proof that, for every natural n, every set of n elements has 2^n subsets. For $n=0$, consider any set A that has 0 elements. Then A is the empty set, and hence it has only one subset, A itself. Since $2^0=1$, a set with 0 elements has 2^0 subsets, and thus the proposition is true for $n=0$.

Next, suppose that n is a number for which the proposition is true; i.e., suppose that every set of n elements has exactly 2^n subsets. We must show that the proposition also holds for $n+1$. Well, consider any set B with $n+1$ elements. Then B has at least one element x; remove x from B and let A be the remaining set ($A = B - \{x\}$). Then $B = A \cup \{x\}$. Since A has n elements, it follows by the inductive hypothesis that A has 2^n subsets. Then B has twice as many subsets as A, so the number of subsets of B is 2×2^n, which is 2^{n+1}. Thus the proposition holds for $n+1$ (assuming it holds for n), and this concludes the induction.

15.4. It is not immediately obvious that from the given conditions, we can infer that if P holds for n, then P holds for $n+1$, so it is not immediately obvious how we can use mathematical induction. But now comes a cute trick: Define $Q(n)$ to mean that P holds for n *and for all natural numbers less than* n.

Does Q hold for 0? Well, P holds for 0, and since there are no natural numbers less than 0, P vacuously holds for all natural numbers less than 0. (If you doubt this, just try to find a natural number less than 0 for which P *doesn't* hold! Let us recall that if there are no Frenchmen in a given club, then all Frenchmen in the club wear hats!) Thus P holds for 0 and all natural numbers less than 0, so Q holds for 0.

Now suppose Q holds for n. Thus P holds for all numbers from 0 to n. Then, as given, P holds for $n+1$. Hence P holds for $n+1$ and all numbers less than $n+1$, which means that Q holds for $n+1$. Thus $Q(n)$ *does* imply $Q(n+1)$, and since Q holds for 0, it follows that, by ordinary mathematical induction, Q holds for all natural numbers. Since $Q(n)$ obviously implies $P(n)$, P, too, holds for all natural numbers.

NOTE. The "trick" I employed is a special case of the following more general "trick," which is sometimes most useful: It frequently happens that one wants to show by induction that a certain property P holds for all (natural) numbers, but one cannot directly show that $P(n)$ implies $P(n+1)$ for all n. One can, however, sometimes find a stronger property Q (stronger in the sense that $Q(n)$ always implies $P(n)$) and show for this stronger property Q that $Q(n)$ implies $Q(n+1)$ for all n, and that Q holds for 0. Then, by induction *with respect to Q*, Q holds for all n, hence also so does P.

15.5. The answer is *yes*, and this principle is known as the principle of *complete* induction. It can easily be derived from the last problem, as follows:

Suppose P is such that for all n, if P holds for all natural numbers less than n, then P holds for n. Then P holds for 0, since P vacuously holds for all natural numbers less than 0. Now suppose P holds for n and all numbers less than n. Then P holds for all numbers less than $n+1$, hence P also holds for $n+1$, as we were given. Thus P holds for 0 and it is also the case that for each n, if P holds for all numbers less than or equal to n, then P holds for $n+1$. Then by the principle of Problem 15.4, P holds for all natural numbers.

15.6. First for an informal argument: Suppose P fails for some number. Then it fails for some lesser number. Then it fails for a still lesser number, and then a still lesser number, until finally it must fail for 0. But then it can't fail for any lesser natural number, because there isn't any. This contradicts the given condition, hence the given condition implies that P can't fail for *any* natural number, hence P holds for *all* natural numbers. Thus it is (2) that follows.

Actually, this is nothing more than the principle of complete induction, stated in a different but equivalent form: To say that if P fails for n, then P fails for some number less than n is but another way of saying that if P holds for all numbers less than n, then P holds for n. So this is indeed the principle of complete induction. Stated in this form, however, it is known as the *principle of finite descent*.

15.7. (a) Starting with mathematical induction, we can derive the least number principle as follows: Define $P(n)$ to mean that every set of natural numbers that contains any number no greater than n has a least member. We show by induction that $P(n)$ holds for every n.

BASIS STEP. Any set A that contains a number no greater than 0 contains 0, hence 0 is the least member of A. Thus $P(0)$ holds.

INDUCTION STEP. Suppose $P(n)$. Now let A be any set that contains a number no greater than $n+1$. Either it contains a number no greater than n or it doesn't. If it does, then by the inductive hypothesis $P(n)$ it must contain a least member. If it doesn't, then it must contain $n+1$ (since it contains *some* number no greater than $n+1$) and then $n+1$ must be the least member of A (since A contains no number less than $n+1$). This completes the induction, so $P(n)$ holds for all n.

Now consider any non-empty set A of natural numbers. It must contain some number n, hence some number no greater than n; and since $P(n)$ holds, A must have a least number.

(b) If we go in the other direction and start with the least number principle, we can derive the principle of mathematical induction as follows. Suppose P is a property satisfying the following two conditions:

(1) P holds for 0.

(2) For any number n, if P holds for n, then P holds for $n+1$.

Now suppose there are some numbers for which P fails. Then by the least number principle there must be a least number m for which P fails. By C_1, m cannot be 0. Hence $m = n+1$ for some n (namely, for $m-1$). Since m is the least number for which P fails, and n is less than m, P must hold for n. Thus P holds for n but fails for $n+1$, violating C_2. Therefore there cannot be any number for which P fails.

REMARKS. There is a well-known mathematical joke that proves that all natural numbers are *interesting*. Well, suppose there were uninteresting numbers. Then by the least number principle, there would have to be the *least* uninteresting number, and such a number would be most interesting indeed!

15.8. Intuitively, this is obvious. We are given $P(1)$ outright, and, by (2), we can successively get $P(2), P(3), \ldots$, up to $P(k)$.

Also, this can be more rigorously proved as a consequence of the principle of mathematical induction: Define $Q(n)$ to mean $n \leq k \Rightarrow P(n)$. From the given condition (1), we easily get $Q(1)$. And from the given condition (2), we can get $Q(n) \Rightarrow Q(n+1)$ (I leave the details to the reader); thus, by ordinary mathematical induction, $Q(n)$ holds for all n, and hence $P(n)$ holds for all $n \leq k$.

15.9. We prove this by induction on the number of elements of the set. That is, we show by induction on n that every n-element set (every set with n elements) has property P.

BASIS STEP. The only zero-element set is the empty set, and we are given that it has property P; hence *all* zero-element sets (of which there is only one) have property P.

INDUCTION STEP. Suppose that n is a number such that all n-element sets have property P. Any $(n+1)$-element set B consists of some n-element set A with an element x not in A adjoined to A (that is, if we let x be any element of B and let $A = B - \{x\}$, then $B = A \cup \{x\}$). Then A has property P by our inductive hypothesis about n, therefore so does B, by the second given condition. This completes the induction.

More informally, and perhaps more convincingly, the empty set initially has property P, and every time we add a new element to a set having the property, the resulting set has property P, so of course all finite sets have property P.

15.10. Let us first observe the following more general fact, which is quite useful to know in its own right: One way to show that a given property P cannot hold for any *finite* set is to show that if P holds for some finite set F, then P must hold for some *proper* subset of F (and hence for some proper subset of that subset, and so on down the line, and eventually reach the empty set, which has no proper subsets). The informal argument can be made more rigorous by induction on the number of elements of the set, or, better still, by the method of finite descent.

Now, let $P(S)$ be the property that S can be put into a 1-1 correspondence with a *proper* subset of itself. Suppose a *finite* set A has this property, so that A can be put into a 1-1 correspondence with its proper subset B. Since B is a proper subset of A, A must contain some element x not in B. Under the 1-1 correspondence, x corresponds to some element y of B. Well, remove x from A and y from B, and the resulting set A_1 (i.e., $A-\{x\}$) is in a 1-1 correspondence with the resulting set B_1 (i.e., $B-\{y\}$). Thus the *proper* subset A_1 of A is in a 1-1 correspondence with *its* proper subset B_1. Hence if A has the property P, so does its proper subset A_1. Therefore, no finite set can have this property.

15.11. The induction fails to go from (1) to (2). For any number n greater than 1, there is indeed an overlap between the sets $\{H_1, \ldots, H_n\}$ and $\{H_2, \ldots, H_{n+1}\}$; at the very least, H_n is common to both. But for $n=1$, we have the sets $\{H_1\}$ and $\{H_2\}$, with no overlap. So it is false that for all n, $P(n) \Rightarrow P(n+1)$, because $P(1) \Rightarrow P(2)$ is false. Indeed, if any two horses were of the same color, then *all* horses would be of the same color!

QUESTION. What about the empty set of horses? Are all its members of the same color? The answer is *yes*. If you doubt this, just try to find in the empty set two horses that are *not* of the same color!

15.12. We are to show that for all n, W_n is desirable.

BASIS STEP. We are given that all the men love W_1, hence W_1 is desirable.

INDUCTIVE STEP. Suppose W_n is desirable. We are to show that W_{n+1} is—that is, that every man M_m loves W_{n+1}. We do this by induction on m. (This is where an induction within an induction comes in!) Well, we are given that M_1 loves W_{n+1} (he loves all the women). Now suppose that M_m loves W_{n+1}. Also M_m and M_{m+1} both love W_n (since she is desirable). Thus M_m loves W_n, M_m loves W_{n+1}, and M_{m+1} loves W_n; so, by the given condition (2), M_{m+1} loves W_{n+1}. This proves that if M_m loves W_{n+1}, so does M_{m+1}; and since M_1 does, it follows by induction that every man loves W_{n+1}; hence W_{n+1} is desirable. This proves that if W_n is desirable, so is W_{n+1}; and since W_1 is desirable, it follows by mathematical induction that all the women are desirable. Hence all the men love all the women.

15.13. This proof is more tricky!

STEP 1. We show that for any n, if all the women love M_n, then W_n loves all the men. So suppose that all the women love M_n. We show by induction on m that for all m, W_n loves M_m.

BASIS STEP. It is given that W_n loves M_1 (all the women do).

INDUCTION STEP. Suppose that W_n loves M_m. Also W_m loves M_n (all women love M_n by the inductive hypothesis). Thus W_n loves M_m and W_m loves M_n, hence W_n loves M_{m+1} (by the given condition (2)). This completes the induction.

STEP 2. We now show by induction on n that all the women love M_n (for all n).

BASIS STEP. We are given that all the women love M_1.

INDUCTION STEP. Now suppose that n is a number such that all the women love M_n. Then also W_n loves all the men (by Step 1). Therefore for every number m, W_m loves M_n and W_n loves M_m (she loves all the men) and therefore W_m loves M_{n+1} (by the given condition (2)); hence M_{n+1} is loved by every woman W_m. This completes the induction.

15.14. As you have most likely guessed, or proved, the probability is zero; all the pages must be blank, by the following argument: Let $B(n,k)$ mean that the kth page of Book n is blank, and let $P(n)$ mean that all the pages of Book n are blank. By induction we will prove $P(n)$ for all n.

BASIS STEP. We are given $B(1,1)$. It follows from (2) (the second given condition) that for all k, $B(1,k)$ implies $B(1,k+1)$. Then by induction on k, it follows that $B(1,k)$ holds for all k, and hence $P(1)$. (Here is an induction within an induction!)

INDUCTION STEP. Suppose n is such that $P(n)$. Then $B(n+1,1)$ (by (3)). Also (by (2)) for any k, $B(n+1,k)$ implies $B(n+1,k+1)$. Hence by induction on k, $B(n+1,k)$ must hold for every k, and thus $P(n+1)$. This proves $P(n) \Rightarrow P(n+1)$, which completes the induction.

REMARKS. Isn't that a funny library with infinitely many books, each with infinitely many pages, and all of them blank! A perfect case of much ado about nothing! I am reminded of the Chinese novel *Monkey*, in which the hero, a mischievous but rather lovable monkey, goes to great lengths to get a copy of the scriptures of Buddha. When he finally gets them he angrily finds all the pages blank! He goes to Buddha to complain. Buddha smiles and says: "Actually blank scriptures are the best. But the people of China don't realize this, so to keep them happy, when I give them scriptures, I put some writing in them."

- Chapter 16 -

Generalized Induction, König's Lemma, Compactness

First, for a little problem: There is a strange planet named *Vlam* on which the inhabitants are very much like us, except that they are immortal. The planet had a beginning in time, however. A curious thing about this planet is that, for any inhabitant x, if all children of x have blue eyes, so does x.

Problem 16.1. Suppose that an individual x on this planet has no children. Can it be determined from the above given condition whether x has blue eyes or not?

A more serious question: Can it be determined from the given condition just what percentage of the inhabitants have blue eyes? Yes, it can, and the solution will emerge from the subject to which we now turn.

Generalized Induction

The principle of complete mathematical induction (Theorem 15.3), which is a theorem about the natural numbers, has an important generalization to arbitrary sets, even non-denumerable ones! We now consider an arbitrary set A of any size and a relation $C(x,y)$ between elements of A, which we read "x is a *component* of y." (The notion of *component* has many applications in set theory, number theory and logic. In set theory, the components of a set are the elements of the set. In some applications to number theory, m is said to be a component of n if $m<n$. In other applications, m is said to be a component of n if $m+1=n$. For logic, the notion of *components* of signed formulas (as defined in earlier chapters) will be seen to play a key role.)

Descending Chains

By a *descending chain* (for the component relation $C(x,y)$) is meant a finite sequence $(x_1, x_2, \ldots, x_n)$ or a denumerable sequence $(x_1, x_2, \ldots, x_n, \ldots)$ such that each term of the sequence other than the first is a component of the preceding term (x_2 is a component of x_1, x_3 is a component of x_2, etc.).

We will be mainly concerned with component relations in which there are no infinite descending chains, because such relations will be seen to obey a very important induction principle. But, first:

PROBLEM 16.2. Suppose that all descending chains are finite.

(a) Is it possible for an element x to be a component of itself?

(b) Is it possible for there to be two elements x and y such that x is a component of y and y is a component of x?

PROBLEM 16.3.

(a) Consider the natural numbers $0, 1, 2, \ldots, n, \ldots$, and define $C(x,y)$ to be "y is the successor of x (i.e., $x+1=y$)." Are there any infinite descending chains?

(b) Suppose that instead of the natural numbers we consider the set of *all* whole numbers, positive and negative $(\ldots, -3, -2, -1, 0, 1, 2, 3, 4, \ldots)$. Are there any infinite descending chains?

Generalized Induction

We continue to consider a component relation $C(x,y)$ on the elements of a set A. A property P of elements of A will be said to be *inductive* (with respect to the component relation, understood) if for every element x of A, if all components of x have the property P, so does x. This is understood to imply that if x has no components at all, then x must have the property, since it is vacuously true that *all* components of x (of which there are none) have the property. Again, I must remind the reader that if a set S has no elements at all, then anything we say about *all* elements of S must be true.

We now say that the component relation obeys the *Generalized Induction Principle* if, for every inductive property P, P must hold for all elements of A. Thus if the Generalized Induction Principle holds—that is, if $C(x,y)$ obeys the Generalized Induction Principle—then to show that a given property P holds for all elements of A, it suffices to show that for each element x of A, if all components of x have property P, so does x.

Let us note that the principle of complete mathematical induction (Theorem 15.3) is but a special case of generalized induction, in which A is the set of natural numbers and the component relation $C(x,y)$ is $x+1=y$.

The following result is basic:

THEOREM 16.1 (GENERALIZED INDUCTION THEOREM). *A sufficient condition for a component relation $C(x,y)$ to obey the Generalized Induction Principle is that there be no infinite descending chains (all descending chains are finite). In other words, if all descending chains are finite, then $C(x,y)$ obeys the Generalized Induction Principle.*

PROBLEM 16.4. Prove Theorem 16.1. *Hint:* If a property P is inductive, then for any element x of A, if P fails to hold for x, then it must fail for at least one component of x.

We can now answer the question we raised about the planet Vlam. We take A to be the set of inhabitants of Vlam and the component relation $C(x,y)$ to be "x is a child of y." Thus the components of x are the children of x. The given condition about blue eyes (x has blue eyes provided all children of x do) is thus that the property of having blue eyes is inductive. A chain $(x_1,\ldots,x_n)$ now is simply a sequence in which for each $i < n$, x_{i+1} is a child of x_i. Obviously all such chains are finite, so, by Theorem 16.1, the Generalized Induction Principle holds, and since the property of having blue eyes is inductive, it follows that *all* inhabitants of Vlam have blue eyes!

Theorem 16.1 has a very important application to formulas of propositional and first-order logic: We recall that we defined the *components* of an α to be α_1 and α_2; those of β are β_1 and β_2; those of γ are all formulas $\gamma(a)$, where a is any parameter, and the components of δ are all formulas $\delta(a)$. Obviously for any formula X, all components of X have fewer logical connectives or quantifiers than X, hence every descending chain starting with X must after finitely many steps eventuate in an atomic formula, so there are no infinite chains. Indeed, we define the *degree* of a formula as the number of occurrences of logical connectives or quantifiers, so no descending chain starting with a formula X can be longer than n, where n is the degree of X.

Since there are no infinite descending chains for formulas under our component relation, we have the following vital induction principles for formulas.

COROLLARY 16.1 (FORMULA INDUCTION). *To show that a given property of formulas holds for all formulas, it suffices to show that, for any formula X, if the property holds for all components of X, then it holds for X too.*

The converse of Theorem 16.1 also holds, namely:

THEOREM 16.2. *If the component relation obeys the generalized induction principle, then all descending chains are finite.*

PROBLEM 16.5. Prove Theorem T16.2. *Hint:* Consider the property $P(x)$ "all descending chains beginning with x are finite."

WELL-FOUNDEDNESS

We continue to consider a component relation $C(x,y)$ among elements of a set A. For any subset S of A, an element x of S is called an *initial* element of S if it has no components inside of S (either it has no components at all, or it does, but all of them lie outside of S).

Now we define the relation $C(x,y)$ to be *well-founded* if every non-empty subset S of A contains at least one initial element.

As an example, the relation $x+1=y$ on the natural numbers is well-founded (obviously for any non-empty set S of natural numbers, its least element is an initial element).

The following problem is easy.

PROBLEM 16.6. Show that well-foundedness implies that there are no infinite descending chains.

The following problem is a bit more tricky.

PROBLEM 16.7. Prove that, if there are no infinite descending chains, then the component relation is well-founded.

By virtue of the last two problems and Theorems 16.1 and 16.2, we now have the following lovely result.

THEOREM 16.3. *For any component relation $C(x,y)$ on a set A, the following three conditions are equivalent.*

(1) The Generalized Induction Principle holds.

(2) There are no infinite descending chains.

(3) The relation is well-founded.

Trees and Ball Games

A Ball Game

Imagine that you live in a universe in which everyone is immortal, except for the possibility of being executed. You are required to play the following game:

There is an infinite supply of pool balls available, each bearing a number (a positive integer). There are infinitely many 1's, infinitely many 2's, and, for each n, there are infinitely many n's. In a certain box there are *finitely* many of these balls, but the box has infinite capacity. Each day you are required to throw out a ball and replace it by *any* finite number of *lower numbered* balls—for example, you may throw out a 58 and replace it by a billion 57's. If you throw out a 1, you cannot replace it by anything. If ever the box becomes empty, you get executed!

Problem 16.8. Is there a strategy by which you never get executed, or is execution inevitable sooner or later?

König's Lemma

We have already encountered *trees* in the form of tableaux. More generally, a tree consists of an element a_0 called the *origin*, which is connected to a finite or denumerable set of elements called the *successors* of a_0, each of which in turn is connected to a finite or denumerable set of *its* successors, and so forth. In a tree, if y is a successor of x, then we say that x is a *predecessor* of y. The origin has no predecessor and is the only element of the tree that has no predecessor, and every other element has one and only one predecessor. Elements of trees are also called *points*. An element is called an *end point* if it has no successors, and a *junction point* otherwise. The origin of the tree is said to be of *level* 0, its successors are of level 1, the successors of these successors of level 2, and so forth. Thus for any n, if x is of level n, all successors of x are of level $n+1$. In displaying trees diagrammatically, the origin is placed at the top and the tree grows *downward*, as with tableaux. By a *path* of the tree is meant a finite or denumerable sequence such that the first term is the origin of the tree and each of the other terms is a tree-successor of the preceding term. (Thus a path is obtained by starting at the origin a_0, then taking a successor a_1 of a_0, then a successor a_2 of a_1, and so forth.) Since an element cannot have more than one predecessor, it follows that for any element x, there is one and only one path from the origin down to x. For any finite path, by its *length* is meant the number of terms of the sequence, or, what is the same thing, the level of its last term. Thus a path $(a_0, \ldots, a_n)$ is of length $n+1$. By the *descendants* of a point x are meant the successors of x, together with the successors of the successors of x, and so forth. Thus y is a descendant of x if and only if there is a path through x down to y.

Problem 16.9. Suppose we are given that for each positive integer n, there is at least one path of length n (and thus every level is hit by at least one path). Does it necessarily follow that there must be at least one infinite path? (Most people answer this incorrectly!)

Finitely Generated Trees

A tree is said to be *finitely generated* if each point has only finitely many successors (though it might have infinitely many descendants). Now, suppose that in Problem 16.9 we were given the additional information that the tree is finitely generated. Would that change the answer? It sure would!, as we will soon see. But first for two subsidiary problems.

Problem 16.10. Suppose a tree is finitely generated. Does it necessarily follow that each level contains only finitely many points?

Problem 16.11. Suppose a tree is finitely generated. Prove that the following two statements are equivalent:

(1) For each positive integer n, there is at least one path of length n.

(2) The tree has infinitely many points.

Note. This can be equivalently stated: For any finitely generated tree the following two conditions are equivalent.

(1) There is some n such that no path has length greater than n.

(2) The tree has only finitely many points.

Now, in the solution to Problem 16.9, we saw an example of a tree such that, for each n, there is a path of length n, yet there is no infinite path. Of course, that tree is not finitely generated (the origin has infinitely many successors). Now, to repeat an earlier question, if for each n there is a path of length n *and the tree is finitely generated*, does it then necessarily have at least one infinite path? The answer is *yes*, and this is König's famous Infinity Lemma. We have seen (Problem 16.11) that for a finitely generated tree, to say that for each positive integer n there is a path of length n, is to say nothing more nor less than that the tree has infinitely many points; the following is an equivalent form of König's Lemma.

Theorem 16.4 (König's Lemma). *A finitely generated tree having infinitely many points must have at least one infinite path.*

Problem 16.12. Prove König's Lemma. *Hint:* Call a point of the tree *rich* if it has infinitely many descendants (I mean *descendants* now, not successors!). If a rich point has only finitely many successors, must it necessarily have at least one *rich* successor?

A tree is called *finite* (not to be confused with finitely generated!) if it has only finitely many points; otherwise it is called *infinite*. Thus Theorem 4 says that any infinite but finitely generated tree must have an infinite path.

Now, suppose we have a finitely generated tree such that all its paths are finite. Thus the tree is finitely generated but has no infinite path. Then by Theorem 16.4, the tree must be finite (because if it were infinite, it *would* have an infinite path). Thus Theorem 16.4 implies (and in fact is equivalent to):

Theorem 16.5 (Brouwer's Fan Theorem). *For a finitely generated tree, if all paths are finite, then the tree is finite.*

Discussion. Brouwer's Fan Theorem is in fact the *contrapositive* of König's Lemma. (By the *contrapositive* of an implication $p \Rightarrow q$ is meant the proposition $\sim q \Rightarrow \sim p$.) Now, the logic we are using in this book is known as *classical* logic, a feature of which is that any implication is logically equivalent to its contrapositive. There is a weaker system of logic known as *intuitionistic* logic, which we will discuss in a later chapter, in which not all implications can be shown to be equivalent to their contrapositives. In particular, the Fan Theorem (Theorem 16.5) can be proved in intuitionistic logic, but König's Lemma cannot.

Tree Induction

The proof we have given of König's Lemma largely duplicated some results of the first section of this chapter—that is, we can derive the Fan Theorem (and hence also König's Lemma) as a corollary of Theorem 16.1. Let us now look at trees from the viewpoint of that first section: We take the *components* of a point x to be the *successors* of x. The following, then, is but a special case of Theorem 16.1.

Theorem 16.6 (Principle of Tree Induction). *Suppose that all branches of a tree are finite. Then a sufficient condition for a property P to hold for all points of the tree is that, for each point x, P holds for x provided it holds for all successors of x.*

Theorem 16.6 holds even if the tree is not finitely generated. It may have infinitely many points, but each branch must be finite. Now suppose that, in addition, the tree is finitely generated. Then for any point

x, if each successor of x has only finitely many descendants, so does x. Therefore, by the above theorem, every point of the tree has only finitely many descendants. In particular, the origin has only finitely many descendants, so the tree is finite. This proves the Fan Theorem.

BALL GAMES REVISITED

In the solution to Problem 16.8 we proved

THEOREM 16.7 (BALL GAME THEOREM). *Every ball game of the type described in Problem 16.8 must terminate in finitely many steps.*

This theorem is closely related to the Fan Theorem—indeed, it can be derived as a corollary of it, thus yielding an interesting alternate proof: Without loss of generality, we can assume that there is only one ball initially in the box, because if we start with a finite set S of balls in the box, we could instead start with just one ball x whose label is a greater number than any of those in S, and on the first move, replace x by the balls in S. So we shall assume that we initially have only one ball in the box. Then to any such ball game $\mathfrak{B}$, we associate the following tree, which we will denote $\text{tr}(\mathfrak{B})$ ("the tree of $\mathfrak{B}$"). We take the ball that is initially in the box as the *origin*, and for any ball that is ever in the box, we take its *successors* to be the balls that replace it (balls numbered 1 are of course end points). Since each ball is replaced by only finitely many balls, the tree is finitely generated. Since each ball is replaced by balls bearing lower numbers, each path must be finite. Then, by the Fan Theorem, the whole tree is finite—there are only finitely many points on the tree, which means that only finitely many balls ever enter the box, which wouldn't be the case if the game never terminated. Therefore the game must terminate.

We have now seen two different proofs of the Ball Game Theorem—the first was by mathematical induction on the greatest number assigned to any ball originally in the box, and the second, by the Fan Theorem. But there is a simpler proof yet, which uses neither the induction technique of the first proof nor the Fan Theorem. The proof is simply this: In the tree $\text{tr}(\mathfrak{B})$ associated with a ball game, not only are all paths finite, but no path can be of length greater than n, where n is the number assigned to the original ball! Thus only finitely many levels are hit, and since in a finitely generated tree, each level can have only finitely many points (Problem 16.10), the tree must be finite.

You see, a ball game is really more than just a finitely generated tree whose paths are all finite; it is that together with an assignment of positive integers to all points of the tree such that each end point is assigned the number 1 and each other point x is assigned a greater number than any of its successors. Such a tree, we will say, is *well-numbered*. Can every finite tree be well-numbered? (If so, then every finite tree is the tree of some ball game.) As a matter of fact, even an infinite tree can be well-numbered, provided that there is an upper bound to the lengths of all its branches.

Problem 16.13. How can such a tree be well-numbered?

Compactness

König's Lemma will be seen to have interesting applications to propositional and first-order logic. We shall shortly consider another principle that has equally important such applications, but we first turn to some related problems.

We consider a universe V with denumerably many inhabitants. They have formed various clubs. A club C is called *maximal* if it is not a *proper* subset of any other club. Thus if C is a maximal club, then for any set S that contains all the people in C, if S contains so much as one person who is not in C, then S fails to be a club.

Problem 16.14.

(a) In a denumerable universe V, assuming that there is at least one club, is there necessarily a maximal club?

(b) What about if V is finite instead of denumerable? If there is at least one club, is there necessarily a maximal club?

Now for the *key* problem!

Problem 16.15. Again we consider a universe V of denumerably many people, and we assume there is at least one club. We have seen that it does not follow that there must then be a maximal club, but now, suppose we are given the additional information that for any set S of inhabitants of V, S is a club if and only if every *finite* subset of S is a club. Then it *does* follow that there must be a maximal club! Better yet, it follows that *every* club C is a subset of some maximal club. The problem is to prove this.

Here are the key steps in the proof.

(1) Show that, under the given conditions, every subset of a club must be a club.

(2) Since V is denumerable, its inhabitants can be arranged in a denumerable sequence $x_0, x_1, x_2, \ldots, x_n, \ldots$. Let C be any club. Now

define the following infinite sequence $C_0, C_1, C_2, \ldots, C_n, \ldots$ of sets (of people):

(a) Take C_0 to be C.

(b) If adding x_0 to C yields a club, let C_1 be this club $C_0 \cup \{x_0\}$; otherwise let C_1 be C_0 itself. Then consider x_1; if $C_1 \cup \{x_1\}$ is a club, let C_2 be this club; otherwise take C_2 to be C_1. And continue this throughout the series—that is, once C_n has been defined, take C_{n+1} to be $C_n \cup \{x_n\}$ if this set is a club, and otherwise take C_{n+1} to be C_n. Now show that each C_n is a club.

(3) Now consider the set S of all members of C_0 together with all the people who have been added at any stage (thus S is the set of all people x who belong to at least one of the sets $C_0, C_1, C_2, \ldots, C_n, \ldots$). Then show that S must be a club, and in fact a maximal club.

Consider an arbitrary set A and a property P of subsets of A. The property P is said to be of *finite character*, or *compact*, if, for any subset S of A, S has property P if and only if all *finite* subsets of S have property P. Also, a collection Σ of subsets of A will be said to be *compact* if the property of being a member of Σ is compact—in other words, for any subset S of A, S is a member of Σ if and only if all finite subsets of S are members of Σ. A set S is said be a *maximal* subset of A having property P if S has property P and is not a proper subset of any other subset of A having property P. Likewise, S is said to be a maximal element of Σ if S is an element of Σ and is not a proper subset of any other element of Σ.

In Problem 16.15 we were given a *denumerable* set V and a collection of subsets called clubs, and we were given that this collection was compact, from which we concluded that any club was a subset of a maximal club. Now, there is nothing special about *clubs* that made our argument go through; the same reasoning yields the following:

THEOREM 16.8 (DENUMERABLE COMPACTNESS THEOREM). *For any compact property P of subsets of a* denumerable *set A, any subset of A having property P is a subset of a* maximal *subset of A having property P.*[1]

REMARKS. In mathematical logic, we have various procedures for showing certain sets of formulas to be *inconsistent*. Any derivation of the inconsistency of a set uses only finitely many members of the set, hence a set S is inconsistent if and only if some finite subset of S is inconsistent, and thus S is *consistent* if and only if all of its finite subsets are consistent. Thus consistency is a property of finite character, a fact that will later be seen to be of considerable importance.

[1] Actually, the above result holds even if the set A is non-denumerable, but this more advanced result will not be needed for anything in this book.

More on Compactness

For any property P of sets and any set S, define $P^{\#}(S)$ to mean that all finite subsets of S have property P. Thus, given a property P, we have a new property $P^{\#}$ closely associated with P.

The following result will prove quite useful.

Theorem 16.9. *For any property P, the property $P^{\#}$ is compact.*

Problem 16.16. Prove Theorem 16.9.

From Theorems 16.8 and 16.9 we thus have:

Theorem 16.10. *For any property P of subsets of a denumerable set A, any subset S of A having property $P^{\#}$ (the property that all finite subsets of S have property P) is a subset of some* maximal *subset of A having property $P^{\#}$.*

Solutions

16.1. Yes, it can. If x has no children, then whatever is said about *all* of x's children must be true. (We recall once again that in any club with *no* Frenchmen, *all* Frenchmen in the club wear hats.) Thus it is *vacuously* true that all children of this childless x have blue eyes, and hence so does x. And so every childless person in Vlam has blue eyes.

16.2. (a) No, it is not possible, because if x were a component of itself, we would have the infinite descending chain $(x, x, x, x, \ldots, x, x, \ldots)$.

(b) Also not possible, because if it was, then we would have the infinite descending chain $(x, y, x, y, \ldots, x, y, \ldots)$.

16.3. (a) Of course not!

(b) Of course! For example, the infinite descending chain $(4, 3, 2, 1, 0, -1, -2, -3, \ldots, -n, \ldots)$.

16.4. It is indeed obvious that if a property P is inductive, then if it fails for a given x, it must fail for at least one component of x (because if it failed for no component of x, it would hold for all components, and hence for x, which it doesn't).

Now suppose P is inductive. If it failed for some x, it would have to fail for some component x_1 of x, hence for some component x_2 of x_1, and so forth, and so we would have an infinite descending chain. Thus if there are no infinite descending chains, then P cannot fail for any element, hence must hold for all.

16.5. It is indeed obvious that if no component of x begins an infinite descending chain, neither does x. So the property of not beginning any infinite descending chain is indeed inductive. If the generalized induction principle holds, then every element x has that property, and hence there are then no infinite descending chains.

16.6. If there were an infinite descending chain, the set of terms of the chain would have no initial element, hence the component relation would not be well-founded.

16.7. Suppose that there are no infinite descending chains. Let S be any non-empty subset of A. We will show that S contains an initial element. Take any element x of S. If x is an initial element of S, we are done; otherwise x has a component y in S. If y is an initial element, we are done; otherwise, y has a component z in S—and we generate a descending chain in this manner. Since all descending chains are finite, we must sooner or later reach an initial element of S.

Thus every non-empty subset S of A contains an initial element, and so the component relation is well-founded.

16.8. Sooner or later the box must become empty. Here is one proof.

Call a positive integer n a *losing* number if it is the case that every ball game in which every ball originally in the box is numbered n or less—every such ball game must terminate. We show by mathematical induction that every n is a losing number.

Obviously 1 is a losing number. (If only 1's are originally in the box, the game clearly must terminate.) Now suppose that n is a losing number. Consider any box in which every ball is numbered $n+1$ or less. You can't keep throwing out balls numbered n or less forever (because by hypothesis, n is a losing number), hence sooner or later you must throw out a ball numbered $n+1$ (assuming there is at least one in the box). Then, sooner or later, you must throw out another ball numbered $n+1$ (if there are some left). And so, continuing this way, sooner or later you must get rid of all balls numbered $n+1$. Then you are down to a box in which the highest numbered ball is n or less, and from then on, the process must terminate (since n is a losing number). This completes the induction.

We have thus proved:

THEOREM 16.11. *Every ball game of the type described in Problem 16.8 must terminate in a finite number of steps.*

Another (and, I believe, more elegant) proof of the above will be given later.

REMARKS. The curious thing about this ball game is that there is no finite limit to how long the player can live (assuming that at least one initial ball is numbered 2 or higher), yet he cannot live forever. Given any finite number n, he can be sure to live n or more days, yet there is no way he can live forever!

16.9. The answer is *no*, as the following tree illustrates:

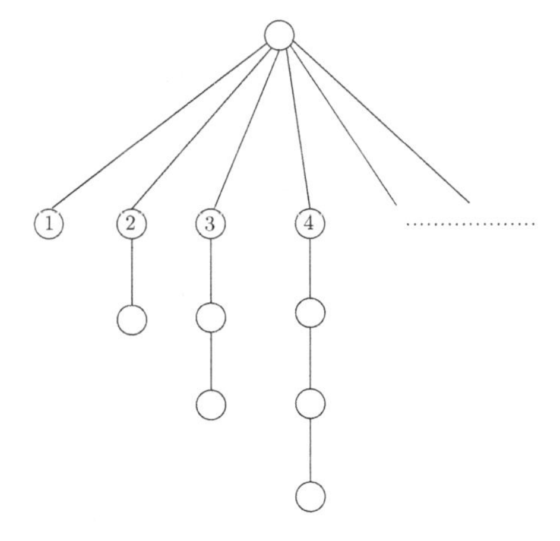

That is, the origin has denumerably many successor numbered $1, 2, \ldots, n, \ldots$. Every other point of the tree has at most one successor. The path through (1) stops at level 1. The path through (2) goes down only two levels, and so forth (for each n, the path through (n) stops at level n). Thus for each n, there is a path of length n (the path through (n)), but the tree has no infinite path.

Someone once objected: "What about the point at infinity?" Well, there is no such point on *this* tree; he was talking about a possible tree, but not the one I just described.

Some of you may still be unconvinced. For those of you who are, maybe this will help: Suppose I have a collection of sticks. One of them is 1 foot long, another is 2 feet long, and so on—for each positive n, I have a stick n feet long. Does it necessarily follow that one of the sticks must be infinitely long? Of course not! True, the collection itself is infinite—there must be infinitely many sticks in it—but that does not mean that one of the sticks itself must be infinite in length. And so a tree can have infinitely many paths, with no finite upper bound to the lengths of the paths, yet each path itself could be finite. In short, there is all the difference between a set having infinitely many members and having a member that itself is infinite. Indeed, the set of all natural numbers is infinite, but each of its members is a *finite* number.

16.10. Yes; for each positive integer n, the nth level contains only finitely many points, as is easily seen by induction on n: For $n=1$, there is only one point of level n, the origin. Now suppose n is such that there are only finitely many points $x_1, \ldots, x_k$ of level n; let n_1 be the number of successors of x_1, n_2 the number of successors of x_2, ..., n_k the number of successors of x_k. Then the number of points of level $n+1$ is $n_1 + \cdots + n_k$, a finite number.

16.11. Statement (1) obviously implies (2), since (1) implies that every level has at least one point. Next, suppose that (2) holds, that the tree has infinitely many points. If there was some n such that no path was of length n, then only finitely many levels would be hit, and hence the tree would have only finitely many points (as we have seen in the solution to the last problem), contrary to the given condition. Therefore for every n, there must be a path of length n.

16.12. The proof of König's Lemma (also called König's Infinity Lemma) is both remarkably simple and beautifully elegant!

We are calling a point *rich* if it has infinitely many descendants (I mean *descendants* now, not successors!), and let us call it *poor* if it has only finitely many descendants. Now, the key idea behind König's proof is that if a point x is rich and if x has only finitely many successors, then at least one of its successors must be rich, because if all its successors were poor, the point x itself would be poor. (More specifically, if $x_1, \ldots, x_k$ are all the successors of x and if x_1 has n_1 descendants, ..., x_k has n_k descendants, then x has exactly $k+n_1 + \cdots + n_k$ descendants, and hence is poor.) Thus every rich point with only finitely many successors must have at least one rich successor. (A rich point with infinitely many successors doesn't necessarily have to have a rich successor—for example, in the tree of Problem 16.9, the origin is rich, but all its successors are poor.)

Now suppose that the tree is finitely generated. Then each point has only finitely many successors, hence each *rich* point must have at least one rich successor. Supposing furthermore that the tree is infinite, the origin is obviously rich (since all points of the tree other than the origin are descendants of the origin). Since the origin (call it a_0) is rich, it must have at least one rich successor a_1, which in turn has at least one rich successor a_2, and so forth. In this way we get an infinite path $(a_0, a_1, a_2, \ldots, a_n, \ldots)$.

16.13. Very simply! Just assign to each point x the number n, where n is the length of the longest descending chain beginning with x.

16.14. (a) The answer is *no*, there doesn't necessarily have to be a maximal club. As an example, it could be that just the finite sets are clubs, and no others. There certainly is no *maximal* finite subset of V!

(b) If V were finite instead of denumerable, then the answer would be *yes*: Given a club C, either it is already maximal or it isn't. If it is, then no more need be said. If it isn't, then we can add some people to C to get a larger club C_1. If C_1 is maximal, we are done. Otherwise, we extend C_1 to a larger club C_2, and so on. Since V is finite, the process must terminate in some maximal club (which might or might not happen to be V itself). And so for a finite universe V the answer is *yes*, but for a denumerable universe, the answer is *no*.

16.15. (1) Suppose C is a club and S is a subset of C. Let F be any finite subset of S. Then F is a finite subset of C (because $F \subseteq S \subseteq C$). Therefore F must be a club (C is a club if and only if all finite subsets of C are; this is our hypothesis). Thus every finite subset of S is a club. Hence S is a club. This proves that every subset of a club is a club.

(2) In the sequence $C_0, C_1, \ldots, C_n, \ldots$, each set C_m is a subset of C_{m+1}, from which it follows by induction on n that for all n greater than m, C_m is a subset of C_n. Thus each term of the sequence is a subset of all later terms. From this it follows that for any two sets C_m and C_n of the sequence, one of them is a subset of the other, from which it further follows that given any finite bunch of sets of the sequence, one of them includes all the others.

Also, C_0 is a club, and our definition of the sequence is such that, if C_n is a club, so is C_{n+1}, from which it follows by induction that every C_n is a club.

(3) We are letting S consist of all the elements of all the C_n's. We first must show that S is a club. To do this it suffices to show that every finite subset of S is a club. Well, consider any finite subset $\{y_1, \ldots, y_n\}$ of S. Each y_i belongs to some set C_{m_i}, and if we let m be the greatest of the numbers $m_1, \ldots, m_n$, then *all* the elements $y_1, \ldots, y_m$ are members of the club C_m, and therefore $\{y_1, \ldots, y_m\}$ is a club (every finite subset of a club is a club).

This proves that every finite subset of S is a club, and therefore S is a club.

As to maximality, we first show that for any person x_n, if $S \cup \{x_n\}$ is a club, then x_n must already be in S, and therefore for any person x outside S, the set $S \cup \{x\}$ is not a club. So suppose that $S \cup \{x_n\}$ is a club. Now, $C_n \cup \{x_n\}$ is a subset of $S \cup \{x_n\}$, hence $C_n \cup \{x_n\}$ is a club. Therefore, $C_{n+1} = C_n \cup \{x_n\}$, hence $x_n \in C_{n+1}$, and so $x_n \in S$ (because $C_{n+1} \subseteq S$). This proves that for any person x outside of S, the set $S \cup \{x\}$ is not a club. Now consider any set A of people such that S is a *proper* subset of A. Then A contains some element x not in S. Then $S \cup \{x\}$ is not a club, hence A is not a club, for if it were, its subset $S \cup \{x\}$ would be a club, which it isn't. Thus if S is a proper subset of A, then A is not a club. Thus S is a maximal club.

16.16. We are to show that for any set S, S has property $P^{\#}$ if and only if all finite subsets of S have property $P^{\#}$.

(1) Suppose S has property $P^{\#}$. Then not only do all finite subsets of S have property $P^{\#}$, but *all* subsets of S must have property $P^{\#}$. To see this, let A be any subset of S. Then every finite subset of A is also a finite subset of S, hence has property P (since S has property $P^{\#}$). Thus all finite subsets of A have property P, which means that A has property $P^{\#}$. Thus all subsets of S have property $P^{\#}$, and, in particular, all finite subsets of S have property $P^{\#}$.

(2) Conversely, suppose all finite subsets of S have property $P^{\#}$. Let F be any finite subset of S. Then F has property $P^{\#}$, which means that all finite subsets of F have property P. Well, F is a finite subset of itself, hence F has property P. Thus all finite subsets of S have property P, which means that S has property $P^{\#}$. Thus if all finite subsets of S have property $P^{\#}$, so does S.

By (1) and (2), S has property $P^{\#}$ if and only if all finite subsets of S have property $P^{\#}$, and thus $P^{\#}$ is compact.

- Part IV -

Fundamental Results in First-Order Logic

- Chapter 17 -

Fundamental Results in Propositional Logic

We are now in a position to answer the questions raised in Chapter 11:

(1) Is the tableaux method for propositional logic *correct*, in the sense that every formula provable is really a tautology?

(2) Is the method *complete*, in the sense that *all* tautologies are provable by tableaux?

(3) The same questions for synthetic tableaux.

(4) Can more formulas be proved by synthetic tableaux than by analytic tableaux?

First, for some preliminaries: The reader should recall or review the unifying α, β notation of Chapter 11, which enables us to collapse eight cases into two. We recall that under any interpretation, α is true if and only if both its components α_1 and α_2 are true, and that β is true if and only if at least one of its components β_1, β_2 is true. We also recall that the eight tableau rules collapse to two:

Rule A. $\dfrac{\alpha}{\begin{matrix}\alpha_1\\ \alpha_2\end{matrix}}$ Rule B. $\dfrac{\beta}{\beta_1 \quad \beta_2}$

SUBFORMULAS

If Z is a conjunction $X \wedge Y$ or a disjunction $X \vee Y$ or a conditional $X \Rightarrow Y$, by the *immediate subformulas* of Z we mean the formulas X and Y. By the immediate subformula of a negation $\sim X$ we mean X.

We now define a formula Y to be a *subformula* of a formula Z iff (if and only if) there is a finite sequence of formulas, beginning with Z and ending with Y, such that each term of the sequence, except the first, is an immediate subformula of the preceding term. Thus

(1) The only subformula of a propositional variable p is p itself.

(2) If Z is of the form $X \wedge Y$ or $X \vee Y$ or $X \Rightarrow Y$, then the subformulas of Z are Z, X, Y and the subformulas of X and the subformulas of Y.

(3) The subformulas of a negation $\sim X$ are $\sim X$, X, and the subformulas of X.

Every formula is a subformula of itself. By a *proper* subformula X is meant any subformula of X other than X itself.

DEGREES

To facilitate proofs by mathematical induction, we define the *degree* of an unsigned formula as the number of occurrences of logical connectives. Thus

(1) Every propositional variable is of degree 0.

(2) For any formula X of degree d, the formula $\sim X$ is of degree $d+1$.

(3) For any formulas X_1 and X_2 with respective degrees d_1 and d_2, each of the formulas $X \wedge Y$, $X \vee Y$, $X \Rightarrow Y$, $X \equiv Y$ has degrees d_1+d_2+1.

By the degree of a *signed* formula TX or FX, we mean the degree of X. Obviously any α has a higher degree than each of its components α_1, α_2, and the same with β.

With respect to the subformula relation, any descending chain starting with any formula must be finite; indeed, its length cannot be more than the degree of the formula.

We often prove various theorems about sets of formulas by induction on degrees. For example, suppose we have a set S of formulas (either all signed, or all unsigned, it doesn't matter) and we wish to show that a certain property P holds for all members of the set. Then it suffices to show both: (1) All formulas of degree 0 in S have the property; (2) For any positive n, if all elements of S of degree less than n have the property, so do all formulas in S of degree n.

We call a set S (whether finite or infinite) of formulas *true* under an interpretation I if every member of S is true under I, and we call S *satisfiable* if it is true under at least one interpretation. (In the literature, such a set is usually called *simultaneously satisfiable*, but we prefer the simpler term *satisfiable*.) For any tableau $\mathfrak{T}$, by an *immediate* extension of $\mathfrak{T}$ we mean any extension of $\mathfrak{T}$ resulting from using just one application of Rule A or Rule B.

Some Notation

For any set S and any element x, by $S{:}x$ (usually written $S \cup \{x\}$) we shall mean the result of adjoining x to S—i.e., the set whose elements are those of S together with the element x. By $S{:}x,y$ (usually written $S \cup \{x,y\}$) we shall mean the set whose elements are x, y and the elements of S. And we will use an analogous notation for tableau branches: If Θ is a branch of a tableau and X is a formula, we will denote by $(\Theta{:}X)$ the branch resulting from adjoining X to the end of Θ.

If Θ is any finite sequence $(x_1, \ldots, x_n)$, by (Θ, x) we shall mean the sequence $(x_1, \ldots, x_n, x)$ and by (Θ, x, y) we shall mean $(x_1, \ldots, x_n, x, y)$.

Now that we have our annoying preliminaries out of the way, we can finally get down to more interesting business.

Correctness and Completeness

Correctness

It should be intuitively obvious that any formula provable by the tableau method must really be a tautology, or, equivalently, the fact that a tableau closes must mean that the origin is not satisfiable (not true under any interpretation). This intuition can be justified by the following argument.

Consider a tableau $\mathfrak{T}$ and an interpretation I of all the propositional variables that appear in any formula on the tree. We shall say that a branch Θ of the tree is *true* under I if all formulas on the branch are true under I, and we shall say that the whole tree $\mathfrak{T}$ is true under I if at least one branch is true under I. Finally we shall say that $\mathfrak{T}$ is *satisfiable* iff it is true under at least one interpretation I. Obviously, no closed tableau is satisfiable.

The key point, now, is that if a tableau $\mathfrak{T}$ is satisfiable, then any immediate extension of it is also satisfiable—more specifically, if $\mathfrak{T}$ is true under an interpretation I, then any immediate extension of $\mathfrak{T}$ is true under that same interpretation I.

Problem 17.1. Prove the above assertion.

Thus, if a tableau $\mathfrak{T}$ is satisfiable, then any immediate extension $\mathfrak{T}_1$ of $\mathfrak{T}$ is satisfiable, hence any immediate extension $\mathfrak{T}_2$ of the extension $\mathfrak{T}_1$ is satisfiable, and so forth. Thus every extension of $\mathfrak{T}$ is satisfiable, and hence no extension of $\mathfrak{T}$ can close. In particular, if a signed formula FX is satisfiable, then no tableau for FX can close. Stated otherwise, if there is a closed tableau for FX, then FX is not satisfiable, hence X must be a tautology. And so we have proved

THEOREM 17.1. *The tableau method is* correct, *in the sense that every formula provable by this method is really a tautology.*

COMPLETENESS

This is a more delicate matter. It should not be so obvious that every tautology is provable by a tableau, but it is nevertheless true. In fact, we will show something even better.

Call a *branch* Θ of a tableau *completed* iff for every α on the branch, both α_1 and α_2 are on the branch, and for any β on the branch, at least one of its components β_1, β_2 is on the branch; and call a *tableau* $\mathfrak{T}$ *completed* iff every open branch of $\mathfrak{T}$ is completed. Of course, after a tableau has been run to completion, there is no sense in extending it any further, since this will yield only duplications.

What we will now show is that if X is a tautology, then not only is it the case that there is a closed tableau for FX, but every *completed* tableau for FX must be closed. Equivalently, for any *completed* tableau $\mathfrak{T}$, if it is not closed, then the origin is satisfiable. Indeed, we will show that in a *completed* tableau, *every* open branch is satisfiable. In fact, we will prove something more general.

Suppose Θ is a *completed open* branch of a tableau. Then the set S of terms of Θ satisfies the following three conditions:

H_0. For no *variable* p is it the case that Tp and Fp are both in S. (Indeed, no signed formula and its conjugate are both in S, but we don't need this stronger fact.)

H_1. For any α in S, both α_1 and α_2 are in S.

H_2. For any β in S, at least one of its components β_1, β_2 is in S.

Such sets S, whether finite or infinite, which obey conditions H_0, H_1, and H_2, are of fundamental importance and are called *Hintikka sets* (after the logician Jaakko Hintikka, who realized their significance). We will show

LEMMA 17.1 (HINTIKKA'S LEMMA). *Every Hintikka set (whether finite or infinite) is satisfiable.*

To prove Hintikka's Lemma, consider a Hintikka set S. We wish to find an interpretation I in which all elements of S are true. Well, each variable p that occurs in at least one formula in S is assigned a truth value as follows:

(1) If Tp is in S, give p the value *truth.*

(2) If Fp is in S, give p the value *falsehood.*

(3) If neither Tp nor Fp are in S, then it makes no difference which value you give p (for definiteness, give it, say, *truth*).

We note that directives (1) and (2) are not incompatible, since no Tp and Fp both occur in S (by condition H_0).

We assert that every element of S is true under this interpretation I.

PROBLEM 17.2. Prove the above assertion. *Hint:* Use mathematical induction on the degrees of elements of S.

Having proved Hintikka's Lemma, we have now completed the proof of the next theorem:

THEOREM 17.2. *Every tautology is provable by the tableau method. In fact, if X is a tautology, then* every *completed tableau starting with FX must be closed.*

DISCUSSION. The tableau method, like the truth-table method, yields a *decision procedure* for whether a given formula is or is not a tautology: One runs a tableau starting with FX to completion. If it is closed, then X is a tautology. If it is not closed, then X is not a tautology. Moreover, any open branch provides an interpretation in which X is false; just assign *truth* to a variable p if Tp is on the branch, and *falsehood* to p if Fp is on the branch. Under that interpretation, all formulas on the branch are true, hence FX is true, hence X is false.

TABLEAUX FOR SETS

Let S be a set of formulas (whether finite or infinite). By a *tableau for S* is meant a tableau in which the origin is some element of S. At any stage of the construction, we can either use Rule A or Rule B, or we can take any element of S and tack it on to the end of every open branch.

If S is a finite set, then we call a tableau for S *completed* if every open branch Θ contains all elements of S and is also such that for every α on the branch, α_1 and α_2 are both on the branch, and for every β on the branch, either β_1 or β_2 is on the branch. By obvious modifications of previous arguments, we have

THEOREM 17.2A. *A finite set S is unsatisfiable if and only if there is a closed tableau for S. Moreover, any completed tableau for S is closed if and only if S is unsatisfiable.*

REMARK. In practice, to construct a tableau for a finite set $X_1, \ldots, X_n$, the most economical thing to do is simply to *start* the tableau with the elements

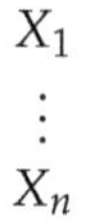

$$X_1 \\ \vdots \\ X_n$$

and then use Rules A and B. This avoids having each X_i written more than once.

REMARK. To test whether a set $\{X_1, \ldots, X_n\}$ of unsigned formulas is or is not satisfiable, one constructs a tableau, not for the set $\{FX_1, \ldots, FX_n\}$ but for the set $\{TX_1, \ldots, TX_n\}$. Alternatively, one can construct a tableau for the single signed formula $T(X_1 \wedge \cdots \wedge X_n)$, but this is less economical.

To test whether an unsigned formula Y is a logical consequence of a finite set $\{X_1, \ldots, X_n\}$ of (unsigned) formulas, one can, of course, construct a tableau for the signed formula $F((X_1 \wedge \cdots \wedge X_n) \Rightarrow Y)$, but it is more economical to construct a tableau for the set $\{TX_1, \ldots, TX_n, FY\}$.

ATOMIC CLOSURE

Let us call a branch Θ of a tableau *atomically* closed iff it contains Tp and Fp for some propositional variable p (or, if we are unhappy with signed formulas, Θ contains p and $\sim p$), and let us call a *tableau* atomically closed if every branch is atomically closed. Now, suppose that we construct a completed tableau $\mathfrak{T}$ for S and declare a branch "closed" only if it is atomically closed. Suppose further that $\mathfrak{T}$ contains an *atomically open* branch Θ (i.e., a branch that is not *atomically* closed). Then the set of elements of Θ is still a Hintikka set (because condition H_0 requires only that for no *variable* p is it the case that Tp and Fp are both in the set), and hence is satisfiable (by Hintikka's Lemma). Thus we have

THEOREM 17.2B.

(a) *S is unsatisfiable if and only if there exists an* atomically *closed tableau for S.*

(b) *X is a tautology if and only if there is an* atomically *closed tableau for FX.*

COROLLARY 17.1. *If there exists a closed tableau for S then there exists an* atomically *closed tableau for S.*

We remark that the above corollary can be proved directly (i.e., without appeal to any completeness theorem such as the above) by the method of the following exercise.

EXERCISE 17.1. Show directly by induction on the degree of X that there exists an atomically closed tableau for any set S that contains both X and its conjugate $\overline{X}$.

SYNTHETIC TABLEAUX

The correctness of synthetic tableaux can be established by an obvious modification of the proof for analytic tableaux. Of course, the method of synthetic tableaux is complete, since the method of analytic tableaux is. Thus a formula is provable by a synthetic tableau if and only if it is provable by an analytic tableau, if and only if it is a tautology. Thus synthetic tableaux cannot prove any more formulas than analytic tableaux, though they may yield shorter proofs. It should be noted, though, that, unlike proofs by synthetic tableaux, any proof of a tautology X by an analytic tableau uses only subformulas of X—or, rather, that this is true for tableaux using *signed* formulas. For tableaux using unsigned formulas, *negations* of subformulas can also appear.

COMPACTNESS

Consider a denumerable set S of formulas arranged in some infinite sequence $X_1, X_2, \dots, X_n, \dots$. Suppose that there is an interpretation I_1 in which X_1 is true, and for each n there is an interpretation I_n in which the first n terms $X_1, \dots, X_n$ are all true. Does it follow that there must be an interpretation in which all the infinitely many of the terms $X_1, \dots, X_n, \dots$ are true? The question is really equivalent to the following question: If all finite subsets of S are satisfiable, is the whole set S necessarily satisfiable?

PROBLEM 17.3. Why are the two questions equivalent? That is, why is it the case that the hypothesis that every finite subset of S is satisfiable is equivalent to the hypothesis that, for every n, the set $\{x_1, \dots, x_n\}$ is satisfiable?

We shall now prove that it is indeed the case that if all finite subsets of a denumerable set S are satisfiable, then S is satisfiable. This is known as the *Compactness Theorem for Propositional Logic,* and will be quite important when we come later to some topics in first-order logic. There are two important proofs of this, each of which reveals certain features not revealed by the other.

Our first proof uses tableaux and König's Lemma (Theorem 16.4).

We note that for a denumerable set S of formulas, if all finite subsets of S are satisfiable, then no tableau for S can close. For suppose that some tableau for S closes. Then all branches of the tableau are finite, hence, by König's Lemma, the whole tree is finite (since the tree is finitely generated—each point has at most two successors), hence only finitely many elements of S were used, hence the tableau is a tableau for that finite set, which must then be unsatisfiable. Thus if all subsets of S are satisfiable, then no tableau for S can close.

Now consider any denumerable set S such that all finite subset of S are satisfiable. Construct a tableau for S as follows: First arrange all elements of S in some infinite sequence $X_1, X_2, \ldots, X_n, \ldots$. Run a completed tableau for X_1. The tableau cannot close, since X_1 is satisfiable. Now tack X_2 on to the end of every open branch and continue the branch to completion (thus getting a completed tableau for the set $\{X_1, X_2\}$). This tableau cannot close, since $\{X_1, X_2\}$ is satisfiable. Then tack X_3 on to the end of every open branch and run the tableau to completion, and continue this process indefinitely (successively tacking on $X_4, X_5, \ldots$). At no stage can the tableau close, for otherwise for some n the finite set $\{X_1, \ldots, X_n\}$ would be unsatisfiable, which it isn't. We thus obtain an *infinite* tree. The tree is obviously finitely generated (each point has at most two successors), and by König's Lemma the tree has an infinite branch Θ, which clearly must be open; the set of terms of Θ is obviously a Hintikka set and contains all elements of S. By Hintikka's Lemma, this set is satisfiable, hence so is its subset S. We have thus proved

THEOREM 17.3 (COMPACTNESS THEOREM FOR PROPOSITIONAL LOGIC). *A denumerable set S of signed formulas is satisfiable if (and only if) all its finite subsets are satisfiable.*

Of course the above theorem also holds for sets of *unsigned* formulas.

TRUTH SETS

Preparatory to the second proof of the Compactness Theorem, we must introduce the notion of *truth sets*: A set S of signed formulas is called a *truth set* if it obeys the following three conditions (for every α and β):

T_0. For every signed formula X, either X or its conjugate $\overline{X}$ is in S, but not both.

T_1. $\alpha \in S$ iff $\alpha_1 \in S$ and $\alpha_2 \in S$.

T_2. $\beta \in S$ iff $\beta_1 \in S$ *or* $\beta_2 \in S$.

EXERCISE 17.2. There is redundancy in the conditions T_0, T_1 and T_2.

(a) Show that T_2 follows from just T_0 and T_1.

(b) Also, T_1 follows from T_0 and T_2. Why?

(c) Given both T_1 and T_2, condition T_0 then holds for the case when X is a signed *variable*. Prove this.

For a set S of *unsigned* formulas, we call S a truth set if it satisfies the above three conditions, but with $\overline{X}$ replaced by $\sim X$ in T_0.

EXERCISE 17.3. If S is a set of unsigned formulas, show that S is a truth set if and only if it satisfies the following conditions:

(1) For every X, either X or $\sim X$ is in S, but not both.

(2) A conjunction $X \wedge Y$ is in S iff X and Y are both in S.

(3) A disjunction $X \vee Y$ is in S iff either X is in S or Y is in S.

(4) A conditional $X \Rightarrow Y$ is in S iff X is not in S or Y is in S.

(5) $\sim\sim X$ is in S iff X is in S.

Returning to signed formulas, it is obvious that, for any interpretation I, the set of all signed formulas that are true under I is a truth set, and we call it the *truth set of* I. Moreover, any truth set S is the truth set of some interpretation; that is, for any truth set S there is an interpretation I (and only one, in fact) such that S consists of all and only those formulas that are true under I (cf. Problem 17.4 below). Of course, every truth set is also a Hintikka set. To say that a set is satisfiable is thus equivalent to saying that it is a subset of some truth set. Thus Hintikka's Lemma is equivalent to the statement that every Hintikka set can be extended to (i.e., is a subset of) some truth set.

PROBLEM 17.4. Given a truth set S, assign *truth* to each variable p for which $Tp \in S$, and *falsehood* to each p for which $Fp \in S$. We already know from the proof of Hintikka's Lemma that all elements of S are true under that interpretation. Now prove by induction on degrees that all formulas true under that interpretation must be in S.

We shall call a set S (of signed formulas) *full* if for every signed formula X, either X or $\overline{X}$ is in S. (For unsigned formulas, replace $\overline{X}$ by $\sim X$ in this definition.) Obviously, every truth set is both full and a Hintikka set. But, conversely, we have the useful

FACT 1. Every full Hintikka set is a truth set.

PROBLEM 17.5. Prove the above fact.

The first proof we gave of the Compactness Theorem showed how an infinite set S whose finite subsets are all satisfiable can be extended to a Hintikka set, which in turn can be extended to a truth set. The proof to which we now turn shows how to extend such a set S to a truth set directly.

Let us define a set S of signed formulas to be *consistent* if all finite subsets of S are satisfiable. By Theorem 16.8, this property of consistency is of finite character (a set is consistent iff all its finite subsets are consistent), hence by Theorem 16.7, any consistent set S can be extended to a maximally consistent set. The key point now is

THEOREM 17.4. *Every maximally consistent set is a truth set.*

PROBLEM 17.6. Prove Theorem 17.4 by successively showing

(a) If a set S is consistent, then for any X, either $S{:}X$ or $S{:}\overline{X}$ is consistent.

(b) Therefore every maximally consistent set is full.

(c) Any full set S in which all finite subsets are satisfiable must be a Hintikka set, and hence (by Fact 1) a truth set.

This completes the second proof of Theorem 17.3.

Of course Theorem 17.3 holds for sets of unsigned formulas as well. It has two important corollaries. Let us call a formula X a *logical consequence* of a set S if X is true under all interpretations that satisfy S.

COROLLARY 17.2 (COMPACTNESS THEOREM OF DEDUCIBILITY). *If X is a logical consequence of a set S, then X is a logical consequence of some finite subset of S*

PROBLEM 17.7. Prove the above corollary.

NOTE. For unsigned formulas, the Corollary 17.2 can be equivalently stated thus: If X is true under all interpretations under which all elements of S are true, then there are finitely many elements $X_1, \ldots, X_n$ such that the formula $(X_1 \wedge \cdots \wedge X_n) \Rightarrow X$ is a tautology.

Call a set S of formulas *disjunctively valid* if, for every interpretation I, at least one element of S is true under I. It is obvious that if some subset of S is disjunctively valid, so is S. But now we have

COROLLARY 17.3. *For any denumerable set S, if S is disjunctively valid then some finite subset of S is disjunctively valid.*

To prove Corollary 17.3, for any set S of signed (unsigned) formulas, let $\overline{S}$ be the set of conjugates (negations) of all the elements of S. We note that a set S is disjunctively valid iff the set $\overline{S}$ is unsatisfiable (why?). Using this fact, and the Compactness Theorem, the above corollary follows quite easily.

PROBLEM 17.8. Prove Corollary 17.3.

NOTE. For unsigned formulas, Corollary 17.3 can be equivalently stated thus: If X is true under every interpretation under which at least one element of S is true, then there are finitely many elements $X_1, \ldots, X_n$ of S such that $(X_1 \vee \ldots \vee X_n) \Rightarrow X$ is a tautology.

SOLUTIONS

17.1. Suppose that $\mathfrak{T}$ is true under I and that $\mathfrak{T}_1$ is an immediate extension of $\mathfrak{T}$. Let Θ be a branch of $\mathfrak{T}$ that is true under I. In extending $\mathfrak{T}$ to $\mathfrak{T}_1$, we have extended either the branch Θ or some other branch. If it was some other branch, then Θ is unaltered, so $\mathfrak{T}_1$ contains the true branch Θ.

On the other hand, suppose it is Θ that was extended: If it was extended by using Rule A, then for some α in Θ, the elements of the extended branch Θ_1 consist of the elements of Θ together with either α_1 or α_2. But since α is true under I, so are both α_1 and α_2, so all elements of Θ_1 are true under I. If, on the other hand, Θ was extended using Rule B, then Θ has been split into two branches $(\Theta{:}\beta_1)$ and $(\Theta{:}\beta_2)$ for some element β on Θ. Since β is true under I, at least one of β_1, β_2 is true under I, and hence at least one of the branches $(\Theta{:}\beta_1)$, $(\Theta{:}\beta_2)$ is true under I. Thus $\mathfrak{T}_1$ contains a branch that is true under I.

17.2. It is immediate from the definition of the interpretation I that all signed *variables* (all signed formulas of degree 0) that are in S must be true under I. Now suppose n is a positive integer such that all elements of S of degree less than n are true under I. We are to show that all elements of S of degree n are true under I. Well, consider any α in S of degree n. Then α_1, α_2 are both in S (by H_1), and since they are both of degree less than n, they are both true under I (by the inductive hypothesis), and hence α is also true under I. Next, consider any β in S of degree n. Then either β_1 is in S or β_2 is in S. Whichever one it is, it is of lower degree than n, hence is true under I. Thus all elements in S of degree n are true under I, which completes the induction.

Note. This induction argument is correct for sets of *signed* formulas, but for unsigned formulas, a modification is necessary: The point is that in dealing with unsigned formulas, the degree of an α or β is not always higher than that of each of its components. For example, consider the formula $p \Rightarrow q$ (where p and q are propositional variables). It has degree 1. Its components are $\sim p$ and q, and $\sim p$ also has degree 1. So what is needed is a notion close to that of degree, but not quite the same—namely, a notion called *rank*. We assign *ranks* to all formulas by the following inductive scheme: We assign rank 0 to propositional variables *and their negations*! (Thus p and $\sim p$ both have rank 0.) Next, for any α, having assigned ranks r_1 and r_2 to α_1 and α_2 respectively, we assign to α the rank r_1+r_2+1. Similarly, to any β, we assign rank r_1+r_2+1, where r_1 and r_2 are the respective ranks of β_1 and β_2. Then, in the above inductive proof, if we deal with unsigned formulas, the proof should be by induction on ranks, instead of degrees.

17.3. Obviously, if every finite subset of S is satisfiable, then, for each n, the set $x_1, \dots, x_n$ is satisfiable. Conversely, suppose that for each n the set $\{x_1, \dots, x_n\}$ is satisfiable. Now consider any finite subset A of S. Then for some n, A must be a subset of $\{x_1, \dots, x_n\}$, and since $\{x_1, \dots, x_n\}$ is satisfiable, so is A.

17.4. Let I be the interpretation so described. It is immediate that every signed *variable* that is true under I must be in S. Now suppose n is any positive integer such that every signed formula of degree less than n that is true under I is in S. Consider any α of degree n that is true under I. Then α_1 and α_2 are both true under I, hence both are in S (by the inductive hypothesis), hence α is in S (by condition T_1). Next, consider any β of degree n that is true under I. Then either β_1 or β_2 is true under I, and, whichever one is true, it is in S (by the inductive hypothesis), and hence β is in S (by condition T_2). This completes the induction.

17.5. Suppose S is a full Hintikka set.

(1) Since S is a Hintikka set, we know that if α is in S, so are α_1 and α_2. We must now show the converse—i.e., that if α_1 and α_2 are in

S, so is α. So suppose α_1 and α_2 are both in S. Then neither $\overline{\alpha_1}$ nor $\overline{\alpha_2}$ can be in S (since a Hintikka set cannot contain both an element and its conjugate). Hence $\overline{\alpha}$ cannot be in S, because if it was, then either $\overline{\alpha_1}$ or $\overline{\alpha_2}$ would have to be in S, since $\overline{\alpha}$ is some β. Since $\overline{\alpha}$ is not in S, it follows that α must be in S (since S is full).

(2) If β is in S, then either β_1 or β_2 is in S (since S is a Hintikka set). We must now show the converse. So suppose either β_1 or β_2 is in S—say β_1 is in S (the argument will be analogous if β_2 is in S). Since β_1 is in S, it follows that $\overline{\beta_1}$ is not in S, and therefore $\overline{\beta}$ cannot be in S, because $\overline{\beta}$ is some α, and hence if it was in S, $\overline{\beta_1}$ (as well as $\overline{\beta_2}$) would have to be in S. Since $\overline{\beta}$ is not in S, it follows that β is in S (since S is full). This completes the proof.

17.6. (a) It suffices to show that if $S{:}X$ and $S{:}\overline{X}$ are both inconsistent, so is S. Well, suppose $S{:}X$ and $S{:}\overline{X}$ are both inconsistent. Since $S{:}X$ is inconsistent, it follows that for some finite subset S_1 of S, the set $S_1{:}X$ is not satisfiable. Likewise, for some finite subset S_2 of S, the set $S_2{:}\overline{X}$ is not satisfiable. Let S_3 be the union $S_1 \cup S_2$ of S_1 and S_2 (the set whose elements are those of S_1 together with those of S_2). Then both $S_3{:}X$ and $S_3{:}\overline{X}$ are unsatisfiable, which means that S_3 is unsatisfiable (because any interpretation making all elements of S_3 true would have to make either X or $\overline{X}$ true). Thus the finite subset S_3 of S is unsatisfiable, and S is therefore inconsistent.

(b) We first note that if S is *maximally* consistent, then any formula X that is consistent with S (i.e., such that $S{:}X$ is consistent) must already be in S (because otherwise S would be a *proper* subset of the consistent set $\{S, X\}$). Now, by (a), if S is consistent then either $S{:}X$ or $S{:}\overline{X}$ is consistent, so if S is *maximally* consistent, then either X or $\overline{X}$ must be in S. Thus S is full.

(c) Suppose that S is full and that every finite subset of S is satisfiable.

For a given α, suppose α_1 and α_2 are both in S. Then $\overline{\alpha}$ is not in S (because the set $\{\alpha_1, \alpha_2, \overline{\alpha}\}$ is not satisfiable), hence $\alpha \in S$ (since S is full).

For a given β, suppose β_1 is in S. Then $\overline{\beta} \notin S$ (since $\{\overline{\beta}, \beta_1\}$ is not satisfiable), hence β is in S. Similarly if β_2 is in S, so is β. Thus S is a Hintikka set.

17.7. Suppose that X is a logical consequence of S. Then $S{:}\overline{X}$ (or $S{:}{\sim}X$, if we are working with unsigned formulas) is not satisfiable. Hence for some finite subset S_0 of S, the set $S_0{:}\overline{X}$ is not satisfiable. Hence X is a logical consequence of S_0.

17.8. We have noted that a set S is disjunctively valid iff $\overline{S}$ is unsatisfiable. Equivalently, S fails to be disjunctively valid iff $\overline{S}$ is *satisfiable*.

Suppose now that no finite subset S_0 of S is disjunctively valid. Then every finite subset $\overline{S}_0$ of $\overline{S}$ is satisfiable. Then by the Compactness Theorem, the whole set $\overline{S}$ is satisfiable, and hence S is not disjunctively valid. Thus if no finite subset of S is disjunctively valid, then neither is S. Hence if S *is* disjunctively valid, then some finite subset of S is disjunctively valid.

- Chapter 18 -

First-Order Logic: Completeness, Compactness, Skolem-Löwenheim Theorem

The first thing we will prove in this chapter is one of the major results in first-order logic—the *completeness theorem* for first-order tableaux, which is that every valid formula of first-order logic is provable by the tableau method.

This result ultimately stems from Gödel, who proved the completeness of an *axiom system* for first-order logic. Later in this book we shall consider an axiom system of the more conventional type and derive its completeness as a consequence of the completeness of the tableau method. We will prove many other related things as well.

We recall that in Chapter 13 (Problem 18.10) we found a formula that is not satisfiable in any finite domain but is satisfiable in a denumerable domain. Now, what about a formula that is not satisfiable in any finite or denumerable domain but is satisfiable in a non-denumerable domain? Can you find one? The answer is *no*! Löwenheim [10] proved the remarkable result that if a formula is satisfiable at all, then it is satisfiable in a denumerable domain! Later, Skolem [16] proved the even more celebrated and important result that, for any denumerable set S of formulas, if S is satisfiable at all (if there is, in some domain, an interpretation under which all elements of S are true) then S is satisfiable in a denumerable domain. This result, the Skolem-Löwenheim Theorem, is of fundamental importance for the entire foundation of mathematics. It means that any axiom system that is intended to apply to a non-denumerable domain can be re-interpreted to apply to a denumerable domain; it cannot force the domain of interpretation to be non-denumerable. We prove this important result in this chapter.

Some Preliminaries

We now consider *sentences* (closed formulas) with or without parameters. By the *degree* of a formula is meant the number of occurrences of the symbols $\sim$, $\wedge$, $\vee$, $\Rightarrow$, $\forall$, $\exists$. We recall the unifying α, β, γ, δ notation introduced in Chapter 11. We have defined the *components* of α to be α_1 and α_2, and the components of β to be β_1 and β_2, We now define the components of γ to be all sentences $\gamma(a)$, where a is any member of the denumerable set of parameters. Similarly, we define the components of δ to be all sentences $\delta(a)$, where a is any parameter. Unlike α's and β's, each γ and each δ has infinitely many components.

Formulas with Constants in V

We consider a non-empty set V of any size (finite, denumerable or non-denumerable), which we call a *domain of individuals* or, more briefly, a *domain*. By a *formula with constants in V* (more briefly, a *V-formula*), we mean an expression like a formula with parameters, except that it has elements of V in place of some or more of the parameters. Thus an *atomic* V-formula is an expression $Pe_1,\ldots,e_n$, where each e_i is either a parameter or an element of V (or a symbol that is the *name* of an element of V, if you prefer[1]). We define *substitution* of elements of V for free occurrences of variables in the same way as we did substitution of parameters, only using elements of V in place of parameters. We will assume that our domain V is disjoint from the set of parameters. (Two sets are called *disjoint* if they have no elements in common.) We shall be dealing now only with *closed* formulas—formulas with no free occurrences of variables, though they may contain parameters.

Let us first consider closed V-formulas without parameters. An *interpretation* of such a formula is specified by assigning to each predicate P a relation P^* of elements of V, the relation being of the same degree as the predicate. An *atomic* V-formula (with no parameters) $Pe_1,\ldots,e_n$ is called *true* under an interpretation I iff the elements $e_1,\ldots,e_n$ (in that order) stand in the relation P^* assigned to P under the interpretation I.

[1] I am thinking of a *formula* as a finite sequence of things called *symbols*, but I see no reason why things other than symbols cannot be members of the sequence, despite the objections of some purists.

Once the atomic sentences receive a truth value, all other closed sentences receive truth values under the following rules:

(1) A conjunction $X \wedge Y$ is true iff X and Y are both true (as in propositional logic).

(2) A disjunction $X \vee Y$ is true iff either X is true or Y is true (as in propositional logic).

(3) A conditional $X \Rightarrow Y$ is true iff either X is not true or Y is true (as in propositional logic).

(4) A negation $\sim X$ is true iff X is not true (as in propositional logic).

(5) A universal quantification $\forall x \phi(x)$ is true iff $\phi(e)$ is true for *every* element e of V.

(6) An existential quantification $\exists x \phi(x)$ is true iff $\phi(e)$ is true for *at least one* element e of V.

A closed formula A *with parameters* is called *true* under an interpretation I (of the predicates and parameters) in a domain V iff the formula obtained by replacing each parameter by its designated element of V is true (thus, e.g., a formula $\phi(a_1, a_2)$ with parameters a_1 and a_2 is true under I iff the formula $\phi(e_1, e_2)$ is true, where e_1 is the element of V assigned to a_1, and e_2 is the element assigned to a_2).

Closed formulas will also be called *sentences*. As with propositional logic, a signed sentence TX is called true under an interpretation iff X is; and FX is called true iff X is not true. We recall that a sentence is called *valid* if it is true in all interpretations and *satisfiable* if it is true under at least one interpretation. As before, we call a *set* S of sentences *satisfiable* if there is at least one interpretation under which all elements of S are true. For any set S and element x we again use the notation $S{:}x$ to mean the result of adjoining x to S, i.e., the set whose elements are x together with all elements of S; and by $S{:}x, y$ we mean the set whose members are x, y, and those of S.

For any set S of sentences (whether signed sentences or unsigned) the following four facts are basic:

F_1. If $S{:}\alpha$ is satisfiable, so is $S{:}\alpha_1, \alpha_2$.

F_2. If $S{:}\beta$ is satisfiable, then either $S{:}\beta_1$ or $S{:}\beta_2$ is satisfiable.

F_3. If $S{:}\gamma$ is satisfiable, then, for every parameter a, the set $S{:}\gamma, \gamma(a)$ is satisfiable (and so is its subset $S{:}\gamma(a)$).

F_4. If $S{:}\delta$ is satisfiable, then, for every parameter a that does not occur in δ or in any element of S, the set $S{:}\delta, \delta(a)$ is satisfiable.

The facts F_1, F_2 and F_3 are obvious. As to F_4, suppose $S{:}\delta$ is satisfiable. Then all elements of $S{:}\delta$ are true under some interpretation I in some domain V—an interpretation of all predicates and parameters that occur in any of the elements of S or in δ. Then δ is true under I; hence, for some element e in V, the V-sentence $\delta(e)$ is true under I. Now, if a is any parameter that does not occur in any formula of S or in δ, then a has not yet been assigned any value under I; so simply assign to a some element e such that $\delta(e)$ is true, and $\delta(a)$ is now also true under this extended interpretation.

Actually, our proof has shown something stronger that will be needed in a later chapter: Consider a set B of predicates and parameters and a subset B_0 of B. Consider an interpretation I of the predicates and parameters of B and an interpretation I_0 of the predicates and parameters of B_0. We call I an *extension* of I_0 iff for every element (predicate or parameter) x of the smaller set B_0, the value of x under I_0 (i.e., the element assigned to x by I_0) is the same as the value of x under I. For any set S of formulas, by an interpretation I of S is meant an interpretation of the predicates and parameters of the elements of S, and we say that I *satisfies* S iff all elements of S are true under I. Now, fact F_4 says that if a does not occur in S nor in δ and there is an interpretation that satisfies $S{:}\delta$, then there is an interpretation of $S{:}\delta, \delta(a)$ that satisfies $S{:}\delta, \delta(a)$. However, the proof given above has shown the following stronger fact:

F_4^*. If a does not occur in S nor in δ, then any interpretation that satisfies $S{:}\delta$ can be *extended* to an interpretation that satisfies $S{:}\delta, \delta(a)$.

In other words, if a is new to S and δ, then for any interpretation I that satisfies $S{:}\delta$, there is an interpretation I' that satisfies $S{:}\delta, \delta(a)$ and is an extension of I.

It easily follows from the above facts F_1–F_4 that if a formula X is satisfiable then no tableau starting with X can close. The proof is quite similar to that for propositional logic and is left to the reader. It then follows that if a tableau for X does close, then X is unsatisfiable, and therefore, for any unsigned formula X, a closed tableau for FX means that FX is unsatisfiable and hence X is valid. Thus the tableau method for first-order logic is *correct* in that every provable formula is valid.

We now turn to the question of *completeness*, which is particularly interesting!

Completeness

Hintikka Sets

By a *Hintikka set* for first-order logic is meant a set S of closed formulas satisfying the five conditions below (where the first three are the same as for propositional logic).

H_0. No atomic sentence X and its conjugate $\overline{X}$ are both in S.

H_1. If $\alpha \in S$, so are α_1 and α_2.

H_2. If $\beta \in S$, then either $\beta_1 \in S$ or $\beta_2 \in S$.

H_3. If $\gamma \in S$, then for every parameter a, the sentence $\gamma(a)$ is in S.

H_4. If $\delta \in S$, then for at least one parameter a, the sentence $\delta(a)$ is in S.

We shall collectively call α's and γ's *conjunctive* elements, and β's and δ's *disjunctive* elements, so conditions H_1–H_4 can be more briefly and neatly stated: For every conjunctive element in S, all its components are in S, and for every disjunctive element in S, at least one of its components is in S.

Lemma 18.1 (Hintikka's Lemma for First-Order Logic). *Every Hintikka set is satisfiable—indeed, satisfiable in the denumerable domain of the parameters.*

The proof of the above is not very different from that of Hintikka's Lemma for propositional logic. Consider a Hintikka set S. We interpret the predicates as follows: For each predicate P of degree n we let P^* be the relation defined by the condition "$P^*(a_1, \ldots, a_n)$ if and only if the sentence $TPa_1, \ldots, a_n$ is in S." Obviously, $TPa_1, \ldots, a_n$, if in S, is true under this interpretation. Now consider a formula $FPa_1, \ldots, a_n$ in S. We know that $TPa_1, \ldots, a_n$ is not in S (by H_0), hence the unsigned formula $Pa_1, \ldots, a_n$ is not true, and therefore $FPa_1, \ldots, a_n$ is true. Thus every atomic element of S, whether signed with T or with F, is true under the interpretation. Then by induction on degrees of elements of S, it can be shown that all elements of S are true under the interpretation.

Problem 18.1. Complete the proof. Show by induction that all elements of S are true under the above interpretation.

Note. Suppose we consider a *finite* set D of parameters and let S be a set of closed formulas all of whose parameters are in D. We define such a set S to be a Hintikka set *for the domain* D as we did above, only in H_3 replacing "every parameter a" by "every parameter a in D." The same proof shows, of course, that every Hintikka set for the domain D is satisfiable *in the finite domain* D.

The Completeness Theorem

Now we wish to show the fundamental result that every valid formula is provable by the tableau method. The proof is more remarkable than the proof for the case of propositional logic. In propositional logic, tableaux terminate after finitely many steps. But a tableau for first-order logic may run on infinitely without ever closing, in which case there must be an infinite branch Θ (by König's Lemma), but the set of formulas on Θ is *not* necessarily a Hintikka set! For example, we might have some γ on a branch, and we might successively adjoin infinitely many components $\gamma(a_1), \gamma(a_2), \ldots, \gamma(a_n), \ldots$ (for all the parameters $a_1, a_2, \ldots, a_n, \ldots$) and totally neglect some other α, β or γ on the branch. The crucial thing now is to devise some *systematic* procedure that will guarantee that any tableau constructed according to the procedure is such that, if it runs on infinitely, then for every open branch Θ the set of elements of Θ will be a Hintikka set. There are many such procedures in the literature, and the reader might try working out one before reading further.

For any non-atomic formula X on an open branch Θ, define X to be *fulfilled* on Θ if

(1) X is an α and both α_1 and α_2 are on Θ;

(2) X is a β and at least one of β_1, β_2 is on Θ;

(3) X is a γ and for *every* parameter a, the sentence $\gamma(a)$ is on Θ; or

(4) X is a δ and $\delta(a)$ is on Θ for at least one parameter a.

To say that the set of points on an open branch Θ is a Hintikka set is to say that every non-atomic formula on Θ has been fulfilled.

As I have said, many *systematic* procedures for constructing tableaux have been considered in the literature, and the following one (published in [17]) seems to be as simple and direct as any. In this procedure for generating a tree, at each stage certain points of the tree are declared to have been "used." (As a practical book-keeping device, we can put a checkmark to the right of the formula as soon as we have used it.) We start the tableau by placing the formula whose satisfiability we are testing at the origin. This concludes the first stage. Now suppose we have concluded the nth stage. Then our next act is determined as follows. If the tableau at hand is already closed, then we stop. If not, then we pick an unused point X as high up in the tree as possible (say, the leftmost such one, if the reader wants the procedure to be completely deterministic). Then we take *every* open branch Θ passing through point X and proceed as follows:

(1) If X is an α, we adjoin α_1 and α_2 to the end of Θ.

(2) If X is a β, we split Θ into the two branches $(\Theta{:}\beta_1)$, $(\Theta{:}\beta_2)$.

(3) If X is a δ, then we take the first parameter a (in some pre-arranged sequence of all parameters) that does not occur in any formula on the tree, and adjoin $\delta(a)$ to the end of Θ.

(4) If X is a γ (and here is the delicate case!), we take the first parameter a such that $\gamma(a)$ does not appear on Θ, and then extend Θ to $(\Theta{:}\gamma(a),\gamma)$ (in other words, we add $\gamma(a)$ as an endpoint to Θ and then repeat γ!).

Having performed one of the above four acts, we then declare the point X to have been *used*, and this concludes stage $n+1$ of the procedure.

In the above procedure, we are systematically working our way down the tree, fulfilling all α, β and δ formulas that come our way. As to the γ-formulas, when we use an occurrence of γ on a branch Θ to subjoin an instance $\gamma(a)$, the purpose of *repeating* an occurrence of γ is that we must sooner or later come down the branch Θ and use this repeated occurrence, from which we adjoin another instance $\gamma(b)$ and again repeat an occurrence of γ, which we in turn later use again, and so forth. In this way we are sure of fulfilling all the γ-formulas (as well as the α, β and δ formulas). Thus if the tableau runs on infinitely without closing, it has an infinite open branch Θ (by König's Lemma) and the set of elements of Θ is a Hintikka set (for the denumerable set of parameters) and is thus satisfiable—in fact, in a denumerable domain (by Hintikka's Lemma).

By a *systematic* tableau we shall mean one constructed by the above procedure. We thus see that if a systematic tableau doesn't close, then the origin is satisfiable, in fact, in a denumerable domain. Therefore, if the origin is not satisfiable, a systematic tableau (unlike a non-systematic tableau) is bound to close. Suppose now that X is a valid unsigned formula. Then FX is not satisfiable, hence a systematic tableau for FX must close, and thus X is provable (by the tableau method). And we have proved

Theorem 18.1 (Completeness Theorem). *Every valid formula is provable by the tableau method. Indeed, if X is valid then a* systematic *tableau for FX must close.*

Löwenheim's Theorem

Suppose that X is satisfiable. Then a systematic tableau for X cannot close, hence, as we have seen, X is satisfiable in a denumerable domain. We therefore have

Theorem 18.2 (Löwenheim). *If X is satisfiable, then X is satisfiable in a denumerable domain.*

Discussion. Strictly speaking, the tree generated by our systematic procedure is not literally a tableau, since there are no tableau rules allowing for arbitrary repetition of formulas; but if we simply delete all the repetitions of γ's, the resulting tree will literally be a tableau. Alternatively, in the process of constructing the tree, we could delete repeated γ's as soon as they have been used. The finished infinite tree will then literally be a tableau. But all this is really a trivial matter.

Next it should be pointed out that, in general, a systematic tableau is usually longer than a tableau constructed using some ingenuity. One can easily program a computer to construct *systematic* tableaux. But a clever human will often obtain *more efficient* proofs of a valid formula than such a computer. If the reader has worked some of the tableaux exercises of Chapter 13, it is highly unlikely that any of them were systematic. It would be profitable for the reader to construct *systematic* tableaux for some of these formulas and compare their lengths with those of the previously constructed tableaux.

In short, a systematic tableau is bound to close, if any closure at all is possible, and a mindless computer can perform this. But a clever human can do the job more efficiently, in considerably fewer steps. At this point I'd like to quote the logician Paul Rosenbloom, who said, about a certain theorem: "This means that man can never eliminate the necessity of using his own intelligence, regardless of how cleverly he tries."

The particular systematic procedure that we have given is by no means the quickest. The following procedure, though a bit more difficult to justify, will in general give shorter proofs: At any given stage, first use up all unused α and δ points (we recall that a δ need be used only once). This clearly must terminate in finitely many steps. Then use up all unused β's (again this terminates after finitely many steps). Then use a γ-point of maximal height in the manner indicated previously.

A few working examples should convince the reader of the practical superiority of this procedure. But many improvements are possible—such a study is a subject in itself and is known as "mechanical theorem proving."

Satisfiability in a Finite Domain

It is possible that in the construction of a tableau, a stage is reached at which the tableau is not closed, yet there is at least one open branch Θ that is a Hintikka set *for a finite domain of parameters* (the parameters that occur in the terms of Θ). In this case, it is pointless to continue further, for, by Hintikka's Lemma, the set of elements of Θ (and hence the origin in particular) is satisfiable in the finite domain of those parameters.

As an example, the formula

$$(\forall x Px \lor \forall x Qx) \Rightarrow \forall x(Px \lor Qx)$$

is valid, but its converse

$$\forall x(Px \lor Qx) \Rightarrow (\forall x Px \lor \forall x Qx)$$

is not. Thus the signed formula

$$F \forall x(Px \lor Qx) \Rightarrow (\forall x Px \lor \forall x Qx)$$

is satisfiable. It happens to be satisfiable in a finite domain, indeed, a domain of only two elements. A tableau will reveal this.

EXERCISE 18.1. Construct a tableau that reveals this, and then give such an interpretation in a two-element domain.

DISCUSSION. We thus see how a tableau can be used not only to show certain formulas to be unsatisfiable—or, equivalently, to show certain formulas to be valid—but also sometimes to show certain formulas to be satisfiable, if they happen to be satisfiable in a finite domain. The real "mystery class" consists of those formulas that are neither unsatisfiable nor satisfiable in a finite domain. If we construct a systematic tableau for such a formula, it will run on infinitely without ever closing, and at no stage can we know whether or not it will close at some later stage. Very frustrating!

THE SKOLEM-LÖWENHEIM AND COMPACTNESS THEOREMS

Consider now a denumerable set S of formulas with no parameters. As in propositional logic, by a tableau for S is meant a tableau starting with some element of S, and where at any stage we may append any element of S to the end of any open branch. As with propositional logic, if a tableau for S closes, then all branches must be finite, hence the whole tableau must be finite (by König's Lemma) and contain only finitely many elements of S; thus the tableau is for a finite subset of S. So, as with propositional logic, if all finite subsets of S are satisfiable, then no tableau for S can close.

We now construct a *systematic* tableau for S as follows: We arrange all elements of S in some denumerable sequence $X_1, X_2, \ldots, X_n, \ldots$. We start the tableau by placing X_1 at the origin. This concludes the first stage. We then proceed as in the construction of a systematic tableau for a single formula as described earlier, except that, for each n, we tack an X_n to the end of each open branch. Now, if all finite subsets of S are satisfiable, then the systematic tableau will not close, hence there will be an infinite branch (by König's Lemma) which must be open; hence the set of its terms is a Hintikka set and contains all elements of S, and is satisfiable in a denumerable domain (by Hintikka's Lemma). So we have proved the following theorem, which yields both the Skolem-Löwenheim Theorem and the Compactness Theorem for first-order logic.

Theorem 18.3. *If all finite subsets of S are satisfiable, then the entire set S is satisfiable in a denumerable domain.*

Remarks. We have so far proved Theorem 18.3 only for sets S of *pure* formulas (formulas with no parameters). If some of the elements of S contain parameters, we might slightly modify the above systematic construction by requiring, when using the δ-rule (from δ infer $\delta(a)$) that the parameter a must occur neither in any formula on the tableau, *nor in any element of* S! This is fine if there are infinitely many parameters that do not occur in any element of S, but what do we do if this is not the case—for example, if *every* parameter occurs in some element of S? Then we could simply add another denumerable group of symbols to our set of parameters. Alternatively, we could arrange all parameters in a denumerable sequence $a_1, a_2, \ldots, a_n, \ldots$ and in each formula X in S, we could replace each parameter a_n in X by, say, a_{2n}. The resulting set S' will obviously have all its finite subsets satisfiable if and only if S does, and the parameters that occur in elements of S' will be $a_2, a_4, \ldots, a_{2n}, \ldots$, and hence leave out all the odd-numbered ones. Thus Theorem 18.3 holds for every set S of closed formulas, with or without parameters.

We say that a formula X is a *logical consequence* of a set S of formulas, or that X is logically *implied* by S, if X is true in all interpretations that satisfy (all elements of) S. As an immediate corollary of Theorem 18.2, we have

Corollary 18.1. *If X is a logical consequence of S, then X is a logical consequence of some finite subset of S.*

Problem 18.2. Prove the above corollary.

Solutions

18.1. Suppose n is a positive integer such that all elements of S of degree less than n are true under the interpretation. Now consider any element in S of degree n. If it is a conjunctive element c (an α or a γ), then all its components are in S (by H_1 or H_3), and since they are all of lower degree than n, they are all true (by the inductive hypothesis), which makes c true. Now consider a disjunctive element d (a β or δ) of degree n. At least one of its components must be in S (by H_2 or H_4), and, being of lower degree than n, must be true (again by the inductive hypothesis), which makes d true. This completes the induction.

18.2. Suppose that X is a logical consequence of S. Then $S \cup \{\sim X\}$ is unsatisfiable. Hence for some finite subset S_0 of S, the set $S_0 \cup \{\sim X\}$ is unsatisfiable. Therefore X is a logical consequence of S_0.

- CHAPTER 19 -

THE REGULARITY THEOREM

In this chapter, we state and prove a basic result of first-order logic that not only provides an elegant treatment of first-order axiom systems (which we present in a later chapter) but is of interest in its own right.

PRELIMINARIES

Let us recall the difference between a sentence of first-order logic being *valid* and being a *tautology* (which is a much stronger condition): Consider an intelligent individual who is learning mathematical logic and who understands the meaning of the logical connectives $\sim$, $\wedge$, $\vee$, $\Rightarrow$ and knows propositional logic but who has not yet been told the meaning of the quantifiers $\forall$ and $\exists$. Suppose you show this individual the formula $\forall xPx \Rightarrow \forall xPx$ and ask whether he has any reason to believe that it is true. He will probably reply something like "Of course it is true! I don't know what $\forall xPx$ means, but I do know that for any proposition p, the proposition $p \Rightarrow p$ must be true, so, in particular, it is true when p is the proposition $\forall xPx$, whatever that proposition means." Similarly, he will recognize the truth of $((\forall xPx \Rightarrow \forall xQx) \wedge \forall xPx) \Rightarrow \forall xQx$, since it is of the form $((p \Rightarrow q) \wedge p) \Rightarrow q$; but he will have no way of recognizing the truth of the valid formula $(\forall x(Px \Rightarrow Qx) \wedge \forall xPx) \Rightarrow \forall xQx$, whose propositional form is the non-tautological $(p \wedge q) \Rightarrow r$. In the first two cases we were dealing with formulas that are *tautologies*, whereas in the third case the formula, though valid, is not a tautology. We recall that a formula X of first-order logic is a *tautology* if it is an instance of a tautology of *propositional* logic—i.e., if there is a tautology Y of propositional logic such that

X is obtained from Y by substituting first-order formulas for the propositional variables of Y. Equivalently, X is a tautology iff there is a closed tableau for FX which uses only the propositional Rules A and B (the rules for the α's and β's). If a formula X is valid but not a tautology, then it cannot be proved by a tableau using only the α and β rules; it must require at least one use of the γ or δ rules.

Here is another way of characterizing the important difference between the notions of being *valid* and being a *tautology*. By a *valuation* v of all closed formulas (with or without parameters) is meant an assignment of a truth value t or f to each formula X. By $v(X)$ is meant the value (t or f) assigned to X by v. We say that X is *true under* v if $v(X)=t$, and *false under* v if $v(X)=f$. A valuation v is called a *Boolean valuation* if it satisfies the following three conditions (for all formulas X, α, β):

B_0. X and $\sim X$ are given opposite values under v.

B_1. α is true under v iff α_1 and α_2 are both true under v.

B_2. β is true under v iff at least one of β_1, β_2 is true under v.

A valuation v is called a *first-order valuation* if it is a Boolean valuation and also satisfies the following two conditions:

F_1. γ is true under v iff for every parameter a, the sentence $\gamma(a)$ is true under v.

F_2. δ is true under v iff there is at least one parameter a such that $\delta(a)$ is true under v.

Now, to say that a formula X is a *tautology* is to say that X is true under all *Boolean* valuations, whereas to say that X is *valid* is to say that X is true under all *first-order* valuations. Obviously all tautologies are valid, but not all valid formulas are tautologies.

First-order valuations are very closely related to *interpretations*, in the following sense: Let us say that a valuation v *agrees* with an interpretation I, or that I *agrees* with v, if, for each formula X, X is true under I if and only if $v(X)=t$. Obviously, if v agrees with some interpretation I, then v is a first-order valuation. Also, for any interpretation I, if for each formula X we define $v(X)$ to be t if and only if X is true under I, then this v agrees with I, and this v is said to be the valuation *induced* by I. Going in the other direction, if we start with a first-order valuation v, there is one and only one interpretation I that agrees with v—namely, for each predicate P of degree n, I assigns to P the relation R such that $R(a_1,\ldots,a_n)$ holds if and only if the sentence $v(Pa_1,\ldots,a_n)=t$.

In short, for any interpretation I there is one and only one first-order valuation v that agrees with I, and for any first-order valuation v there is one and only one interpretation I that agrees with it.

Truth-Functional Satisfiability versus First-Order Satisfiability

We shall say that a set S is *truth-functionally satisfiable* if there is some *Boolean* valuation in which all elements of S are true, and *first-order satisfiable* if there is some *first-order* valuation in which all elements of S are true. For example, $\{\forall xPx, \exists x{\sim}Px\}$ is obviously not first-order satisfiable, but it is truth-functionally satisfiable (you can't get a closed tableau for this set using only Rules A and B. We also say that a formula X is *truth-functionally implied*, or *tautologically implied*, by a set S of formulas if X is true in all *Boolean* valuations that satisfy (all elements of) S. If S is a *finite* set $\{X_1, \ldots, X_n\}$, then X is truth-functionally implied by S iff the sentence $(X_1 \wedge \cdots \wedge X_n) \Rightarrow X$ is a tautology. We say that X is *validly* implied by S if X is true in all interpretations that satisfy S, and, if S is a finite set $\{X_1, \ldots, X_n\}$, this is equivalent to saying that $(X_1 \wedge \cdots \wedge X_n) \Rightarrow X$ is valid. Every tautology is of course valid, but a valid formula is not necessarily a tautology, so if X is truth-functionally implied by S, it is of course validly implied by S, but if X is validly implied by S, it is not necessarily truth-functionally implied by S. Thus truth-functional implication is a much stronger condition than valid implication.

We say that X *tautologically implies* Y, or that Y is tautologically implied by X, if $X \Rightarrow Y$ is a tautology, and we say that sentences X and Y are *tautologically equivalent* if $X \equiv Y$ is a tautology.

Boolean Atoms

By a *Boolean atom* we shall mean a first-order sentence that is neither a negation of some sentence, nor a conjunction, disjunction or conditional of other sentences—in other words, it is either an atomic sentence $Pa_1, \ldots, a_n$ or of the form $\forall x\phi(x)$ or $\exists x\phi(x)$.

What we shall do in this chapter is show how validity is related to tautological truth in a very interesting and useful way.

Regular Sets

We are working now with *unsigned* formulas. By a *regular* formula we shall mean one of the form $\gamma \Rightarrow \gamma(a)$ or of the form $\delta \Rightarrow \delta(a)$, where a does not occur in δ. Regular formulas of type $\gamma \Rightarrow \gamma(a)$ shall be called regular formulas of *type C*, and regular formulas of type $\delta \Rightarrow \delta(a)$ shall be called regular formulas of *type D*.

We shall henceforth use the symbol "Q" to denote any γ or δ, and by $Q(a)$ we shall mean the corresponding $\gamma(a)$ or $\delta(a)$.

By a *regular sequence* we shall mean a finite (possibly empty) sequence

$$(Q_1 \Rightarrow Q_1(a_1), \ldots, Q_n \Rightarrow Q_n(a_n))$$

such that every term is regular, and, furthermore, for each $i<n$, if $Q_{i+1} \Rightarrow Q_{i+1}(a_{i+1})$ is of type D, then a_{i+1} does not occur in any of the earlier terms

$$Q_1 \Rightarrow Q_1(a_1), \ldots, Q_i \Rightarrow Q_i(a_i).$$

By a *regular set* R we shall mean a finite set whose members can be arranged in a regular sequence. Alternatively, a regular set can be characterized as any finite set constituted according to the following rules:

R_0. The empty set $\emptyset$ is regular.

R_1. If R is regular, so is $R{:}\gamma \Rightarrow \gamma(a)$.

R_2. If R is regular, so is $R{:}\delta \Rightarrow \delta(a)$, provided a does not occur in δ or in any element of R.

We aim to show that for any sentence X without parameters, X is valid if and only if X is truth-functionally implied by some regular set R. Equivalently, X is valid iff there is a regular set $\{X_1, \ldots, X_n\}$ such that $(X_1 \wedge \cdots \wedge X_n) \Rightarrow X$ is a tautology. We shall, in fact, show something stronger; but before we state the stronger result, observe that every regular formula is of one of the following four forms:

(1) $\forall x \phi(x) \Rightarrow \phi(a)$.

(2) $\sim\forall x \phi(x) \Rightarrow \sim\phi(a)$.

(3) $\exists x \phi(x) \Rightarrow \phi(a)$.

(4) $\sim\exists x \phi(x) \Rightarrow \sim\phi(a)$.

In each of the four cases, we shall refer to $\phi(x)$ as the *principal part* of the formula. The stronger result that we will show is that for any pure sentence X (i.e., a sentence with no parameters), X is valid if and only if X is truth-functionally implied by a regular set R such that, for each element of R, its principal part is a subformula of X. In fact, we will show even more—but first for some further preliminaries (recall for any set S and formula X we are using the notation $S{:}X$ to abbreviate $S \cup \{X\}$, i.e., the set whose elements are X together with the elements of S).

FACT 1. For any satisfiable set S,

(a) if X is valid then $S{:}X$ is satisfiable;

(b) if $S{:}X$ is not satisfiable, then for any formula Y, the set $S{:}X \Rightarrow Y$ is satisfiable.

PROBLEM 19.1. Prove Fact 1.

Next, we need

LEMMA 19.1. *Let S be a satisfiable set of sentences (maybe with parameters).*

(a) For every parameter a, the set $S{:}\gamma \Rightarrow \gamma(a)$ is satisfiable;

(b) For every parameter a that occurs neither in δ nor in any element of S, the set $S{:}\delta \Rightarrow \delta(a)$ is satisfiable.

PROBLEM 19.2. Prove Lemma 19.1. *Hint:* For (b), consider these two cases: $S{:}\delta$ is satisfiable, $S{:}\delta$ is not satisfiable. In one of these cases, use Fact F_4 of Chapter 18.

We shall call a parameter a a *critical* parameter of a regular set R if, for some δ, the sentence $\delta \Rightarrow \delta(a)$ is in R. Let us recall that for any sets A and B, by their *union* $A \cup B$ is meant the set whose elements are those of A together with those of B. In what follows, we shall always use "R" to denote some regular set.

THEOREM 19.1.

(a) If S is satisfiable and R is a regular set such that no critical parameter of R occurs in any element of S, then $R \cup S$ is satisfiable.

(b) Every regular set is satisfiable.

(c) If X is validly implied by a regular set R and if no critical parameter of R occurs in X, then X is valid. In particular, any pure *sentence X validly implied by a regular set R is valid.*

(d) If $(\gamma \Rightarrow \gamma(a)) \Rightarrow X$ is valid, so is X.
If $(\delta \Rightarrow \delta(a)) \Rightarrow X$ is valid and a does not occur in δ nor in X, then X is valid.

PROBLEM 19.3. Prove Theorem 19.1. *Hint:* (b) and (c) are easy consequences of (a), and (d) is an easy consequence of (c). As to (a), use Lemma 19.1.

Concerning (b), a regular set R is not only satisfiable but has an even stronger property midway in strength between satisfiability and validity. Consider first a single formula $\phi(a_1, \ldots, a_n)$ whose parameters are $a_1, \ldots, a_n$. We call the formula *sound* if, for every interpretation I of the *predicates* of the formula in some universe V, there are elements $e_1, \ldots, e_n$

of V such that $\phi(e_1, \ldots, e_n)$ is true under I. This condition is equivalent to the following: Let $\phi(x_1, \ldots, x_n)$ be the result of substituting the variables $x_1, \ldots, x_n$ for the parameters $a_1, \ldots, a_n$, respectively. Then $\phi(a_1, \ldots, a_n)$ is sound if and only if the sentence

$$\exists x_1 \cdots \exists x_n \phi(x_1, \ldots, x_n)$$

is *valid*. More generally, we call a set S of sentences *sound* iff for every interpretation of the predicates occurring in S, there is a choice of values for the parameters that makes all elements of S true. Now, a regular set R is not only satisfiable, but even sound—in fact, it has the even stronger property that, for any interpretation of the predicates of R and any choice of values for the non-critical parameters of R, there exists a choice of values for the critical parameters of R that makes all the elements of R true.

EXERCISE 19.1. Prove the preceding statement.

A FUNDAMENTAL THEOREM

THEOREM 19.2 (THE REGULARITY THEOREM). *Every valid sentence X is truth-functionally implied by a regular set R—in fact, by one in which no critical parameter of R occurs in X, and, moreover, the principal part of each member of R is a subformula of X.*

We will prove the Regularity Theorem by showing how, from a closed tableau $\mathfrak{T}$ for $\sim X$, we can effectively find a regular set that implies truth-functionally X. Can the reader guess how? The answer is beautifully simple! Just take for R the set of all formulas $Q \Rightarrow Q(a)$ such that $Q(a)$ was inferred from Q by Rule C or Rule D (see Chapter 13)! We will call this the *associated* regular set for $\mathfrak{T}$.

PROBLEM 19.4. Prove that this set R works. *Hint:* Consider a tableau starting with $\sim X$ and the elements of R and show that it can be made to close using only Rules A and B.

Let us consider an example. Take the valid formula

$$\forall x(Px \Rightarrow Qx) \Rightarrow (\exists x Px \Rightarrow \exists x Qx).$$

Here is a closed tableau for the negation of this formula:

$$
\begin{array}{ll}
(1)\ \sim[\forall x(Px \Rightarrow Qx) \Rightarrow (\exists x Px \Rightarrow \exists x Qx)] & \\
(2)\ \forall x(Px \Rightarrow Qx) & (1) \\
(3)\ \sim(\exists x Px \Rightarrow \exists x Qx) & (1) \\
(4)\ \exists x Px & (3) \\
(5)\ \sim\exists x Qx & (3) \\
(6)\ Pa & (4) \\
(7)\ \sim Qa & (5) \\
(8)\ Pa \Rightarrow Qa & (2)
\end{array}
$$

$\sim Pa$ (8) $\quad$ Qa (8)

We have used the quantification rules (C and D) to infer (6) from (4), (7) from (5), and (8) from (2). Accordingly, the associated regular set R is

$$\left\{ \begin{array}{l} (4) \Rightarrow (6) \\ (5) \Rightarrow (7) \\ (2) \Rightarrow (8) \end{array} \right\}$$

That is,

$$R = \left\{ \begin{array}{l} \exists x Px \Rightarrow Pa \\ \sim\exists x Qx \Rightarrow \sim Qa \\ \forall x(Px \Rightarrow Qx) \Rightarrow (Pa \Rightarrow Qa) \end{array} \right\}$$

According to our theorem, this set should truth-functionally imply X, i.e., the formula

$$\forall x(Px \Rightarrow Qx) \Rightarrow (\exists x Px \Rightarrow \exists x Qx).$$

To see this more clearly, let us abbreviate by propositional variables all *Boolean atoms* involved. We let

$$
\begin{aligned}
p &= Pa, \\
q &= Qa, \\
r &= \exists x Px, \\
s &= \exists x Qx, \\
m &= \forall x(Px \Rightarrow Qx).
\end{aligned}
$$

Then R is the set

$$\left\{ \begin{array}{l} r \Rightarrow p \\ \sim s \Rightarrow \sim q \\ m \Rightarrow (p \Rightarrow q) \end{array} \right\}$$

and $X = m \Rightarrow (r \Rightarrow s)$.

It is now easy to see that X is tautologically implied by R—in other words, that

$$[(r \Rightarrow p) \wedge (\sim s \Rightarrow \sim q) \wedge (m \Rightarrow (p \Rightarrow q))] \Rightarrow (m \Rightarrow (r \Rightarrow s))$$

is a tautology.

One obtains some very curious tautologies in this manner. The reader might have a bit of fun by taking some of the valid formulas proved in exercises of earlier chapters and finding the regular sets that tautologically imply them.

Theorem 19.2 has a generalization worth considering: For any finite set S of closed formulas (maybe with parameters), call a regular set R an *associate* of S if the following three conditions hold:

(1) $R \cup S$ is not truth-functionally satisfiable.

(2) The principal part of each element of R is a subformula of some element of S.

(3) No critical parameter of R occurs in any element of S.

THEOREM 19.2*. *Every unsatisfiable set S has an associate.*

EXERCISE 19.2. Prove Theorem 19.2* and explain why Theorem 19.2 is only a special case of it.

SIMPLIFIED REGULAR FORMULAS

In unified notation, every regular formula is of one of two forms: $\gamma \Rightarrow \gamma(a)$, $\delta \Rightarrow \delta(a)$. In non-unified notation, every regular formula is of one of *four* forms: $\forall x \phi(x) \Rightarrow \phi(a)$, $\sim \exists x \phi(x) \Rightarrow \sim \phi(a)$, $\exists x \phi(x) \Rightarrow \phi(a)$, $\sim \forall x \phi(x) \Rightarrow \sim \phi(a)$. Let us define the *simplification* of $\sim \exists x \phi(x) \Rightarrow \sim \phi(a)$ to be the truth-functionally equivalent formula $\phi(a) \Rightarrow \exists x \phi(x)$ and the *simplification* of $\sim \forall x \phi(x) \Rightarrow \sim \phi(a)$ to be the truth-functionally equivalent formula $\phi(a) \Rightarrow \forall x \phi(x)$. Thus if X is of the form $\sim \exists x \phi(x) \Rightarrow \sim \phi(a)$ or of the form $\sim \forall x \phi(x) \Rightarrow \sim \phi(a)$, then X is the *contrapositive* of its simplification. (The *contrapositive* of a formula $X \Rightarrow Y$ is $\sim Y \Rightarrow \sim X$.) For any X of the form $\forall x \phi(x) \Rightarrow \phi(a)$ or of the form $\exists x \phi(x) \Rightarrow \phi(a)$, define the simplification of X to be X itself.

Now, given any regular set R, define its *simplification* R to be the result of replacing each element of R by its simplification. Such a set R we will call a *simplified* regular set. It is obvious that a formula X is truth-functionally implied by R if and only if X is truth-functionally implied by R, so, by the Regularity Theorem, every valid formula is truth-functionally implied by a simplified regular set.

Simplified regular formulas strike me as neater and more natural to work with than regular formulas, and my only purpose in working initially with the latter was to allow the unified γ, δ-notation. For *simplified* regular formulas, however, a unified notation is also possible, as follows: Define $\gamma \rightarrow \gamma(a)$ to be the *simplification* of $\gamma \Rightarrow \gamma(a)$, and $\delta \rightarrow \delta(a)$ to be the simplification of $\delta \Rightarrow \delta(a)$. (Thus, if γ is of the form $\forall x \phi(x)$, then $\gamma \rightarrow \gamma(a)$ is the same thing as $\gamma \Rightarrow \gamma(a)$; but if γ is of the form $\sim \exists x \phi(x)$, then $\gamma \rightarrow \gamma(a)$ is not $\gamma \Rightarrow \gamma(a)$ but the formula $\phi(a) \Rightarrow \exists x \phi(x)$, whose *contrapositive* is $\gamma \Rightarrow \gamma(a)$. Similarly with δ.)

In this notation, a simplified regular set is any set constructed by the following rules:

(1) The empty set $\emptyset$ is a simplified regular set.

(2) For any simplified regular set S, the set $S : Q \rightarrow Q(a)$ is a simplified regular set, provided that either Q is some γ or Q is some δ, and a does not occur in either Q or any element of S.

The reader should note that Lemma 19.1 and item (d) of Theorem 19.1 both still hold if one replaces $\gamma \Rightarrow \gamma(a)$ by $\gamma \rightarrow \gamma(a)$, and $\delta \Rightarrow \delta(a)$ by $\delta \rightarrow \delta(a)$.

SOLUTIONS

19.1. (a) If X is valid, then it is true in *all* interpretations, hence, in particular, true under any interpretation that satisfies S. Thus, if S is satisfiable, it is satisfied by some interpretation I that also satisfies X.

(b) Let I be an interpretation that satisfies S. Since $S{:}X$ is not satisfiable, X is false under I; hence for any Y, the sentence $X \Rightarrow Y$ is true under I. Thus I satisfies S and also $X \Rightarrow Y$, so $S{:}X \Rightarrow Y$ is satisfiable.

19.2. We are given that S is satisfiable.

(a) Since $\gamma \Rightarrow \gamma(a)$ is valid, $S{:}\gamma \Rightarrow \gamma(a)$ is satisfiable, by (a) of Fact 1.

(b) Either $S{:}\delta$ is satisfiable or it isn't. If it isn't, then $S{:}\delta \Rightarrow \delta(a)$ is satisfiable, by (b) of Fact 1. Now suppose that $S{:}\delta$ is satisfiable. Let a be any parameter that occurs neither in δ nor in any element of S. Then by Fact F_4 of Chapter 18, the set $S{:}\delta, \delta(a)$ is satisfiable. Let I be any interpretation that satisfies it. Since $\delta(a)$ is true under I, so is $\delta \Rightarrow \delta(a)$, and thus I satisfies $\delta \Rightarrow \delta(a)$ as well as S. Thus $S{:}\delta \Rightarrow \delta(a)$ is satisfiable.

19.3. (a) Suppose S is satisfiable, R is regular, and no critical parameter of R occurs in any element of S. Then if we successively adjoin elements of R to S, at no stage do we destroy satisfiability (by Lemma 19.1). Strictly speaking, this argument uses mathematical induction on the number of elements of R.

(b) This is immediate from (a), taking S to be the empty set $\emptyset$.

(c) Suppose X is validly implied by R and no critical parameter of R occurs in X. Then no critical parameter of R occurs in $\sim X$. Also, since X is validly implied by R, the set $R \cup \{\sim X\}$ is not satisfiable. Hence $\{\sim X\}$ is not satisfiable (by (a), taking $\{\sim X\}$ for S), hence X is valid.

(d) If $(\gamma \Rightarrow \gamma(a)) \Rightarrow X$ is valid, then so is X, since $\gamma \Rightarrow \gamma(a)$ is itself valid.

Now suppose $(\delta \Rightarrow \delta(a)) \Rightarrow X$ is valid and a does not occur in δ nor in X. Then X is validly implied by the regular set $\{\delta \Rightarrow \delta(a)\}$, and the critical parameter occurs neither in δ nor in X, hence X is valid by (c).

19.4. Let us first note that if we add the *Modus Ponens* rule (from X and $X \Rightarrow Y$ to infer Y) to our tableau rules, we do not increase the class of provable formulas. That is, suppose we add to our tableau rules the rule "Given a branch Θ containing X and $X \Rightarrow Y$, we may adjoin Y as an endpoint of Θ—i.e., we may extend Θ to $(\Theta{:}Y)$." This does not increase the class of provable formulas, because, given a branch Θ containing X and $X \Rightarrow Y$, instead of using Modus Ponens to extend Θ to $(\Theta{:}Y)$, we could have split Θ into the two branches $(\Theta{:}\sim X)$ and $(\Theta{:}Y)$ (by Rule B, since Θ contains $X \Rightarrow Y$), but the branch $(\Theta{:}\sim X)$ is immediately closed (since X is also on the branch), and thus we can effectively extend Θ to $(\Theta{:}Y)$.

Now for the problem at hand: We are given a closed tableau $\mathfrak{T}$ for $\sim X$, and we now construct another tableau $\mathfrak{T}_1$ starting with $\sim X$ and all formulas $Q \Rightarrow Q(a)$ such that $Q(a)$ was inferred from Q in $\mathfrak{T}$ by Rule C or Rule D. Whereas we inferred $Q(a)$ from Q on some branch Θ in $\mathfrak{T}$ by Rule C or D, since in our new tableau $\mathfrak{T}_1$ the formula $Q \Rightarrow Q(a)$ is already on Θ (it is on all branches of $\mathfrak{T}_1$), we can now infer $Q(a)$ from Q and $Q \Rightarrow Q(a)$ by Modus Ponens (which is reducible to truth-functional rules). Thus we need only Rules A and B to close $\mathfrak{T}_1$.

- Part V -

Axiom Systems

- CHAPTER 20 -

BEGINNING AXIOMATICS

AXIOM SYSTEMS IN GENERAL

One purpose of the field known as *mathematical logic* is to make precise the notion of a *proof*. I like to illustrate the need for this with the following example: Suppose that, in a high-school course in Euclidean geometry, a student is asked on an examination to give a proof of, say, the Pythagorean Theorem. The teacher returns the paper with a grade of zero and the comment "This is no proof!" Now, a really sophisticated student could well retort: "How do you know that what I handed you is not a proof? You have never once in this course defined what you mean by a *proof*. You have defined with admirable precision various geometric notions, such as *angle, triangle, congruence*, etc., but never have you given a definition of the word *proof*. So on what basis can you say that what I gave you is not a proof? How would you *prove* that what I wrote is not a proof?"

Well, in modern rigorous axiom systems for fields like geometry, or number theory, or set theory, the underlying logic is made quite explicit, the notion of *proof* in the system is well defined, and there is no question of whether a purported proof is really a proof or not—indeed, even a computer could check whether a purported proof is a genuine proof.

Axiom systems, in the Greek sense of the term, consisted of a set of propositions considered *self-evident* and a set of logical rules that enabled one to derive from these self-evident propositions, other propositions, some of which were far from being self-evident. An axiom system $\mathfrak{A}$ in the *modern* sense of the term consists of a domain (set) D of elements called *formal objects*, together with a subset A of D whose elements are called the *axioms* of the system, together with certain relations called *inference rules*, each being of the form "From (objects) $X_1, \ldots, X_n$, one may infer (object) X." (Formally, an inference rule is simply a set $\{X_1, \ldots, X_n\}$ together with a single object X.)

What the formal objects are varies from system to system. In so-called *Hilbert-type* axiom systems, the formal objects are formulas of propositional logic or first-order logic. In this chapter, we shall consider only Hilbert-type systems. We do not require that the set of axioms be finite. One sometimes displays an infinite set of axioms by means of an *axiom schema* which specifies the general form of a set of formulas—an example of an axiom schema for propositional logic could be the schema $((X \Rightarrow Y) \vee (Y \Rightarrow Z)) \Rightarrow (X \Rightarrow Z)$, which, if adopted as part of an axiom system, would specify that any formula of that logical form is an axiom. An example of an axiom schema for first-order logic could be $\forall x \phi(x) \Rightarrow \phi(a)$ (which we will use in one of our systems). A typical inference rule for both propositional and first-order logic is the rule known as *Modus Ponens*, which is "From X and $X \Rightarrow Y$, to infer Y." In an inference rule "From $X_1, \ldots, X_n$ to infer X," the objects $X_1, \ldots, X_n$ are called the *premises* of the rule, and X is called the *conclusion*. An inference rule is usually displayed in the form of a figure in which a horizontal line is drawn: the premises are written above the line and the conclusion is drawn below the line. For example, the rule of Modus Ponens is displayed thus:

$$\frac{X, X \Rightarrow Y}{Y}.$$

By a *proof* in an axiom system is meant a finite sequence $X_1, \ldots, X_n$ such that each term of the sequence is either an axiom of the system or is inferable from earlier terms by one of the inference rules. A proof $X_1, \ldots, X_n$ is also called *a proof of* its last term X_n, and an element X is called *provable* in the system, or a *theorem* of the system, if there exists a proof of it. The terms $X_1, \ldots, X_n$ of a proof are usually called the *lines* of the proof and are displayed vertically.

Here is an important general fact about axiom systems: Consider an axiom system $\mathfrak{A}$. Given a property P of the formal objects of $\mathfrak{A}$, we will say that an inference rule R of $\mathfrak{A}$ *preserves* the property P to mean that if each of the premises of R has the property, so does the conclusion. Now, to show that every provable object of $\mathfrak{A}$ has a given property, it suffices to show that each axiom of $\mathfrak{A}$ has the property and that each inference rule of $\mathfrak{A}$ preserves the property—because, if these conditions hold, then in any proof in $\mathfrak{A}$, for each line of the proof, the line has the property provided all earlier lines do (since each line is either an axiom, and hence has the property, or is inferable from earlier lines, all of which have the property, by an inference rule that preserves the property). It then follows by mathematical induction that every line of the proof has the property—in particular, the last line does. Thus, every provable object has the property.

This general principle has several important applications: Consider an axiom system $\mathfrak{A}$ for propositional logic (the formal objects are thus *formulas* of propositional logic). The system as called *correct* if every provable formula is a tautology. Applying our principle to the property of being a tautology, we see that, to show a system is correct, it suffices to show that all axioms are tautologies and that, for each inference rule, if all its premises are tautologies, so is its conclusion. Similarly, an axiom system for first-order logic is called *correct* if only valid formulas are provable in it. To show that a first-order axiom system is correct, it suffices to show that all the axioms are valid and that each inference rule preserves validity.

Here is another application: Consider two axiom systems $\mathfrak{A}_1$ and $\mathfrak{A}_2$. It is sometimes desirable to show that everything provable in $\mathfrak{A}_1$ is also provable in $\mathfrak{A}_2$. To do this, it suffices to show that all axioms of $\mathfrak{A}_1$ are provable in $\mathfrak{A}_2$ and that, for each inference rule R of $\mathfrak{A}_1$, R preserves the property of being provable in $\mathfrak{A}_2$—i.e., if the premises of R are provable in $\mathfrak{A}_2$, so is the conclusion. If this latter condition holds, we say that the inference rule *holds* in the system $\mathfrak{A}_2$, or that R is a *derived* rule of $\mathfrak{A}_2$.

Let us record these facts thus:

Fact 1. To show that an axiom system $\mathfrak{A}$ for propositional logic (first-order logic) is *correct*, it suffices to show that all axioms of $\mathfrak{A}$ are tautologies (respectively, valid formulas) and that for each inference rule R of $\mathfrak{A}$, if all premises of R are tautologies (valid formulas), so is the conclusion.

Fact 2. To show that everything provable in an axiom system $\mathfrak{A}_1$ is also provable in another axiom system $\mathfrak{A}_2$, it suffices to show that all axioms of $\mathfrak{A}_1$ are provable in $\mathfrak{A}_2$ (not necessarily *axioms* of $\mathfrak{A}_2$, but provable in $\mathfrak{A}_2$) and that each inference rule of $\mathfrak{A}_1$ holds in $\mathfrak{A}_2$.

Remark. Suppose that all axioms of $\mathfrak{A}_1$ are provable in $\mathfrak{A}_2$ and that all inference rules of $\mathfrak{A}_1$ hold in $\mathfrak{A}_2$. Although this implies that everything provable in $\mathfrak{A}_1$ is provable in $\mathfrak{A}_2$, it does *not* necessarily imply that every *proof* in $\mathfrak{A}_1$ is also a proof in $\mathfrak{A}_2$. In general, this is not the case: Given a proof in $\mathfrak{A}_1$, one generally must intersperse other lines between the lines of that proof in order to obtain a proof in $\mathfrak{A}_2$.

In this chapter and the next, we will be considering axiom systems for propositional logic only. Axiom systems for first-order logic will be dealt with in a later chapter.

Some History

The earlier treatment of propositional logic was through axiom systems; truth tables came much later, and tableaux came later still. The old axiom systems for propositional logic took a *finite* number of tautologies as axioms (usually six or less) and used two inference rules—Modus Ponens and the *Rule of Substitution*, which I will now explain: By an *instance* (sometimes called a *substitution instance*) of a formula X is meant any formula obtained by substituting formulas for some or all of the propositional variables in X. Examples: An instance of $p \Rightarrow p$ is $(p \vee q) \Rightarrow (p \vee q)$, another is $(p \wedge (q \Rightarrow r)) \Rightarrow (p \wedge (q \Rightarrow r))$; an instance of $p \Rightarrow (q \Rightarrow r)$ is $(q \Rightarrow p) \Rightarrow ((p \wedge q) \Rightarrow (r \vee s))$. The Rule of Substitution is "From X to infer any instance of X." This rule is obviously *correct* in the sense that any instance of a tautology is also a tautology.

I find it remarkable that *all* tautologies are derivable from only a finite number of tautologies using only these two rules. In setting up these systems, one specifies certain of the logical connectives as *primitive*, or *undefined*, and defines other connectives in terms of them. The system used by Russell and Whitehead in *Principia Mathematica* [28] was based on $\sim$ and $\vee$ as primitives (and took $X \Rightarrow Y$ merely as an *abbreviation* for $\sim X \vee Y$). This system had the following five axioms:

(a) $(p \vee p) \Rightarrow p$.

(b) $p \Rightarrow (p \vee q)$.

(c) $(p \vee q) \Rightarrow (q \vee p)$.

(d) $(p \Rightarrow q) \Rightarrow ((r \vee p) \Rightarrow (r \vee q))$.

(e) $(p \vee (q \vee r)) \Rightarrow ((p \vee q) \vee r)$.

It later turned out that axiom (e) was redundant—it was derivable from the other four.

There are many axiom systems on the market that take $\sim$ and $\Rightarrow$ as primitive. The earliest such system goes back as far as Frege, in 1879 [5], and consisted of six axioms. Later, J. ukasiewicz [11] replaced those six by a simpler system that uses the following three axioms:

(1) $p \Rightarrow (q \Rightarrow p)$.

(2) $(p \Rightarrow (q \Rightarrow r)) \Rightarrow ((p \Rightarrow q) \Rightarrow (p \Rightarrow r))$.

(3) $(\sim p \Rightarrow \sim q) \Rightarrow (q \Rightarrow p)$.

A more modern version of, say, the above system would take as axioms all *instances* of the formulas (1), (2), (3) above and use only Modus Ponens as an inference rule. Thus the axioms of a modernized version of the above system consist of all formulas of any of the following three forms:

(i) $X \Rightarrow (Y \Rightarrow X)$.

(ii) $(X \Rightarrow (Y \Rightarrow Z)) \Rightarrow ((X \Rightarrow Y) \Rightarrow (X \Rightarrow Z))$.

(iii) $(\sim X \Rightarrow \sim Y) \Rightarrow (Y \Rightarrow X)$.

Each of (i), (ii), (iii) above is an example of what is termed an *axiom schema*, or, more briefly, a *schema*. A schema is the set of all *instances* of a formula.

Incidentally, a variant of the above system replaces schema (iii) by the following schema:

(iii′) $(\sim X \Rightarrow \sim Y) \Rightarrow ((\sim X \Rightarrow Y) \Rightarrow X)$.

The completeness of this variant is a bit quicker to establish.

Church, in [2], gives an axiom system whose primitives are $\Rightarrow$ and f (for *falsehood*). In it, $\sim X$ is defined as $X \Rightarrow f$. This system uses schemata (i) and (ii) above, and in place of (iii) uses the schema

(iii″) $((X \Rightarrow f) \Rightarrow f) \Rightarrow X$.

Rosser [15] takes $\sim$ and $\vee$ as primitive and uses the following three axiom schemata:

(a) $X \Rightarrow (X \vee X)$.

(b) $(X \wedge Y) \Rightarrow X$.

(c) $(X \Rightarrow Y) \Rightarrow (\sim(X \vee Z) \Rightarrow \sim(Z \vee X))$.

Kleene [9] gives an axiom system that uses 11 schemata and takes all four connectives $\sim$, $\vee$, $\wedge$, $\Rightarrow$ as primitive. Most proofs in this system are quite natural. At the other extreme, Nicod [12] takes only the Sheffer stroke as primitive and uses only one axiom and one rule of inference other than substitution.

Uniform Systems

I will take a different approach from the more conventional ones: I find it of particular interest to consider axiom systems for propositional logic that are *uniform*, in the sense that the choice of logical connectives is immaterial. My uniform α, β notation is what suggested this to me.

In most of the axiom systems for propositional logic in the literature, a good deal of work is required to prove their completeness. By contrast, in the first uniform system $\mathfrak{U}_0$, to which we now turn, the completeness proof is almost immediate—given a truth table for a tautology X, it is relatively easy to convert it into a proof in the axiom system $\mathfrak{U}_0$.

First, let us recall that by $X_1 \wedge X_2 \wedge X_3$ we mean $(X_1 \wedge X_2) \wedge X_3$; $X_1 \wedge X_2 \wedge X_3 \wedge X_4$ is an abbreviation for $((X_1 \wedge X_2) \wedge X_3) \wedge X_4$, and, in general, $X_1 \wedge \cdots \wedge X_{n+1}$ is an abbreviation for $(X_1 \wedge \cdots \wedge X_n) \wedge X_{n+1}$.

Now for the postulates of $\mathfrak{U}_0$.

AXIOMS. All sentences of the form $(X_1 \wedge \cdots \wedge X_n) \Rightarrow X_i$ $(i \leq n)$

INFERENCE RULES.

I. $$\frac{X \Rightarrow \alpha_1 \quad X \Rightarrow \alpha_2}{X \Rightarrow \alpha}.$$

II. (a) $\dfrac{X \Rightarrow \beta_1}{X \Rightarrow \beta}$; (b) $\dfrac{X \Rightarrow \beta_2}{X \Rightarrow \beta}$.

III. $\dfrac{(X \wedge Y) \Rightarrow Z}{X \Rightarrow (Y \Rightarrow Z)}$ (known as *Exportation*).

IV. $$\frac{X \Rightarrow Y \quad \sim X \Rightarrow Y}{Y}.$$

REMARK. The symbols $\Rightarrow$ and $\wedge$ do appear in the above presentation, but I must emphasize that they do not have to be primitive symbols of the system—they can just as well be defined from others, or even from the single primitive of joint denial or the Sheffer stroke. Whatever connectives we take as primitive, the α's and β's are to be understood accordingly. It is in this sense that the system $\mathfrak{U}_0$ is *uniform*.

REMARK. Our presentation of the axioms is, strictly speaking, not an *axiom schema*, because of the variable n. For each *particular* n, the expression $(X_1 \wedge \cdots \wedge X_n) \Rightarrow X_i$ $(i \leq n)$ can be broken up into n schemas—for example, $(X_1 \wedge X_2 \wedge X_3) \Rightarrow X_i$ $(i \leq n)$ is really a condensation of

$$(X_1 \wedge X_2 \wedge X_3) \Rightarrow X_1, (X_1 \wedge X_2 \wedge X_3) \Rightarrow X_2, (X_1 \wedge X_2 \wedge X_3) \Rightarrow X_3.$$

The expression $(X_1 \wedge \cdots \wedge X_n) \Rightarrow X_i$ $(i \leq n)$ might aptly be called a *meta-schema*. Later we will reduce this "meta-schema" to honest-to-goodness axiom schemas, and also replace the inference rules by the single rule of Modus Ponens.

CORRECTNESS. It is obvious that all axioms of $\mathfrak{U}_0$ are tautologies and that, for each inference rule, if the premises are all tautologies, so is the conclusion. Thus all provable formulas are tautologies, so the system is correct.

COMPLETENESS. Preparatory to our completeness proof for $\mathfrak{U}_0$, we must introduce the notions of a *basic sequence* and a *basic formula*: Let us enumerate all propositional variables in some order $p_1, p_2, \ldots, p_n, \ldots$, fixed in advance. By a *basic sequence* with respect to the variables $p_1, \ldots, p_n$, we mean a sequence $q_1, \ldots, q_n$, where each q_i is either p_i or $\sim p_i$ (for $i \leq n$). There are thus 2^n basic sequences with respect to $p_1, \ldots, p_n$. By a *basic formula* with respect to (or *in*) $p_1, \ldots, p_n$, we mean a formula $q_1 \wedge \cdots \wedge q_n$, where $q_1, \ldots, q_n$ is a basic sequence with respect to $p_1, \ldots, p_n$. There are exactly 2^n basic formulas with respect to $p_1, \ldots, p_n$. In preparation for our completeness proof, we must note a few general facts about basic formulas.

FACT 1. (a) For any basic formula $q_1 \wedge \cdots \wedge q_n$ with respect to the variables $p_1, \ldots, p_n$, there is one and only one interpretation I of the variables $p_1, \ldots, p_n$ in which $q_1 \wedge \cdots \wedge q_n$ is true.

(b) For any interpretation I of the variables $p_1, \ldots, p_n$, there is one and only one basic formula $q_1 \wedge \cdots \wedge q_n$ with respect to the variables $p_1, \ldots, p_n$ that is true under I.

PROBLEM 20.1. Prove Fact 1.

Let us note that if $X \Rightarrow \alpha$ is a tautology, then $X \Rightarrow \alpha_1$ and $X \Rightarrow \alpha_2$ are both tautologies (because $\alpha \Rightarrow \alpha_1$ and $\alpha \Rightarrow \alpha_2$ are themselves both tautologies). If $X \Rightarrow \beta$ is a tautology, however, it does *not* necessarily follow that either $X \Rightarrow \beta_1$ or $X \Rightarrow \beta_2$ is a tautology (for example, $(p \vee q) \Rightarrow (p \vee q)$ is a tautology, but neither $(p \vee q) \Rightarrow p$ nor $(p \vee q) \Rightarrow q$ is a tautology). But we do know the following important fact:

FACT 2. For any *basic* formula $q_1 \wedge \cdots \wedge q_n$ with respect to variables $p_1, \ldots, p_n$ and any formula β of type B whose variables are all in the set $\{p_1, \ldots, p_n\}$, if $(q_1 \wedge \cdots \wedge q_n) \Rightarrow \beta$ is a tautology, then either $(q_1 \wedge \cdots \wedge q_n) \Rightarrow \beta_1$ or $(q_1 \wedge \cdots \wedge q_n) \Rightarrow \beta_2$ is a tautology.

PROBLEM 20.2. Prove Fact 2.

The reader should find the following exercise helpful.

EXERCISE 20.1.

(a) Show that for any formula X and any basic formula B in the variables of X, either $B \Rightarrow X$ or $B \Rightarrow \sim X$ is a tautology.

(b) Show that X is a tautology if and only if for every basic formula B in the variables of X, the formula $B \Rightarrow X$ is a tautology.

(c) Suppose B is a basic formula that is true under an interpretation I. Show that, for any formula X, X is true under I if and only if $B \Rightarrow X$ is a tautology.

As we will see from the completeness proof that follows, the only axioms we need are those in which $X_1 \wedge \cdots \wedge X_n$ is a basic formula; the only applications we need of the inference rules I and II are those in which X is a basic conjunction; the only application we need of rule III is that in which X is a basic conjunction and $X \wedge Y$ is a basic formula; and the only application we need of rule IV that we need is that in which X is a propositional variable.

LEMMA 20.1. *For any basic formula B in the variables $p_1, \ldots, p_n$, and any formula X whose variables are all in the set $\{p_1, \ldots, p_n\}$, if $B \Rightarrow X$ is a tautology then it is provable in $\mathfrak{U}_0$ (in fact, using only Rules I and II).*

PROBLEM 20.3. Prove Lemma 20.1. *Hint:* Use induction on the degree of X.

At this point, it will be useful to use the following abbreviations: By $X_1 \Rightarrow X_2 \Rightarrow X_3$ we shall mean $X_1 \Rightarrow (X_2 \Rightarrow X_3)$, by $X_1 \Rightarrow X_2 \Rightarrow X_3 \Rightarrow X_4$ we shall mean $X_1 \Rightarrow (X_2 \Rightarrow (X_3 \Rightarrow X_4))$, etc. Thus $X_1 \Rightarrow \cdots \Rightarrow X_n$ abbreviates $X_1 \Rightarrow (\cdots (X_{n-1} \Rightarrow X_n) \ldots)$. Note that parentheses here are restored to the *right* (unlike in n-fold conjunctions $X_1 \wedge \cdots \wedge X_n$, in which parentheses are restored to the *left*).

Let us now note that for any $n>1$, if $(X_1 \wedge \cdots \wedge X_n) \Rightarrow Y$ is provable in $\mathfrak{U}_0$, so is $(X_1 \wedge \cdots \wedge X_{n-1}) \Rightarrow (X_n \Rightarrow Y)$, by Rule III. If, furthermore, $n-1>1$, then

$$(X_1 \wedge \cdots \wedge X_{n-2}) \Rightarrow (X_{n-1} \Rightarrow (X_n \Rightarrow Y))$$

is provable in $\mathfrak{U}_0$, again, by Rule III. After $n-1$ rounds of this reasoning, we see that $X_1 \Rightarrow X_2 \Rightarrow \cdots \Rightarrow X_n \Rightarrow Y$ is provable in $\mathfrak{U}_0$. From this and Lemma 20.1 follows

PROPOSITION 20.1. *For any basic sequence $q_1, \ldots, q_n$ in the variables $p_1, \ldots, p_n$, and for any formula X whose variables are all in the set $\{p_1, \ldots, p_n\}$, if*

$$(q_1 \wedge \cdots \wedge q_n) \Rightarrow X$$

is a tautology, then $q_1 \Rightarrow \cdots \Rightarrow q_n \Rightarrow X$ is provable in $\mathfrak{U}_0$.

Next, from repeated applications of Rule IV, we have:

PROPOSITION 20.2. *Suppose that for each of the 2^n basic sequences $q_1, \ldots, q_n$ with respect to the variables $p_1, \ldots, p_n$, the formula $q_1 \Rightarrow \cdots \Rightarrow q_n \Rightarrow X$ is provable in $\mathfrak{U}_0$. Then X is provable in $\mathfrak{U}_0$.*

PROBLEM 20.4. Prove Proposition 20.2.

From Propositions 20.1 and 20.2 we easily get

THEOREM 20.1. *The system $\mathfrak{U}_0$ is complete; all tautologies are provable in $\mathfrak{U}_0$.*

PROBLEM 20.5. Prove Theorem 20.1.

DISCUSSION. As mentioned before, it is easy to convert a truth table for a tautology X into a proof of X in $\mathfrak{U}_0$. Here is how: Suppose X is a tautology in the variables $p_1, \ldots, p_n$.

STEP 1. Construct a truth table for X and number the rows 1 to 2^n.

STEP 2. Replace each entry T by the formula at the top of the column in which it appears, and each entry F by the negation of the formula at the top of the column.

EXAMPLE. Here is a truth table for $\sim(p\wedge q)\Rightarrow(\sim p\vee\sim q)$.

	p	q	$\sim p$	$\sim q$	$p\wedge q$	$\sim(p\wedge q)$	$\sim p\vee\sim q$	$\sim(p\wedge q)\Rightarrow(\sim p\vee\sim q)$
(1)	T	T	F	F	T	F	F	T
(2)	T	F	F	T	F	T	T	T
(3)	F	T	T	F	F	T	T	T
(4)	F	F	T	T	F	T	T	T

Making the replacements indicated above, we get

	p	q	$\sim p$	$\sim q$	$p\wedge q$	$\sim(p\wedge q)$	$\sim p\vee\sim q$	$\sim(p\wedge q)\Rightarrow(\sim p\vee\sim q)$
(1)	p	q	$\sim\sim p$	$\sim\sim q$	$p\wedge q$	$\sim\sim(p\wedge q)$	$\sim(\sim p\vee\sim q)$	$\sim(p\wedge q)\Rightarrow(\sim p\vee\sim q)$
(2)	p	$\sim q$	$\sim\sim p$	$\sim q$	$\sim(p\wedge q)$	$\sim(p\wedge q)$	$\sim p\vee\sim q$	$\sim(p\wedge q)\Rightarrow(\sim p\vee\sim q)$
(3)	$\sim p$	q	$\sim p$	$\sim\sim q$	$\sim(p\wedge q)$	$\sim(p\wedge q)$	$\sim p\vee\sim q$	$\sim(p\wedge q)\Rightarrow(\sim p\vee\sim q)$
(4)	$\sim p$	$\sim q$	$\sim p$	$\sim q$	$\sim(p\wedge q)$	$\sim(p\wedge q)$	$\sim p\vee\sim q$	$\sim(p\wedge q)\Rightarrow(\sim p\vee\sim q)$

STEP 3. Let B be a basic formula $q_1\wedge\cdots\wedge q_n$ in the variables $p_1, \ldots, p_n$ of X. It corresponds to the row where the first n entries are $q_1, \ldots, q_n$. (In the above example, for instance, row 2 corresponds to the basic formula $p\wedge\sim q$.) Let $q_1, \ldots, q_n, X_1, \ldots, X_k, X$ be the row corresponding to B. Then the following is a proof of $B\Rightarrow X$ in $\mathfrak{U}_0$:

$$B\Rightarrow q_1$$
$$\vdots$$
$$B\Rightarrow q_n$$
$$B\Rightarrow X_1$$
$$\vdots$$
$$B\Rightarrow X_k$$
$$B\Rightarrow X.$$

Example (continued). Corresponding to row 2, we get the following proof of $(p\wedge\sim q)\Rightarrow(\sim(p\wedge q)\Rightarrow(\sim p\vee\sim q))$:

$$(p\wedge\sim q)\Rightarrow p$$
$$(p\wedge\sim q)\Rightarrow\sim q$$
$$(p\wedge\sim q)\Rightarrow\sim\sim p$$
$$(p\wedge\sim q)\Rightarrow\sim q$$
$$(p\wedge\sim q)\Rightarrow\sim(p\wedge q)$$
$$(p\wedge\sim q)\Rightarrow\sim(p\wedge q)$$
$$(p\wedge\sim q)\Rightarrow(\sim p\vee\sim q)$$
$$(p\wedge\sim q)\Rightarrow(\sim(p\wedge q)\Rightarrow(\sim p\vee\sim q)).$$

Continuing with the general case, we see how, from a truth table for a tautology X in the variables $p_1,\ldots,p_n$, we can get proofs of $B_1\Rightarrow X,\ldots,$ $B_{2^n}\Rightarrow X$, where $B_1,\ldots,B_{2^n}$ are the 2^n basic formulas in $p_1,\ldots,p_n$. Then, from these 2^n formulas $B_i\Rightarrow X$ $(i\leq 2^n)$, we can eliminate the basic formulas B_i in the manner described in the solution to Problem 20.4.

Reduction to Some More Standard-Type Systems

We now wish to reduce the system $\mathfrak{U}_0$ to some more standard-type systems in which Modus Ponens is the only rule of inference. We will introduce some intermediate systems along the way.

To begin with, in place of the axioms of $\mathfrak{U}_0$ we could take the following axioms, and add the following rules.

Axioms. All sentences $X\Rightarrow X$.

Inference Rules.

$$S_1.\ \frac{X\Rightarrow Y}{(Z\wedge X)\Rightarrow Y}.$$

$$S_2.\ \frac{X\Rightarrow Y}{(X\wedge Z)\Rightarrow Y}.$$

Problem 20.6. Prove that from the above axioms $X\Rightarrow X$ we can derive all axioms of $\mathfrak{U}_0$ (all formulas $(X_1\wedge\cdots\wedge X_n)\Rightarrow X_i$, $i\leq n$) by using the above inference rules S_1 and S_2.

We let $\mathfrak{U}_1$ be the system whose axioms are all formulas of the form $X \Rightarrow X$ and whose inference rules are S_1, S_2, and the inference rules I, II, III and IV of the system $\mathfrak{U}_0$. We shall re-number the inference rules, and our system $\mathfrak{U}_1$ is thus as follows:

AXIOM SCHEMA. $X \Rightarrow X$.

INFERENCE RULES.

S_1. $\dfrac{X \Rightarrow Y}{(Z \wedge X) \Rightarrow Y}$.

S_2. $\dfrac{X \Rightarrow Y}{(X \wedge Z) \Rightarrow Y}$.

S_3. $\dfrac{X \Rightarrow \alpha_1 \quad X \Rightarrow \alpha_2}{X \Rightarrow \alpha}$.

S_4. (a) $\dfrac{X \Rightarrow \beta_1}{X \Rightarrow \beta}$; (b) $\dfrac{X \Rightarrow \beta_2}{X \Rightarrow \beta}$.

S_5. $\dfrac{(X \wedge Y) \Rightarrow Z}{X \Rightarrow (Y \Rightarrow Z)}$ (Exportation).

S_6. $\dfrac{X \Rightarrow Y \quad \sim X \Rightarrow Y}{Y}$.

We see that the system $\mathfrak{U}_1$ is complete. At this point, let us note that in any axiom system in which Modus Ponens is an inference rule, any other inference rule R can be replaced by an axiom schema that yields the rule as a derived rule. Indeed, suppose R is of the form "From X to derive Y." If we adopt the axiom schema $X \Rightarrow Y$, we get R as a derived rule, because if X is provable, so is Y (by Modus Ponens, since $X \Rightarrow Y$ is provable). Similarly, if R is a two-premise rule of the form "From X and Y to infer Z," R can be replaced by the axiom schema $X \Rightarrow (Y \Rightarrow Z)$, which will yield R by two applications of Modus Ponens, because, if X and Y are both provable, from X and $X \Rightarrow (Y \Rightarrow Z)$ we get $Y \Rightarrow Z$ (by Modus Ponens), and then from Y and $Y \Rightarrow Z$ we

We can convert the system $\mathfrak{U}_1$ to a Modus Ponens system $\mathfrak{U}_2$ (a system in which Modus Ponens is the only inference rule) by replacing the inference rules S_1 to S_6 by the respective axiom schemas below:

AXIOM SCHEMAS OF $\mathfrak{U}_2$.

A_0. $X \Rightarrow X$

A_1. $(X \Rightarrow Y) \Rightarrow ((Z \wedge X) \Rightarrow Y)$.

A_2. $(X \Rightarrow Y) \Rightarrow ((X \wedge Z) \Rightarrow Y)$.

A_3. $(X \Rightarrow \alpha_1) \Rightarrow ((X \Rightarrow \alpha_2) \Rightarrow (X \Rightarrow \alpha))$.

A_4. (a) $(X \Rightarrow \beta_1) \Rightarrow (X \Rightarrow \beta)$;

(b) $(X \Rightarrow \beta_2) \Rightarrow (X \Rightarrow \beta)$.

A_5. $((X \wedge Y) \Rightarrow Z) \Rightarrow (X \Rightarrow (Y \Rightarrow Z))$.

A_6. $(X \Rightarrow Y) \Rightarrow ((\sim X \Rightarrow Y) \Rightarrow Y)$.

INFERENCE RULE. $\dfrac{X \quad X \Rightarrow Y}{Y}$ (Modus Ponens)

From A_1, A_2, A_4, and A_5, we get rules S_1, S_2, S_4, and S_5, respectively, by one application of Modus Ponens. From A_3 and A_6, we get rules S_3 and S_6, respectively, by two applications of Modus Ponens.

Next, we consider a variant $\mathfrak{U}_3$ of $\mathfrak{U}_2$ that comes closer to some of the systems we will consider in Chapter 21.

AXIOM SCHEMAS OF $\mathfrak{U}_3$.

B_0. $X \Rightarrow X$.

B_1. $(Z \wedge X) \Rightarrow X$.

B_2. $(X \wedge Z) \Rightarrow X$.

B_3. $((X \wedge Y) \Rightarrow Z) \Rightarrow (X \Rightarrow (Y \Rightarrow Z))$ (Exportation).

B_4. $((X \Rightarrow Y) \wedge (Y \Rightarrow Z)) \Rightarrow (X \Rightarrow Z)$ (Syllogism).

B_5. $((X \Rightarrow Y) \wedge (X \Rightarrow Z)) \Rightarrow (X \Rightarrow (Y \wedge Z))$.

B_6. $((X \Rightarrow Z) \wedge (Y \Rightarrow Z)) \Rightarrow ((X \vee Y) \Rightarrow Z)$.

B_7. $(\alpha_1 \wedge \alpha_2) \Rightarrow \alpha$.

B_8. (a) $\beta_1 \Rightarrow \beta$;

(b) $\beta_2 \Rightarrow \beta$.

B_9. $X \vee \sim X$.

INFERENCE RULE. $\dfrac{X \quad X \Rightarrow Y}{Y}$ (Modus Ponens).

PROBLEM 20.7. Prove that $\mathfrak{U}_3$ is complete by appealing to the completeness of $\mathfrak{U}_1$. (The axioms of $\mathfrak{U}_1$ are among the axioms of $\mathfrak{U}_3$, hence it suffices to show that the inference rules of $\mathfrak{U}_1$ all hold in $\mathfrak{U}_3$.)

SOLUTIONS

20.1. (a) The one and only interpretation I under which $q_1 \wedge \cdots \wedge q_n$ is true is the one that, for each $i \leq n$, assigns to p_i *truth* if $q_i = p_i$, and *falsehood* if $q_i = \sim p_i$.

(b) The one and only basic formula $q_1 \wedge \cdots \wedge q_n$ in the sequence $p_1, \ldots, p_n$ that is true under I is the one in which for each $i \leq n$, q_i is p_i if p_i is true under I, and $\sim p_i$ if p_i is false under I.

20.2. Since $q_1 \wedge \cdots \wedge q_n \Rightarrow \beta$ is a tautology, β is true under *all* interpretations of $p_1, \ldots, p_n$ under which $q_1 \wedge \cdots \wedge q_n$ is true, but there is only one such interpretation (Fact 1). Thus β is true under the interpretation I under which $q_1 \wedge \cdots \wedge q_n$ is true. Then either β_1 or β_2 is true under I, hence true under *all* interpretations under which $q_1 \wedge \cdots \wedge q_n$ is true—hence, either $(q_1 \wedge \cdots \wedge q_n) \Rightarrow \beta_1$ or $(q_1 \wedge \cdots \wedge q_n) \Rightarrow \beta_2$ is true under all interpretations of $p_1, \ldots, p_n$, and hence is a tautology.

20.3. Let B be a basic formula $q_1 \wedge \cdots \wedge q_n$ in the variables $p_1, \ldots, p_n$, fixed for the discussion. To avoid verbosity, let us say that a formula X obeys condition H if all variables of X are in the set $\{p_1, \ldots, p_n\}$ and $B \Rightarrow X$ is a tautology. We are to show that, for any formula X, if X obeys condition H, then $B \Rightarrow X$ is provable in $\mathfrak{U}_0$. We do this by induction on the degree of X. So suppose that X obeys condition H and that for every formula Y of degree less than that of X, if Y obeys condition H then $B \Rightarrow Y$ is provable in $\mathfrak{U}_0$. We are to show that $B \Rightarrow X$ is provable in $\mathfrak{U}_0$.

If X is a propositional variable, it must be one of the variables p_i in the set $\{p_1, \ldots, p_n\}$, and since $(q_1 \wedge \cdots \wedge q_n) \Rightarrow X$ is a tautology, q_i must be p_i (it cannot be $\sim p_i$), and thus X is q_i, so $B \Rightarrow X$ is an axiom of $\mathfrak{U}_0$, hence provable in $\mathfrak{U}_0$. Similarly, if X is the negation of a propositional variable, X is $\sim p_i$ for some $i \leq n$, in which case q_i must be $\sim p_i$ (it cannot be p_i, since $(q_1 \wedge \cdots \wedge q_n) \Rightarrow \sim p_i$ is a tautology). Thus X is q_i, and again $B \Rightarrow X$ is an axiom of $\mathfrak{U}_0$.

If X is neither a variable nor the negation of a variable, it must be some α or some β. Suppose it is some α. Then $B \Rightarrow \alpha$ is a tautology, and all variables of α are in the set $\{p_1, \ldots, p_n\}$. Since $B \Rightarrow \alpha$ is a tautology, $B \Rightarrow \alpha_1$ and $B \Rightarrow \alpha_2$ are both tautologies, and of course all variables of α_1 and α_2 are variables of α, hence are all in the set $\{p_1, \ldots, p_n\}$. Also, α_1 and α_2 are of lower degree than α, and, by the inductive hypothesis, $B \Rightarrow \alpha_1$ and $B \Rightarrow \alpha_2$ are both provable in $\mathfrak{U}_0$. Hence $B \Rightarrow \alpha$ is provable in $\mathfrak{U}_0$ by inference rule I.

Now suppose that X is some β. Thus $B \Rightarrow \beta$ is a tautology, and all variables of β are in the set $\{p_1, \ldots, p_n\}$. Since $B \Rightarrow \beta$ is a tautology, either $B \Rightarrow \beta_1$ or $B \Rightarrow \beta_2$ is a tautology (Fact 2). β_1 and β_2 are of lower degree than β, and their variables are all variables of β, hence are all in the set $\{p_1, \ldots, p_n\}$. Then by the inductive hypothesis, either $B \Rightarrow \beta_1$ or $B \Rightarrow \beta_2$ (whichever one is a tautology) is provable in $\mathfrak{U}_0$, hence $B \Rightarrow \beta$ is provable in $\mathfrak{U}_0$ by inference rule II (a) or (b).

This concludes the proof.

20.4. Suppose that for each of the 2^n basic sequences $q_1, \ldots, q_n$, with respect to the variables $p_1, \ldots, p_n$, the formula $q_1 \Rightarrow q_2 \Rightarrow \cdots \Rightarrow q_n \Rightarrow X$ is provable in $\mathfrak{U}_0$. Then for each basic sequence $q_1, \ldots, q_n$, both formulas

$$p_1 \Rightarrow q_2 \Rightarrow \cdots \Rightarrow q_n \Rightarrow X \quad \text{and} \quad {\sim} p_1 \Rightarrow q_2 \Rightarrow \cdots \Rightarrow q_n \Rightarrow X$$

are provable, hence, by Rule IV, the formula $q_2 \Rightarrow \cdots \Rightarrow q_n \Rightarrow X$ is provable. Similarly, we can prove $q_3 \Rightarrow \cdots \Rightarrow q_n \Rightarrow X$, and after $n-1$ rounds of this reasoning, we get $q_n \Rightarrow X$, for each of the two possibilities for q_n. Thus $p_n \Rightarrow X$ and ${\sim} p_n \Rightarrow X$ are both provable, hence so is X (again by Rule IV). Strictly speaking, the proof is by induction on n.

As an example, for $n = 2$, suppose that, for each of the four basic sequences q_1, q_2 with respect to p_1, p_2, the formula $q_1 \Rightarrow (q_2 \Rightarrow X)$ is provable (in $\mathfrak{U}_0$). Thus each of the following formulas is provable:

(1) $p_1 \Rightarrow (p_2 \Rightarrow X)$.

(2) ${\sim} p_1 \Rightarrow (p_2 \Rightarrow X)$.

(3) $p_1 \Rightarrow ({\sim} p_2 \Rightarrow X)$.

(4) ${\sim} p_1 \Rightarrow ({\sim} p_2 \Rightarrow X)$.

From (1) and (2), we get $p_2 \Rightarrow X$.

From (3) and (4), we get ${\sim} p_2 \Rightarrow X$.

Then, from these last two formulas, we get X.

20.5. Suppose X is a tautology and $p_1, \ldots, p_n$ are the variables that occur in X. Then for every basic sequence $q_1, \ldots, q_n$ with respect to $p_1, \ldots, p_n$, the formula $(q_1 \wedge \cdots \wedge q_n) \Rightarrow X$ is a tautology (every formula $Y \Rightarrow X$ is). Hence, by Proposition 1, each of the 2^n formulas $q_1 \Rightarrow \cdots \Rightarrow q_n \Rightarrow X$ is provable in $\mathfrak{U}_0$, hence so is X (by Proposition 2).

20.6. We show this by induction on n.

For $n=1$, we have $X_1 \Rightarrow X_1$ as an axiom of $\mathfrak{U}_1$.

Now suppose that n is such that, for all formulas $X_1, \ldots, X_n$, the formula $(X_1 \wedge \cdots \wedge X_n) \Rightarrow X_i$ is provable in $\mathfrak{U}_1$, for every $i \leq n$. We must show that, for any formulas $X_1, \ldots, X_n, X_{n+1}$, the formula $(X_1 \wedge \cdots \wedge X_n \wedge X_{n+1}) \Rightarrow X_i$ is provable for each $i \leq n+1$.

Suppose $i = n+1$. Since $X_{n+1} \Rightarrow Xn+1$ is provable in $\mathfrak{U}_1$ (it is an axiom), so is $(X_1 \wedge \cdots \wedge X_n \wedge X_{n+1}) \Rightarrow X_{n+1}$ (by Rule S_1, taking X_{n+1} for X, $Xn+1$ for Y, and $X_1 \wedge \cdots \wedge X_n$ for Z).

Next, suppose $i \neq n+1$. Thus $i \leq n$. Hence $(X_1 \wedge \cdots \wedge X_n) \Rightarrow X_i$ is provable in $\mathfrak{U}_1$ (by the inductive hypothesis), hence so is $(X_1 \wedge \cdots \wedge X_n \wedge X_{n+1}) \Rightarrow X_i$ by Rule S_2, taking Z to be X_{n+1}. This concludes the proof.

Here is another way of looking at it, which comes closer to the way we would actually obtain a proof of $(X_1 \wedge \cdots \wedge X_n) \Rightarrow X_i$ in $\mathfrak{U}_1$: For $n=1$, the case reduces to $X_1 \Rightarrow X_1$, which is an axiom of $\mathfrak{U}_1$. So we consider the case $n>1$. If $i=n$, we have $X_n \Rightarrow X_n$ as an axiom, and thus can infer $(X_1 \wedge \cdots \wedge X_n) \Rightarrow X_n$ by Rule S_1. If $i=1$, we have $X_1 \Rightarrow X_1$ as an axiom, and, by iterations of Rule S_2, we can successively obtain $X_1 \wedge X_2 \Rightarrow X_1, \ldots, X_1 \wedge \cdots \wedge X_n \Rightarrow X_1$. For $i \neq n$ and $i \neq 1$, we have $X_i \Rightarrow X_i$ as an axiom, which yields $X_1 \wedge \cdots \wedge X_i \Rightarrow X_i$ by Rule S_1, and then, by iterated use of Rule S_2 we successively get $X_1 \wedge \cdots \wedge X_i \wedge X_{i+1} \Rightarrow X_i$, $X_1 \wedge \cdots \wedge X_n \Rightarrow X_i$.

20.7. The word "provable" shall now mean "provable in $\mathfrak{U}_3$." It will be helpful first to note that the following four inference rules hold in $\mathfrak{U}_3$:

(1) $\dfrac{(X \wedge Y) \Rightarrow Z}{X \Rightarrow (Y \Rightarrow Z)}$ (Exportation).

(2) $\dfrac{X \Rightarrow Y \quad Y \Rightarrow Z}{X \Rightarrow Z}$ (Syllogism).

(3) $\dfrac{X \Rightarrow Y \quad X \Rightarrow Z}{X \Rightarrow (Y \wedge Z)}$.

(4) $\dfrac{X \Rightarrow Z \quad Y \Rightarrow Z}{(X \vee Y) \Rightarrow Z}$.

Re (1): This follows from Axiom Schema B_3 by Modus Ponens (if $(X \wedge Y) \Rightarrow Z$ is provable, then by Modus Ponens so is $X \Rightarrow (Y \Rightarrow Z)$, since

$$((X \wedge Y) \Rightarrow Z) \Rightarrow (X \Rightarrow (Y \Rightarrow Z))$$

is provable).

Re (2): By B_4 and Exportation, we have $(X \Rightarrow Y) \Rightarrow ((Y \Rightarrow Z) \Rightarrow (X \Rightarrow Z))$ as provable. Then by two applications of Modus Ponens we obtain Rule (2).

Re (3): Similarly, from B_5 by Exportation and two applications of Modus Ponens.

Re (4): Similarly, using B_6 in place of B_5.

Now to show that Rules S_1 to S_6 hold in $\mathfrak{U}_3$:

S_1. Suppose $X \Rightarrow Y$ is provable in $\mathfrak{U}_3$. $(Z \wedge X) \Rightarrow X$ is also provable (B_1), hence so is $(Z \wedge X) \Rightarrow Y$, by the syllogism rule (2).

S_2. Similar proof, using B_2 instead of B_1.

Note. For the purposes of Chapter 21, it is important to note that the only axioms we used to establish rules S_1 and S_2 were the axiom schemas $X \Rightarrow X$, $(Z \wedge X) \Rightarrow X$, $(X \wedge Z) \Rightarrow X$ and $((X \Rightarrow Y) \wedge (Y \Rightarrow Z)) \Rightarrow (X \Rightarrow Z)$ (syllogism).

S_3. Suppose $X \Rightarrow \alpha_1$ and $X \Rightarrow \alpha_2$ are both provable. Then so is $X \Rightarrow (\alpha_1 \wedge \alpha_2)$ (by Rule (3)). Also $(\alpha_1 \wedge \alpha_2) \Rightarrow \alpha$ is provable (B_7), hence so is $X \Rightarrow \alpha$ (by Rule (2)).

S_4. (a) If $X \Rightarrow \beta_1$ is provable, then, since $\beta_1 \Rightarrow \beta$ is provable (B_8(a)), so is $X \Rightarrow \beta$, by the syllogism rule.

(b) Similar, using B_8(b) in place of B_8(a).

S_5. This is Rule (1).

S_6. Suppose $X \Rightarrow Y$ and $\sim X \Rightarrow Y$ are both provable. Then so is $(X \vee \sim X) \Rightarrow Y$ (by Rule (4)). Also, $X \vee \sim X$ is provable (B_9), hence by Modus Ponens, Y is provable.

This concludes the proof.

- Chapter 21 -

More Propositional Axiomatics

Fergusson Surprises McCulloch

My friend Norman McCulloch, a character from some of my puzzle books such as *The Lady or the Tiger?* [23], like my friend Inspector Craig of Scotland Yard—whom I have also written about, in *What is the Name of this Book?* [24] and in *To Mock a Mockingbird* [25]—was very interested in all matters pertaining to logic. At one point, he became particularly interested in propositional logic and invented several axiom systems. By an amazing coincidence, one of his systems was my very system $\mathfrak{U}_3$ of the last chapter! Of course, he had the axiom schemas in a different order and numbered differently than mine, but I will continue with my numbering scheme B_1 to B_8.

My immediate question is this: Who should get the credit for this system, McCulloch or me? Which one of us came up with it first? This is very difficult to tell, since McCulloch lives in an imaginary world, and I live in a real one. How can one determine whether an event in an imaginary world came earlier or later than an event in the real world? I don't know! Perhaps it is best that he and I share the credit for the system.

At any rate, McCulloch proudly showed the system $\mathfrak{U}_3$ to our friend Malcolm Fergusson, a very distinguished logician who invented some remarkable logic machines, which I wrote about in *The Lady or the Tiger?*. Fergusson looked at the system with interest, and at one point a little smile crossed his face.

"Is there something wrong with my system?" asked McCulloch, who was very sensitive and noticed Fergusson's smile.

"Oh, no," replied Fergusson. "Nothing wrong, but I noticed something interesting."

"What was that?" asked McCulloch eagerly.

"It's just that one of your axiom schemas—perhaps the most self-evident one of all—is derivable from the others."

"That's interesting," replied McCulloch. "Which one is that?"

"It's axiom schema B_0: $X \Rightarrow X$," replied Fergusson. "It can be obtained from just B_1 and B_3, using Modus Ponens."

"How?" asked McCulloch.

PROBLEM 21.1. Good question! How? *Hint:* Use the axiom schema $(Z \wedge X) \Rightarrow X$, and then Exportation.

"Now that we have eliminated the schema $X \Rightarrow X$," said Fergusson, "let us see what your axiom system looks like in un-unified notation, if we take $\sim$, $\wedge$, $\vee$, and $\Rightarrow$ as primitives."

"Very good," said McCulloch, who then wrote down the axiom schemas as follows:

AXIOM SCHEMAS.

(1) $(Y \wedge X) \Rightarrow X$.

(2) $(X \wedge Y) \Rightarrow X$.

(3) $((X \wedge Y) \Rightarrow Z) \Rightarrow (X \Rightarrow (Y \Rightarrow Z))$ (Exportation).

(4) $((X \Rightarrow Y) \wedge (Y \Rightarrow Z)) \Rightarrow (X \Rightarrow Z)$ (Syllogism).

(5) $((X \Rightarrow Y) \wedge (X \Rightarrow Z)) \Rightarrow (X \Rightarrow (Y \wedge Z))$.

(6) $((X \Rightarrow Z) \wedge (Y \Rightarrow Z)) \Rightarrow ((X \vee Y) \Rightarrow Z)$.

(7) $X \vee \sim X$.

(8)
$$\left.\begin{array}{ll} \text{(a)} & (X \wedge Y) \Rightarrow (X \wedge Y) \\ \text{(b)} & (\sim X \wedge \sim Y) \Rightarrow \sim (X \vee Y) \\ \text{(c)} & (X \wedge \sim Y) \Rightarrow \sim (X \Rightarrow Y) \end{array}\right\} (\alpha_1 \wedge \alpha_2) \Rightarrow \alpha.$$

(9) (d) $X \Rightarrow (X \vee Y)$
(e) $\sim X \Rightarrow \sim (X \wedge Y)$
(f) $\sim X \Rightarrow (X \Rightarrow Y)$ } $\beta_1 \Rightarrow \beta$.

(g) $Y \Rightarrow (X \vee Y)$
(h) $\sim Y \Rightarrow \sim (X \wedge Y)$
(i) $Y \Rightarrow (X \Rightarrow Y)$
(j) $X \Rightarrow \sim\sim X$ } $\beta_2 \Rightarrow \beta$.

INFERENCE RULE. From X, $X \Rightarrow Y$ to infer Y (Modus Ponens).

"Let's clean up this system a little," said Fergusson. "To begin with, schema 8(a) is a special case of $X \Rightarrow X$, which we have seen to be derivable from (1) and (3), so we can get rid of that. Also, 9(i) ($Y \Rightarrow (X \Rightarrow Y)$ can be derived from other axiom schemata on the list."

PROBLEM 21.2. How? *Hint:* Use Axiom Schema (2).

"Now," said Fergusson, "I will tell you something that I believe will surprise you! It is possible to replace the two schemas 7 and 9(f) by the single schema $\sim\sim X \Rightarrow X$, and, at the same time, replace the five schemas 8(b), 8(c), 9(e), 9(h) and 9(j) by the two schemas $(X \wedge (X \Rightarrow Y)) \Rightarrow Y$ and $((X \Rightarrow Y) \wedge (X \Rightarrow \sim Y)) \Rightarrow \sim X$. Thus we can replace the seven schemas 7, 8(b), 8(c), 9(e), 9(f), 9(h) and 9(j) by just three schemas."

"Interesting!" said McCulloch, "how do you do that?"

"That's a fairly long story," replied Fergusson. "I will give you a sketch of the various things to be proved along the way. Let us first re-organize the system I have in mind."

Fergusson then wrote down his system, which I will call *system* $\mathfrak{F}$, as follows:

AXIOM SCHEMAS.

F_1. $((X \wedge Y) \Rightarrow Z) \Rightarrow (X \Rightarrow (Y \Rightarrow Z))$ (Exportation).

F_2. $((X \Rightarrow Y) \wedge (Y \Rightarrow Z)) \Rightarrow (X \Rightarrow Z)$ (Syllogism).

F_3. $((X \Rightarrow Y) \wedge (X \Rightarrow Z)) \Rightarrow (X \Rightarrow (Y \wedge Z))$.

F_4. $(X \wedge (X \Rightarrow Y)) \Rightarrow Y$.

F_5. $(X \wedge Y) \Rightarrow X$.

F_6. $(X \wedge Y) \Rightarrow Y$.

F_7. $X \Rightarrow (X \vee Y)$.

F_8. $Y \Rightarrow (X \vee Y)$.

F_9. $((X \Rightarrow Z) \wedge (Y \Rightarrow Z)) \Rightarrow ((X \vee Y) \Rightarrow Z)$.

F_{10}. $((X \Rightarrow Y) \wedge (X \Rightarrow \sim Y)) \Rightarrow \sim X$.

$\Box F_{11}$. $\sim\sim X \Rightarrow X$.

INFERENCE RULE. From X, $X \Rightarrow Y$ to infer Y (Modus Ponens).

"Why did you put a box to the left of F_{11}?" asked McCulloch.

"Oh," replied Fergusson, "F_{11} has a very special status that I will explain later. For certain purposes we will later need to know what results require F_{11}. Those formulas and derived inference rules that require use of F_{11} will be displayed with a box before their numbering.

"Now, the first thing to note is that the following five rules hold in this system:

R_1. $\dfrac{(X \wedge Y) \Rightarrow Z}{X \Rightarrow (Y \Rightarrow Z)}$ (Exportation).

R_2. $\dfrac{X \Rightarrow Y \quad Y \Rightarrow Z}{X \Rightarrow Z}$ (Syllogism).

R_3. $\dfrac{X \Rightarrow Y \quad X \Rightarrow Z}{X \Rightarrow (Y \wedge Z)}$.

R_4. $\dfrac{X \Rightarrow Z \quad Y \Rightarrow Z}{(X \vee Y) \Rightarrow Z}$.

R_5. $\dfrac{X \Rightarrow Y \quad X \Rightarrow \sim Y}{\sim X}$.

"These follow rather easily from the axiom schemas F_1, F_2, F_3, F_9, and F_{10}, respectively," continued Fergusson. "More specifically, R_1 follows from F_1 by Modus Ponens. Then, using Exportation, from F_2, F_3, F_9, and F_{10}, respectively, we get

(2) $(Y \Rightarrow Z) \Rightarrow ((X \Rightarrow Y) \Rightarrow (X \Rightarrow Z))$,

(3) $(X \Rightarrow Y) \Rightarrow ((X \Rightarrow Z) \Rightarrow (X \Rightarrow (Y \wedge Z))$,

(4) $(X \Rightarrow Z) \Rightarrow ((Y \Rightarrow Z) \Rightarrow ((X \vee Y) \Rightarrow Z))$,

(5) $(X \Rightarrow Y) \Rightarrow ((X \Rightarrow \sim Y) \Rightarrow \sim X)$.

"Then from, (2)–(5), we get R_1 to R_5, respectively, each by two applications of Modus Ponens. In fact, more generally speaking, whenever you have an axiom schema of the form $X \Rightarrow (Y \Rightarrow Z)$, in a system in which the rule of Modus Ponens holds, the rule From X and Y to infer Z' automatically holds, because if X and Y are both provable, then $Y \Rightarrow Z$ is provable (from X, $X \Rightarrow (Y \Rightarrow Z)$ by Modus Ponens), and then Z is provable (from Y and $Y \Rightarrow Z$ by Modus Ponens)."

"That makes sense," said McCulloch.[1]

"Here are five other useful rules that hold in my system," said Fergusson. "Using them will save you a lot of labor."

R_6. $\dfrac{Y}{X \Rightarrow Y}$.

R_7. $\dfrac{X \Rightarrow Y}{(Y \Rightarrow Z) \Rightarrow (X \Rightarrow Z)}$.

R_8. $\dfrac{X \Rightarrow Y \quad X \Rightarrow (Y \Rightarrow Z)}{X \Rightarrow Z}$.

R_9. $\dfrac{X \Rightarrow Y \quad X \Rightarrow Z \quad (Y \wedge Z) \Rightarrow W}{X \Rightarrow W}$.

R_{10}. $\dfrac{X \Rightarrow (Y \Rightarrow Z) \quad X \Rightarrow (Y \Rightarrow \sim Z)}{X \Rightarrow \sim Y}$.

$\square R_{11}$. $\dfrac{X \Rightarrow Y \quad \sim X \Rightarrow Y}{Y}$.

Problem 21.3. Show that R_6–R_{10} all hold in the system.

"Now," said Fergusson, "our job is to obtain 7, 8(b), 8(c), 9(e), 9(f), 9(h) and 9(j). To this end, I suggest that you successively show that the following schemas are derivable in the system."

Fergusson then wrote down the following schemas:

P_1. $X \Rightarrow (Y \Rightarrow X)$.

P_2. $X \Rightarrow \sim\sim X$ (this is 9(j)).

P_3. $(X \wedge \sim Y) \Rightarrow \sim(X \Rightarrow Y)$ (this is 8(c)).

P_4. $(X \Rightarrow Y) \Rightarrow (\sim Y \Rightarrow \sim X)$.

P_5. $(X \Rightarrow \sim Y) \Rightarrow (Y \Rightarrow \sim X)$.

P_6. $\sim X \Rightarrow \sim(X \wedge Y)$ (this is 9(e)).

[1] Actually, we already derived rules R_1 to R_4 in the last chapter.

P_7. $\sim Y \Rightarrow \sim(X \wedge Y)$ (this is 9(h)).

P_8. $(\sim X \wedge \sim Y) \Rightarrow \sim(X \vee Y)$ (this is 8(b)).

$\square P_9$. $\sim X \Rightarrow (X \Rightarrow Y)$ (this is 9(f)).

$\square P_{10}$. $((\sim X \Rightarrow Y) \wedge (\sim X \Rightarrow \sim Y)) \Rightarrow X$.

$\square P_{11}$. $X \vee \sim X$ (this is 7).

PROBLEM 21.4. Successively derive P_1–P_{11} in the system. Then derive R_{11}.

"Now you see that the system is complete," said Fergusson, after having derived P_1–P_{11}.

"Very good," said McCulloch, "but I would like to know why you singled out axiom schema F_{11} for special attention."

"Ah, that's an interesting business!" replied Fergusson. "One of these days I will tell you all about it."

McCulloch Surprises Fergusson

Some weeks after McCulloch's visit to Fergusson, he paid another visit and said, "You really surprised me with your simplifications, but now I have another simplification that I believe will surprise you!"

"What is that?" asked Fergusson with considerable interest.

"It's just this," said McCulloch. "Consider the following three axiom schemas of your system:

F_2. $((X \Rightarrow Y) \wedge (Y \Rightarrow Z)) \Rightarrow (X \Rightarrow Z)$.

F_3. $((X \Rightarrow Y) \wedge (X \Rightarrow Z)) \Rightarrow (X \Rightarrow (Y \wedge Z))$.

F_4. $(X \wedge (X \Rightarrow Y)) \Rightarrow Y$."

"What about them?" asked Fergusson.

"Well, I have found a way of replacing all three of them by the following single schema:

M. $((X \Rightarrow Y) \wedge (X \Rightarrow (Y \Rightarrow Z)) \Rightarrow (X \Rightarrow Z)$".

"That's interesting," replied Fergusson. "How do you do that?

"It's a bit complicated," said McCulloch, "but it can be done. Moreover, you don't need to use any of the schemas F_7–F_{11} to do it. That is, from the schemas F_1, F_5, F_6, and my schema M, one can derive F_2, F_3, and F_4."

PROBLEM 21.5. How can this be done?

SUGGESTIONS. Call the present system (the system that replaces F_2, F_3 and F_4 by the single schema M) $\mathfrak{F}_M$. Now successively do the following things:

(1) Show that the rules R_6 and R_8 of Fergusson's system $\mathfrak{F}$ hold in $\mathfrak{F}_M$.

(2) Next, show that the syllogism rule R_2 ($X \Rightarrow Y$, $Y \Rightarrow Z$ yields $X \Rightarrow Z$) holds in $\mathfrak{F}_M$.

Now that R_2 is established, get the following schemas S_1 and S_2 as before (using F_5 and F_6):

S_1. $((X \wedge Y) \wedge Z) \Rightarrow X$.

S_2. $((X \wedge Y) \wedge Z) \Rightarrow Y$.

In fact, by induction, we can get the entire meta-schema $X_1 \wedge \cdots \wedge X_n \Rightarrow X_i$ ($i \leq n$), but we don't need this.

(3) From S_1 and S_2 and F_7, get the Syllogism formula

$$((X \Rightarrow Y) \wedge (Y \Rightarrow Z)) \Rightarrow (X \Rightarrow Z)$$

(which is F_2) by using the Syllogism rule R_2. (Before, we derived the rule from the formula, whereas now, we derive the formula from the rule! This is curious.)

(4) We can get the schema $Y \Rightarrow Y$ as before (using the schema $(X \wedge Y) \Rightarrow Y$, by Exportation we get $X \Rightarrow (Y \Rightarrow Y)$, and then we take for X any provable formula and use Modus Ponens to get $Y \Rightarrow Y$). In particular, we then have the schema $(X \wedge Y) \Rightarrow (X \wedge Y)$, hence, by Exportation, we get the following:

S_3. $X \Rightarrow (Y \Rightarrow (X \wedge Y))$.

Now we are ready to get $((X \Rightarrow Y) \wedge (X \Rightarrow Z)) \Rightarrow (X \Rightarrow (Y \wedge Z))$, which is F_3. How do we get this?

(5) Then we get the schema $(X \wedge (X \Rightarrow Y)) \Rightarrow Y$, which is F_4.

At this point, let us re-number McCulloch's axiom schemas (we will call the resulting system $\mathfrak{M}$):

M_1. $((X \wedge Y) \Rightarrow Z) \Rightarrow (X \Rightarrow (Y \Rightarrow Z))$ (Exportation).

M_2. $((X \Rightarrow Y) \wedge (X \Rightarrow (Y \Rightarrow Z)) \Rightarrow (X \Rightarrow Z)$.

M_3. $(X \wedge Y) \Rightarrow X$.

M_4. $(X \wedge Y) \Rightarrow Y$.

M_5. $X \Rightarrow (X \vee Y)$.

M_6. $Y \Rightarrow (X \vee Y)$.

M_7. $((X \Rightarrow Z) \wedge (Y \Rightarrow Z)) \Rightarrow ((X \vee Y) \Rightarrow Z)$.

M_8. $((X \Rightarrow Y) \wedge (X \Rightarrow \sim Y)) \Rightarrow \sim X$.

□M_9. $\sim\sim X \Rightarrow X$.

EXERCISE 21.1. Let $\mathfrak{F}_0$ be Fergusson's system using only the axiom schemas F_1–F_6, and let $\mathfrak{M}_0$ be McCulloch's system using only the axiom schemas M_1–M_4. We have seen that F_1–F_6 are all provable in $\mathfrak{M}_0$. Now show that, conversely, M_1–M_4 are provable in $\mathfrak{F}_0$, and hence that $\mathfrak{M}_0$ and $\mathfrak{F}_0$ are equivalent systems, in the sense that the class of formulas provable in $\mathfrak{F}_0$ is the same as the class of formulas provable in $\mathfrak{M}_0$.

REMARKS. The full system $\mathfrak{F}$ is obviously equivalent to the full system $\mathfrak{M}$, since both systems are complete and correct, and hence the class of provable formulas of each is simply the class of all tautologies. Now, suppose we delete the axiom schema $\sim\sim X \Rightarrow X$ from both systems; does it follow that the remaining systems are equivalent? Certainly not just from the fact that the same axiom schema has been deleted from both, but, nevertheless, the two remaining systems *are* equivalent, by virtue of the above exercise.

Deduction Theorem

"That's a neat system," said Fergusson, after McCulloch had shown him his system $\mathfrak{M}$ and proved its completeness by showing how all the axiom schemas of $\mathfrak{F}$ were derivable in it. "Your system," continued Fergusson, "comes fairly close to a system I call $\mathcal{K}$, which can be found in Kleene's book [9], but I like your system better."

"What is this system $\mathcal{K}$?" asked McCulloch. Fergusson then wrote down the following ten axiom schemas of system $\mathcal{K}$:

K_1. $X \Rightarrow (Y \Rightarrow X)$.

K_2. $(X \Rightarrow Y) \Rightarrow ((X \Rightarrow (Y \Rightarrow Z)) \Rightarrow (X \Rightarrow Z))$.

K_3. $X \Rightarrow (Y \Rightarrow (X \wedge Y))$.

K_4. $(X \wedge Y) \Rightarrow X$.

K_5. $(X \wedge Y) \Rightarrow Y$.

K_6. $X \Rightarrow (X \vee Y)$.

K_7. $Y \Rightarrow (X \vee Y)$.

K_8. $(X \Rightarrow Z) \Rightarrow ((Y \Rightarrow Z) \Rightarrow ((X \vee Y) \Rightarrow Z))$.

K_9. $(X \Rightarrow Y) \Rightarrow ((X \Rightarrow \sim Y) \Rightarrow \sim X)$.

□K_{10}. $\sim\sim X \Rightarrow X$.

"I like your system better for several reasons," continued Fergusson. "For one thing, your system has only nine axioms schemas, whereas system $\mathcal{K}$ has ten, but that is really a minor point. More importantly, your schemas seem to me more natural and less convoluted than those of system $\mathcal{K}$—they come closer to the normal way we logically think. Also, your system gets much quicker results. It is relatively easy to show that the axiom schemas of $\mathcal{K}$ are provable in your system, but going in the other direction is much more involved. For example, the proof of $X \Rightarrow X$ in system $\mathcal{K}$ is more convoluted and unnatural than the proof in your system."

"What is the proof?" asked McCulloch.

"Well, it uses axiom schemas K_1 and K_2 and goes like this: In K_2, we take X for Z, so we have

$$(X \Rightarrow Y) \Rightarrow ((X \Rightarrow (Y \Rightarrow X)) \Rightarrow (X \Rightarrow X)). \tag{21.1}$$

Next, we replace Y by $Y \Rightarrow X$ in (21.1), and we have

$$(X \Rightarrow (Y \Rightarrow X)) \Rightarrow ((X \Rightarrow ((Y \Rightarrow X) \Rightarrow X)) \Rightarrow (X \Rightarrow X)). \tag{21.2}$$

Then, by (21.2) and K_1 and Modus Ponens, we have

$$(X \Rightarrow ((Y \Rightarrow X) \Rightarrow X)) \Rightarrow (X \Rightarrow X). \tag{21.3}$$

Next, in K_1, we replace Y by $Y \Rightarrow X$, and we have

$$X \Rightarrow ((Y \Rightarrow X) \Rightarrow X). \tag{21.4}$$

Finally, we get $X \Rightarrow X$ from (21.3) and (21.4) by Modus Ponens."

"Good grief!" cried McCulloch. "What a roundabout proof!"

"That's the way it is," laughed Fergusson. "I actually broke it up into easy pieces for you. The usual proof goes as follows:

(1) $X \Rightarrow (X \Rightarrow X)$ (K_1).

(2) $(X \Rightarrow (X \Rightarrow X)) \Rightarrow ((X \Rightarrow ((X \Rightarrow X) \Rightarrow X)) \Rightarrow (X \Rightarrow X))$ (K_2).

(3) $(X \Rightarrow ((X \Rightarrow X) \Rightarrow X)) \Rightarrow (X \Rightarrow X)$ (1, 2, and Modus Ponens).

(4) $X \Rightarrow ((X \Rightarrow X) \Rightarrow X)$ (K_1).

(5) $X \Rightarrow X$ (3, 4, and Modus Ponens)."

"That's even worse!" said McCulloch.

"I agree!" replied Fergusson. "Just consider how much neater the proof is in your system, which you recall is simply that, from the axiom $(Y \wedge X) \Rightarrow X$ and Exportation, we get $Y \Rightarrow (X \Rightarrow X)$ and then take for Y any provable formula and then use Modus Ponens.

"As I said," continued Fergusson, "it is relatively easy to prove the axiom schemas of $\mathcal{K}$ in your system, but going in the other direction is more involved. Indeed, it would be quite tedious without the use of a result known as the *deduction theorem,* which, I am pleased to say, you never had to use for your system in order to prove completeness. You did so in a perfectly straightforward way."

"What is the Deduction Theorem?" asked McCulloch.

"It's an interesting theorem in its own right," said Fergusson, "and has applications even to systems that are not necessarily complete. It formalizes a procedure common to ordinary mathematical reasoning. Suppose one wishes to show that a certain proposition p implies a certain proposition q. One can do this by first saying, Suppose p is true.' One then goes through an argument and reaches q as a conclusion. From this it follows that p does imply q. Thus, if q is deducible from the premise p, then $p \Rightarrow q$ is established. More generally, suppose one wishes to show that a proposition $p \Rightarrow q$ is implied by a set S of propositions. Well, one can do this by assuming $S{:}p$ (i.e., $S \cup \{p\}$) as premises and showing that q is a conclusion. It then follows that q is logically implied by the elements of S together with p, and hence that $p \Rightarrow q$ is a logical consequence of S.

"Now, let us consider an axiom system $\mathfrak{A}$ fixed for the discussion, and one that is not necessarily complete. A formula X is said to be *deducible* from a set S of formulas (with respect to the axiom system $\mathfrak{A}$, understood) if there is a finite sequence of formulas whose last member is X and such that each term of the sequence is either an axiom of $\mathfrak{A}$ *or a member of* S, or is derivable from earlier terms by one of the inference rules of $\mathfrak{A}$. Such a sequence is called a *deduction* of X from S (with respect to $\mathfrak{A}$, understood).

"Now that you know what I mean by saying that X is deducible from S, I will define $\mathfrak{A}$ to have the (normal) *deduction property* if, whenever X is deducible from $S{:}Y$, the formula $Y \Rightarrow X$ is deducible from S. In particular, for the case when S is empty, if Y is deducible from X, then $Y \Rightarrow X$ is provable (all with respect to $\mathfrak{A}$, understood). Now consider an axiom system $\mathfrak{A}$ for propositional logic, in which Modus Ponens is the only rule of inference. One form of the Deduction Theorem is that a sufficient condition for $\mathfrak{A}$ to have the deduction property is that the following two schemas are axiom schemas, or at least deducible schemas, of $\mathfrak{A}$:

K_1. $X \Rightarrow (Y \Rightarrow X)$.

K_2. $(X \Rightarrow Y) \Rightarrow ((X \Rightarrow (Y \Rightarrow Z)) \Rightarrow (X \Rightarrow Z))$.

"If K_1 and K_2 hold, then the following facts hold (for all formulas X, Y and Z, and any set S of formulas):

D_1. $X \Rightarrow X$ is provable in $\mathfrak{A}$.

D_2. If Y is deducible from S, so is $X \Rightarrow Y$.

D_3. If $X \Rightarrow Y$ and $X \Rightarrow (Y \Rightarrow Z)$ are deducible from S, so is $X \Rightarrow Z$.

"I have already shown you D_1, and it is an easy exercise to establish D_2 and D_3."

EXERCISE 21.2. Show that D_2 and D_3 follow from K_1 and K_2.

"Now, from D_1, D_2, and D_3 we get the Deduction Theorem as follows: Consider a deduction $X_1, \ldots, X_n = Y$ of Y from a set S:X. Now consider the sequence $X \Rightarrow X_1, \ldots, X \Rightarrow X_n$. An easy induction argument shows that, for each $i \leq n$, the formula $X \Rightarrow X_i$ is deducible from S—in particular, $X \Rightarrow Y$, which is $X \Rightarrow X_n$, is deducible from S."

EXERCISE 21.3. Consider the sequence $X \Rightarrow X_1, \ldots, X \Rightarrow X_n$ above. Suppose $i \leq n$ is such that, for each $j < i$, the formula $X \Rightarrow X_j$ is deducible from S. Show that $X \Rightarrow X_i$ is then deducible from S by considering three possible cases:

(1) $X_i = X$ (in which case, use D_1).

(2) X_i is a member of S (in which case, use D_2).

(3) X_i comes from two earlier terms X_{j_1} and X_{j_2} by Modus Ponens (in which case, use D_3).

It then follows by mathematical induction that for each $i \leq n$, the formula $X \Rightarrow X_i$ is deducible from S.

EXERCISE 21.4. Show that, conversely, if $\mathfrak{A}$ has the deduction property, then K_1 and K_2 both hold.

EXERCISE 21.5. We see now that Kleene's system $\mathcal{K}$ has the deduction property. Using this, show that all axiom schemas of McCulloch's system $\mathfrak{M}$ are derivable in $\mathcal{K}$. I suggest that you successively derive

(1) $((X\wedge Y)\Rightarrow Z)\Rightarrow(X\Rightarrow(Y\Rightarrow Z))$ (Exportation),

(2) $(X\Rightarrow(Y\Rightarrow Z))\Rightarrow((X\wedge Y)\Rightarrow Z)$ (Importation),

(3) $((X\Rightarrow Y)\wedge(Y\Rightarrow Z))\Rightarrow(X\Rightarrow Z)$ (Syllogism),

(4) $((X\Rightarrow Y)\wedge(X\Rightarrow Z))\Rightarrow(X\Rightarrow(Y\wedge Z))$,

(5) $(X\wedge(X\Rightarrow Y))\Rightarrow Y$.

"There is another deduction property," said Fergusson, "that I like even better: Let me say that $\mathfrak{A}$ has the *conjunctive deduction property* if, whenever Y is deducible from $\{X_1,\ldots,X_n\}$ (with respect to $\mathfrak{A}$, understood), the formula $X_1\wedge\cdots\wedge X_n\Rightarrow Y$ is provable in $\mathfrak{A}$. Now, if $\mathfrak{A}$ has the (usual) deduction property, and if the Importation rule holds (from $X\Rightarrow(Y\Rightarrow Z)$ to infer $(X\wedge Y)\Rightarrow Z)$), then $\mathfrak{A}$ also has the conjunctive deduction property—because suppose Y is deducible from $\{X_1,\ldots,X_n\}$. Then by successive applications of the Deduction Theorem, the formula

$$X_1\Rightarrow(X_2\Rightarrow\cdots\Rightarrow(X_{n-1}\Rightarrow X_n)\cdots)$$

is provable in $\mathfrak{A}$. Then by successive applications of Importation we get a proof of $X_1\wedge\cdots\wedge X_n\Rightarrow Y$ in $\mathfrak{A}$. But this process is exceedingly tedious! First, from a deduction of Y from $X_1,\ldots,X_n$ we get a deduction of $X_n\Rightarrow Y$ from $X_1,\ldots,X_{n-1}$. Then we have to make another deduction of $X_{n-1}\Rightarrow(X_n\Rightarrow Y)$ from $X_1,\ldots,X_{n-2}$, and so forth. This involves a lot of writing! The following approach works fine in either your system or mine, does not make use of the axiom schemas involving disjunction or negation, and, in fact, is generalizable to any axiom system satisfying the following three conditions:

C_1. All formulas of the form $(X\wedge\cdots\wedge X_n)\Rightarrow X_i$ $(i\leq n)$ are provable. (This includes $X_1\Rightarrow X_1$.)

C_2. If X is provable, so is $Y\Rightarrow X$.

C_3. If $X\Rightarrow Y$ and $X\Rightarrow(Y\Rightarrow Z)$ are provable, so is $X\Rightarrow Z$.

"The approach is this. Suppose Y is deducible from $X_1,\ldots,X_n$ (with respect to $\mathfrak{A}$); let $Z_1,\ldots,Z_k=Y$ be such a deduction. Then each of the formulas $(X\wedge\cdots\wedge X_n)\Rightarrow Z_1,\ldots,(X\wedge\cdots\wedge X_n)\Rightarrow Z_k$ (which is $(X\wedge\cdots\wedge X_n)\Rightarrow Y)$ can be successively seen to be provable in $\mathfrak{A}$. If fact, if we let $\mathfrak{A}^*$ be the system resulting from $\mathfrak{A}$ by adjoining all formulas

$(X\wedge\cdots\wedge X_n)\Rightarrow X_i$ $(i\leq n)$, and adding C_2 and C_3 as inference rules, then the above sequence

$$(X\wedge\cdots\wedge X_n)\Rightarrow Z_1,\ldots,(X\wedge\cdots\wedge X_n)\Rightarrow Z_k$$

is a proof in $\mathfrak{A}^*$, hence $(X\wedge\cdots\wedge X_n)\Rightarrow Y$ is provable in $\mathfrak{A}$."

EXERCISE 21.6. Verify Fergusson's claims.

EXERCISE 21.7. Show that if $\mathfrak{A}$ has the Exportation property (if $(X\wedge Y)\Rightarrow Z$ is provable, so is $X\Rightarrow(Y\Rightarrow Z))$, then, conversely, the conjunctive deduction property implies that conditions C_1, C_2 and C_3 hold.

A Glimpse of Intuitionistic Logic

"You never did tell me," said McCulloch, "why you singled out the axiom schema $\sim\sim X\Rightarrow X$ for special attention. What's so special about it?"

"Ah, yes!" replied Fergusson. "As I told you, this is a very interesting matter. There is an important school of logic known as *intuitionism*, whose adherents, known as *intuitionists*, do not accept many principles of reasoning belonging to what is known as *classical logic*, which includes all tautologies. The schema $\sim\sim X\Rightarrow X$ is one of the tautologies the intuitionists do not accept. Another is the principle $X\vee\sim X$, known as the *law of the excluded middle.*"

"Why should they have any doubts about them?" asked McCulloch, somewhat puzzled.

"Well," replied Fergusson, "they somehow seem to identify truth with provability, or at least potential provability. Let us first consider the principle of the excluded middle—X or not X. They accept this principle when applied to reasoning about *finite* sets, but not when applied to infinite ones. In intuitionistic logic, the only way a disjunction $X\vee Y$ can be proved is by first either proving X or proving Y. So if X is such that no proof can be found for X and none for $\sim X$, they see no reason to accept $X\vee\sim X$. As an illustration, suppose S is a set and P is a property such that, given any element x of S, there is an effective way of testing whether of not x has the property. Now let X be the proposition that at least one element of S has property P. If S is finite, then we can test each member of S in turn and thus either find an element of S that has the property or verify that none of them do—and thus either verify X or verify $\sim X$, so we can establish $X\vee\sim X$ to the intuitionist's satisfaction. But if S is infinite, then it is not possible to search through the entire set, so in this case, the intuitionist sees no reason to accept $X\vee\sim X$.

"As to the schema $\sim\sim X \Rightarrow X$, in intuitionist logic, as well as in classical logic, if a formula Y leads to a contradiction, then $\sim Y$ is established. Hence if $\sim X$ leads to a contradiction, then $\sim\sim X$ is established—but to the intuitionist, this does not guarantee that X is true! To the intuitionist, $\sim\sim X$ means that X can be disproved, but not necessarily that $\sim X$ can be proved!

"Now," continued Fergusson, "the axiom schemas F_1–F_{10} are all intuitionistically valid, but F_{11}, though classically valid, is not intuitionistically valid."

"All this strikes me as quite curious, though interesting," said McCulloch. "How do you stand with respect to intuitionism?"

"I am certainly not an intuitionist," replied Fergusson. "I definitely accept classical logic, but I also find intuitionistic logic very interesting. Very often an intuitionistic proof of a mathematical theorem yields more information than a classical one, and I find it of interest to know which mathematical theorems can be proved intuitionistically and which not. I therefore am all in favor of continued research in this area. I strongly oppose the *rejection* of classical logic—I think the two should be studied side by side."

McCulloch looked thoughtful.

"What are you thinking?" asked Fergusson.

"Oh, I was just wondering whether M_9 might not be deducible from the axiom schemas M_1 through M_8."

"Impossible," replied Fergusson. "It cannot be done."

"Now, how can you say *impossible*?" cried McCulloch. "Just because no one has yet done it doesn't mean it is impossible!"

"It's not merely that no one has yet *done it*; it can be logically *proved* that M_9 does not follow from M_1 through M_8."

"Really!" said McCulloch in amazement, "how can it be proved?"

Fergusson then gave him the proof, which the reader, if interested, can find below—or, if not interested, can proceed to the next chapter.

The general idea behind the proof is to exhibit a set S of formulas having the following three properties:

P_1. All instances of schemas M_1–M_8 are in S.

P_2. S is closed under Modus Ponens—i.e., if X and $X \Rightarrow Y$ are both in S, so is Y.

P_3. Not all instances of M_9 are in S.

It would then follow from P_1 and P_2 that every formula provable using only the schemas M_1–M_8 is in S. But by P_3, some instance of M_9 is not in S, and therefore not provable using only M_1–M_8.

Now for the construction of such a set S. The construction of this set is *inductive*, in the sense that we add formulas to S in stages, dependent on their degrees—that is, we first decide which formulas of degree 0 (propositional variables) are to be put in S, and then we decide which formulas of degree 1 are to go in S, on the basis of which formulas of degree 0 have already been put in S. Then we decide which formulas of degree 2 are to be put in S, on the basis of which formulas of degree less than 2 have been put in S, and so forth with formulas of degree 3, $4, \ldots, n, n+1, \ldots$.

Preparatory to this, we define a formula to be *positively true* if it is true under the interpretation under which *all* propositional variables are assigned *truth*. (Equivalently, a formula X with propositional variables $p_1, \ldots, p_n$ is positively true iff $(p_1 \wedge \cdots \wedge p_n) \Rightarrow X$ is a tautology.) Obviously every tautology is positively true. Also, if X and $X \Rightarrow Y$ are positively true, then Y must also be.

Now for the conditions defining the set S.

(0) All propositional variables are to be left *outside* S. Thus S contains *no* propositional variables.

(1) A formula $X \wedge Y$ (which is of higher degree than both X and Y) is to be put in S if and only if X and Y are both in S.

(2) $X \vee Y$ is to be put in S iff at least one of X, Y is in S.

(3) (This is a more delicate case.) $X \Rightarrow Y$ goes in S iff either $X \notin S$ or $Y \in S$, *and also* $X \Rightarrow Y$ is positively true!

(4) (Another delicate case) $\sim X$ is put in S iff X is not in S *and also* $\sim X$ is positively true.

Now to see that S satisfies the desired conditions P_1, P_2, P_3 above.

Let's first consider condition P_2—closure under Modus Ponens. Well, suppose X and $X \Rightarrow Y$ are both in S. The only way $X \Rightarrow Y$ can be in S is that either X is not in S or Y is in S. But it cannot be that X is not in S (since we are given that X *is* in S), hence the other alternative must hold—Y must be in S. This proves that if X and $X \Rightarrow Y$ are both in S, then Y must also be in S.

Now let's consider condition P_3—i.e., that not all instances of $\sim\sim X \Rightarrow X$ are in S. Well, in fact, we see that for *no* propositional variable p can it be that $\sim\sim p \Rightarrow p$ is in S: To begin with, p itself is not in S (no propositional variable is). Since $\sim p$ is not positively true (it is obviously *false* under that interpretation that assigns *truth* to p), it cannot be in S. Hence, since $\sim p$ is not in S and $\sim\sim p$ *is* positively true ($p \Rightarrow \sim\sim p$ is a tautology), $\sim\sim p$ must be in S. Thus $\sim\sim p$ is in S but p is not, so it is neither the case that

$\sim\sim p$ is not in S, nor that p is in S; therefore $\sim\sim p \Rightarrow p$ cannot be in S. This takes care of condition P_3.

As for condition P_1, some preliminaries are in order.

LEMMA 21.1. *All elements of S are positively true.*

This is proved by induction on degrees of formulas. Since no propositional variables are in S, it is vacuously true that all propositional variables in S are positively true. (That old trick again! If you don't believe that all propositional variables in S are positively true, just try to find one in S that isn't!) Next, it is obvious that if X and Y are positively true, so is $X \wedge Y$, and also, if either X or Y is positively true, so is $X \vee Y$. As for $X \Rightarrow Y$, it can only be in S if it is positively true, and the same with $\sim X$.

Next, let us note a general observation that should be helpful. Consider formulas X and Y such that $X \Rightarrow Y$ is a tautology, and we wish to show that $X \Rightarrow Y$ is in the set S. Now, either X is in S or it isn't. We really do not need to consider the case where X isn't, because if X is not in S, then $X \Rightarrow Y$ is automatically in S, since $X \Rightarrow Y$ is a tautology and hence obviously positively true. Thus we need to consider only the case when X *is* in S. In short, to show that a *tautology* $X \Rightarrow Y$ is in S, it suffices to show that if X is in S, so is Y. In particular, since each axiom schema M_1–M_8 is tautological (in the sense that all its instances are tautologies), and each instance is of the form $X \Rightarrow Y$, to show that such an instance is in S it suffices to show that if X is in S, so is Y. The reader should be able to do this on his or her own, but I will do a sample case, to help the reader along the way.

Let's do F_1 (Exportation), which is a relatively complex one. We are to show that the formula $((X \wedge Y) \Rightarrow Z) \Rightarrow (X \Rightarrow (Y \Rightarrow Z))$ is in S, for any formulas X, Y and Z. We suppose that $(X \wedge Y) \Rightarrow Z$ is in S, and we are to show that $X \Rightarrow (Y \Rightarrow Z)$ must also be in S. Well, since $(X \wedge Y) \Rightarrow Z$ is in S, it must be positively true, hence so is $X \Rightarrow (Y \Rightarrow Z)$ (since $((X \wedge Y) \Rightarrow Z) \Rightarrow (X \Rightarrow (Y \Rightarrow Z))$ is a tautology, hence trivially positively true). Let us record this as

(*) $(X \wedge Y) \Rightarrow Z$ and $X \Rightarrow (Y \Rightarrow Z)$ are both positively true (under the assumption, of course, that $(X \wedge Y) \Rightarrow Z$ is in S).

Next, since $(X \wedge Y) \Rightarrow Z$ is in S, it follows that either $(X \wedge Y)$ is not in S or Z is in S.

CASE 1. $X \wedge Y$ is not in S. Then either X is not in S or Y is not in S. Suppose X is not in S. Then, since $X \Rightarrow (Y \Rightarrow Z)$ is positively true, it must be in S. Thus if $X \notin S$, then $X \Rightarrow (Y \Rightarrow Z) \in S$. Now suppose $X \in S$. Then $Y \notin S$ (because we are assuming $X \wedge Y$ is not in S).

Since $X \in S$, X is positively true, and since $X \Rightarrow (Y \Rightarrow Z)$ is positively true, so is $Y \Rightarrow Z$. Since $Y \Rightarrow Z$ is positively true and $Y \notin S$, $Y \Rightarrow Z$ must be in S. Then, since $X \Rightarrow (Y \Rightarrow Z)$ is positively true and $Y \Rightarrow Z$ is in S, $X \Rightarrow (Y \Rightarrow Z)$ must be in S. This proves that if $X \wedge Y$ is not in S, then regardless of whether it is X or Y that is not in S, the formula $X \Rightarrow (Y \Rightarrow Z)$ *is* in S.

Case 2. $Z \in S$. This case is simple! Since $Z \in S$, it follows that Z is positively true, hence $Y \Rightarrow Z$ is positively true (since $Z \Rightarrow (Y \Rightarrow Z)$ is a tautology). Then, since $Z \in S$ and $Y \Rightarrow Z$ is positively true, $Y \Rightarrow Z$ must be in S. Then, since $X \Rightarrow (Y \Rightarrow Z)$ is positively true and $Y \Rightarrow Z$ is in S, $X \Rightarrow (Y \Rightarrow Z)$ must be in S. This concludes the proof.

exercise 21.8. Show that all instances of M_2–M_8 are in S.

We have seen that from $\sim\sim X \Rightarrow X$ (together with the other axiom schemas of the system $\mathfrak{M}$) we can derive the schemas $\sim X \Rightarrow (X \Rightarrow Y)$ and $X \vee \sim X$. Now, the schema $\sim X \Rightarrow (X \Rightarrow Y)$ is intuitionistically acceptable, but the schema $X \vee \sim X$ is not.

exercise 21.9. Consider the "test" set S that we have been working with.

(a) Show that all instances of $\sim X \Rightarrow (X \Rightarrow Y)$ are in S.

(b) Show that for any propositional variable p, the formula $p \vee \sim p$ is *not* in S.

It is of interest to note that in the system $\mathfrak{M}$, if we replace the schema $\sim\sim X \Rightarrow X$ by the two schemas $\sim X \Rightarrow (X \Rightarrow Y)$ and $X \vee \sim X$, we obtain an equivalent system—the schema $\sim\sim X \Rightarrow X$ is then derivable.

exercise 21.10. Show how to derive $\sim\sim X \Rightarrow X$ from those other two schemas (together, of course, with M_1–M_8).

Thus a system equivalent to $\mathfrak{M}$ is one that uses schemas M_1–M_8 and the two schemas

M_9a. $\sim X \Rightarrow (X \Rightarrow Y)$.

M_9b. $X \vee \sim X$.

This is a system of *classical* propositional logic. If we delete M_9b, we obtain a system of *intuitionistic* propositional logic. For the rest of this volume, we will be dealing with classical logic only.

SOLUTIONS

21.1. From the axiom schema $((X \wedge Y) \Rightarrow Z) \Rightarrow (X \Rightarrow (Y \Rightarrow Z))$ and Modus Ponens, we get the Exportation rule: From $(X \wedge Y) \Rightarrow Z$ to infer $X \Rightarrow (Y \Rightarrow Z)$. Now, $(Z \wedge X) \Rightarrow X$ is an axiom, hence by Exportation we have $Z \Rightarrow (X \Rightarrow X)$. Now take for Z any provable formula, and then by Modus Ponens we get $X \Rightarrow X$.

21.2. From $(Y \wedge X) \Rightarrow Y$ (axiom schema (2)), we get $Y \Rightarrow (X \Rightarrow Y)$ by Exportation.

21.3. To prove R_6: We have already shown that $X \Rightarrow (Y \Rightarrow X)$ is provable, and only the schema $(X \wedge Y) \Rightarrow X$ and the Exportation rule were used, and these hold also in the present system $\mathfrak{F}$. Thus if X is provable, so is $Y \Rightarrow X$ (by $X \Rightarrow (Y \Rightarrow X)$ and Modus Ponens).

To prove R_7: $(((X \Rightarrow Y) \wedge (Y \Rightarrow Z)) \Rightarrow (X \Rightarrow Z)$ is an axiom (F_2). Hence, by Exportation, we have $(X \Rightarrow Y) \Rightarrow ((Y \Rightarrow Z) \Rightarrow (X \Rightarrow Z))$. Then, by Modus Ponens, if $X \Rightarrow Y$ is provable, so is $(Y \Rightarrow Z) \Rightarrow (X \Rightarrow Z)$.

To prove R_8: Suppose $X \Rightarrow Y$ and $X \Rightarrow (Y \Rightarrow Z)$ are both provable. Then so are the following:

(1) $X \Rightarrow (Y \wedge (Y \Rightarrow Z))$ (by R_3, taking $(Y \Rightarrow Z)$ for Z).

(2) $(Y \wedge (Y \Rightarrow Z)) \Rightarrow Z$ (axiom F_4).

Then $X \Rightarrow Z$ is provable (by (1), (2) and the syllogism rule R_2). Thus, from $X \Rightarrow Y$ and $X \Rightarrow (Y \Rightarrow Z)$, one can infer $X \Rightarrow Z$.

To prove R_9: Suppose $X \Rightarrow Y$, $X \Rightarrow Z$ and $(Y \wedge Z) \Rightarrow W$ are all provable. From $X \Rightarrow Y$ and $X \Rightarrow Z$ we get $X \Rightarrow (Y \wedge Z)$ (by R_3). From this and $(Y \wedge Z) \Rightarrow W$ we get $X \Rightarrow W$ (by R_2, syllogism).

Now for R_{10}: Suppose $X \Rightarrow (Y \Rightarrow Z)$ and $X \Rightarrow (Y \Rightarrow \sim Z)$ are both provable. Also $((Y \Rightarrow Z) \wedge (Y \Rightarrow \sim Z)) \Rightarrow \sim Y$ is provable (axiom schema F_{10}). Hence $X \Rightarrow \sim Y$ by R_9.

21.4. We first note that, to show that a given formula X is provable in the system, it suffices to exhibit a list $X_1, \ldots, X_n = X$ of formulas such that, for each $i \leq n$, the term X_i is either an axiom, or has already been proved, or is derivable from earlier terms either by Modus Ponens or by an inference rule that has already been shown to hold in the system.

Another point: In any axiom schema—or in any provable schema, for that matter—one can substitute for any of the schematic letters any schemas at all to obtain another provable schema. For example, in the provable schema $X \Rightarrow X$, one can substitute $\sim X$ for X to obtain $\sim X \Rightarrow \sim X$. After all, $X \Rightarrow X$ is provable for *every* formula X. Well, $\sim X$ is also a formula, so $\sim X \Rightarrow \sim X$ is subsumed under the schema $X \Rightarrow X$. Another example: For every formula X, the formula $X \vee \sim X$ is an axiom. In particular, we can take $X \Rightarrow Y$ for X, hence $(X \Rightarrow Y) \vee \sim (X \Rightarrow Y)$ is also an axiom.

A third preliminary point: Consider the meta-schema

$$\text{A. } (X_1 \wedge \cdots \wedge X_n) \Rightarrow X_i \quad (i \leq n).$$

We noted in the last chapter that the only axiom schemas necessary to establish all such formulas are $X \Rightarrow X$, $(X \wedge Y) \Rightarrow X$, $(X \wedge Y) \Rightarrow Y$ and

$$((X \Rightarrow Y) \wedge (Y \Rightarrow Z)) \Rightarrow (X \Rightarrow Z) \quad \text{(syllogism)},$$

all of which are provable in the present system $\mathfrak{F}$. So we can use meta-schema A.

Now for the proofs of P_1–P_{11}.

P_0: (already proved) $\quad X \Rightarrow X$.

P_1: $\quad X \Rightarrow (Y \Rightarrow X)$.

PROOF:
1. $(X \wedge Y) \Rightarrow X$ $\quad$ (F_5).
2. $X \Rightarrow (Y \Rightarrow X)$ $\quad$ ((1), Rule R_1).

P_2: $\quad X \Rightarrow \sim\sim X$.

PROOF:
1. $X \Rightarrow (\sim X \Rightarrow X)$ $\quad$ (P_1, taking $\sim X$ for Y).
2. $\sim X \Rightarrow \sim X$ $\quad$ (P_0, taking $\sim X$ for X).
3. $X \Rightarrow (\sim X \Rightarrow \sim X)$ $\quad$ ((2), R_6).
4. $X \Rightarrow \sim\sim X$ $\quad$ ((1), (3), R_{10}).

P_3: $\quad (X \wedge \sim Y) \Rightarrow \sim (X \Rightarrow Y)$.

PROOF:
1. $((X \wedge \sim Y) \wedge (X \Rightarrow Y)) \Rightarrow X$ $\quad$ (Meta-Schema A).
2. $((X \wedge \sim Y) \wedge (X \Rightarrow Y)) \Rightarrow (X \Rightarrow Y)$ $\quad$ (Meta-Schema A).
3. $((X \wedge \sim Y) \wedge (X \Rightarrow Y)) \Rightarrow \sim Y$ $\quad$ (Meta-Schema A).
4. $((X \wedge \sim Y) \wedge (X \Rightarrow Y)) \Rightarrow Y$ $\quad$ ((1), (2), R_8).
5. $(X \wedge \sim Y) \Rightarrow ((X \Rightarrow Y) \Rightarrow Y)$ $\quad$ ((4), Exportation).
6. $(X \wedge \sim Y) \Rightarrow ((X \Rightarrow Y) \Rightarrow \sim Y)$ $\quad$ ((3), Exportation).
7. $(((X \Rightarrow Y) \Rightarrow Y) \wedge ((X \Rightarrow Y) \Rightarrow \sim Y)) \Rightarrow \sim (X \Rightarrow Y)$ $\quad$ (F_{10}).
8. $(X \wedge \sim Y) \Rightarrow \sim (X \Rightarrow Y)$ $\quad$ ((5), (6), (7), R_9).

P_4: $(X \Rightarrow Y) \Rightarrow (\sim Y \Rightarrow \sim X)$.

PROOF:
1. $(((X \Rightarrow Y) \wedge \sim Y) \wedge X) \Rightarrow X$ (Meta-Schema A).
2. $(((X \Rightarrow Y) \wedge \sim Y) \wedge X) \Rightarrow (X \Rightarrow Y)$ (Meta-Schema A).
3. $(((X \Rightarrow Y) \wedge \sim Y) \wedge X) \Rightarrow \sim Y$ (Meta-Schema A).
4. $(((X \Rightarrow Y) \wedge \sim Y) \wedge X) \Rightarrow Y$ ((1), (2), R_8).
5. $((X \Rightarrow Y) \wedge \sim Y) \Rightarrow (X \Rightarrow Y)$ ((4), Exportation).
6. $((X \Rightarrow Y) \wedge \sim Y) \Rightarrow (X \Rightarrow \sim Y)$ ((3), Exportation).
7. $((X \Rightarrow Y) \wedge (X \Rightarrow \sim Y)) \Rightarrow \sim X$ (F_{10}).
8. $((X \Rightarrow Y) \wedge \sim Y) \Rightarrow \sim X$ ((5), (6), R_{10}).
9. $(X \Rightarrow Y) \Rightarrow (\sim Y \Rightarrow \sim X)$ ((8), Exportation).

P_5: $(X \Rightarrow \sim Y) \Rightarrow (Y \Rightarrow \sim X)$.

PROOF:
1. $(X \Rightarrow \sim Y) \Rightarrow (\sim\sim Y \Rightarrow \sim X)$ (P_4, taking $\sim Y$ for Y).
2. $Y \Rightarrow \sim\sim Y$ (P_2).
3. $(\sim\sim Y \Rightarrow \sim X) \Rightarrow (Y \Rightarrow \sim X)$ ((2), R_7).
4. $(X \Rightarrow \sim Y) \Rightarrow (Y \Rightarrow \sim X)$ ((1), (3), R_2 (Syllogism)).

P_6: $\sim X \Rightarrow \sim (X \wedge Y)$.

PROOF:
1. $(X \wedge Y) \Rightarrow X$ (F_5).
2. $((X \wedge Y) \Rightarrow X) \Rightarrow (\sim X \Rightarrow \sim (X \wedge Y))$ (P_4).
3. $\sim X \Rightarrow (\sim (X \wedge Y)$ ((1), (2), Modus Ponens).

P_7: $\sim Y \Rightarrow \sim (X \wedge Y)$.

PROOF: Similar to that of P_6, using axiom $(X \wedge Y) \Rightarrow Y$ in place of $(X \wedge Y) \Rightarrow X$.

P_8: $(\sim X \wedge \sim Y) \Rightarrow \sim (X \vee Y)$.

PROOF:
1. $(\sim X \wedge \sim Y) \Rightarrow \sim X$ (F_5).
2. $X \Rightarrow \sim (\sim X \wedge \sim Y)$ ((1), P_5, Modus Ponens).
3. $(\sim X \wedge \sim Y) \Rightarrow \sim Y$ (F_6).
4. $Y \Rightarrow \sim (\sim X \wedge \sim Y)$ ((3), P_5, Modus Ponens).
5. $(X \vee Y) \Rightarrow \sim (\sim X \wedge \sim Y)$ ((2), (4), R_4).
6. $(\sim X \wedge \sim Y) \Rightarrow \sim (X \vee Y)$ ((5), P_5 (taking $(\sim X \wedge \sim Y)$ for Y and $(X \vee Y)$ for X), Modus Ponens).

$\Box P_9$: $\sim X \Rightarrow (X \Rightarrow Y)$.

PROOF:
1. $(\sim X \wedge X \wedge \sim Y) \Rightarrow X$ (Meta-Schema A).
2. $(\sim X \wedge X \wedge \sim Y) \Rightarrow \sim X$ (Meta-Schema A).
3. $(\sim X \wedge X) \Rightarrow (\sim Y \Rightarrow X)$ ((1), R_1).
4. $(\sim X \wedge X) \Rightarrow (\sim Y \Rightarrow \sim X)$ ((2), R_1).
5. $((\sim Y \Rightarrow X) \wedge (\sim Y \Rightarrow \sim X)) \Rightarrow \sim\sim Y$ (F_{10}).
6. $(\sim X \wedge X) \Rightarrow \sim\sim Y$ ((3), (4), (5), R_9).
7. $(\sim X \wedge X) \Rightarrow Y$ ((6), F_{11}, R_2).
8. $\sim X \Rightarrow (X \Rightarrow Y)$ ((7), R_1).

$\Box P_{10}$: $((\sim X \Rightarrow Y) \wedge (\sim X \Rightarrow \sim Y)) \Rightarrow X$.

PROOF:
1. $((\sim X \Rightarrow Y) \wedge (\sim X \Rightarrow \sim Y)) \Rightarrow \sim\sim X$ (F_{10}).
2. $(\sim X \Rightarrow Y) \wedge (\sim X \Rightarrow \sim Y) \Rightarrow X$ ((1), F_{11}, R_2).

$\Box P_{11}$: $X \vee \sim X$.

PROOF:
1. $X \Rightarrow (X \vee \sim X)$ (F_7).
2. $\sim(X \vee \sim X) \Rightarrow (X \Rightarrow (X \vee \sim X))$ ((1), R_6).
3. $\sim(X \vee \sim X) \Rightarrow (X \Rightarrow \sim(X \vee \sim X))$ (P_1).
4. $\sim(X \vee \sim X) \Rightarrow \sim X$ ((2), (3), R_{10}).
5. $\sim X \Rightarrow (X \vee \sim X)$ (F_8).
6. $\sim(X \vee \sim X) \Rightarrow (\sim X \Rightarrow (X \vee \sim X))$ ((5), R_6).
7. $\sim(X \vee \sim X) \Rightarrow (\sim X \Rightarrow \sim(X \vee \sim X))$ (P_1).
8. $\sim(X \vee \sim X) \Rightarrow \sim\sim X$ ((6), (7), R_{10}).
9. $\sim\sim(X \vee \sim X)$ ((4), (8), R_5).

$\Box$10. $\sim\sim(X \vee \sim X) \Rightarrow (X \vee \sim X)$ (F_{11}).

$\Box$11. $X \vee \sim X$ (9, 10, Modus Ponens).

Having proved P_{11}, we can now easily derive R_{11} thus: Suppose $X \Rightarrow Y$ and $\sim X \Rightarrow Y$ are both provable. Then so is $(X \vee \sim X) \Rightarrow Y$, by R_4. Since $X \vee \sim X$ is provable in the system ($\Box P_{11}$), Y follows by Modus Ponens. (Actually, this is a repetition of an argument from Chapter 20.)

21.5. (a) To begin with, the Exportation rule R_1 follows from the axiom schema F_1 by Modus Ponens (as before). Next, from McCulloch's

$$((X \Rightarrow Y) \wedge (X \Rightarrow (Y \Rightarrow Z)) \Rightarrow (X \Rightarrow Z),$$

we get $(X \Rightarrow Y) \Rightarrow ((X \Rightarrow (Y \Rightarrow Z)) \Rightarrow (X \Rightarrow Z))$ by Exportation (R_1). Therefore, if $X \Rightarrow Y$ and $X \Rightarrow (Y \Rightarrow Z)$ are both provable in $\mathfrak{F}$, then so is $X \Rightarrow Z$, by two applications of Modus Ponens. We thus get rule R_8.

As to R_6, we get this the same way as before: Since $(X \wedge Y) \Rightarrow X$ is an axiom (F_5), it follows that $X \Rightarrow (Y \Rightarrow X)$ is provable (by Exportation); hence if X is provable, so is $Y \Rightarrow X$ (by $X \Rightarrow (Y \Rightarrow X)$ and Modus Ponens).

(b) Suppose $X \Rightarrow Y$ and $Y \Rightarrow Z$ are both provable in $\mathfrak{F}_M$. Since $Y \Rightarrow Z$ is provable, so is $X \Rightarrow (Y \Rightarrow Z)$ (by rule R_6). Now that $X \Rightarrow Y$ and $X \Rightarrow (Y \Rightarrow Z)$ are both provable, so is $X \Rightarrow Z$, by rule R_8.

(c) Now to get $((X \Rightarrow Y) \wedge (Y \Rightarrow Z)) \Rightarrow (X \Rightarrow Z)$ (F_2):

1. $(((X \Rightarrow Y) \wedge (Y \Rightarrow Z)) \wedge X) \Rightarrow X$ (F_6),
2. $(((X \Rightarrow Y) \wedge (Y \Rightarrow Z)) \wedge X) \Rightarrow (X \Rightarrow Y)$ (S_1),
3. $(((X \Rightarrow Y) \wedge (Y \Rightarrow Z)) \wedge X) \Rightarrow Y$ ((1), (2), R_8),
4. $(((X \Rightarrow Y) \wedge (Y \Rightarrow Z)) \wedge X) \Rightarrow (Y \Rightarrow Z)$ (S_2),
5. $(((X \Rightarrow Y) \wedge (Y \Rightarrow Z)) \wedge X) \Rightarrow Z$ ((3), (4), R_8),
6. $((X \Rightarrow Y) \wedge (Y \Rightarrow Z)) \Rightarrow (X \Rightarrow Z)$ ((5), Exportation).

(d) Now to get $((X \Rightarrow Y) \wedge (X \Rightarrow Z)) \Rightarrow (X \Rightarrow (Y \wedge Z))$ (F_3):

1. $(((X \Rightarrow Y) \wedge (X \Rightarrow Z)) \wedge X) \Rightarrow X$ (F_6),
2. $(((X \Rightarrow Y) \wedge (X \Rightarrow Z)) \wedge X) \Rightarrow (X \Rightarrow Y)$ (S_1),
3. $(((X \Rightarrow Y) \wedge (X \Rightarrow Z)) \wedge X) \Rightarrow Y$ ((1), (2), R_8),
4. $(((X \Rightarrow Y) \wedge (X \Rightarrow Z)) \wedge X) \Rightarrow (X \Rightarrow Z)$ (S_2),
5. $(((X \Rightarrow Y) \wedge (X \Rightarrow Z)) \wedge X) \Rightarrow Z$ ((1), (4), R_8),
6. $Y \Rightarrow (Z \Rightarrow (Y \wedge Z))$ (S_3),
7. $(((X \Rightarrow Y) \wedge (X \Rightarrow Z)) \wedge X) \Rightarrow (Z \Rightarrow (Y \wedge Z))$ ((3), (6), R_2),
8. $(((X \Rightarrow Y) \wedge (X \Rightarrow Z)) \wedge X) \Rightarrow (Y \wedge Z)$ ((5), (7), R_8),
9. $((X \Rightarrow Y) \wedge (X \Rightarrow Z)) \Rightarrow (X \Rightarrow (Y \wedge Z))$ ((8), Exportation).

(e) Now to get $(X \wedge (X \Rightarrow Y)) \Rightarrow Y$ (F_4). This is easy:

1. $(X \wedge (X \Rightarrow Y)) \Rightarrow X$ (F_5),
2. $(X \wedge (X \Rightarrow Y)) \Rightarrow (X \Rightarrow Y)$ (F_6),
3. $(X \wedge (X \Rightarrow Y)) \Rightarrow Y$ ((1), (2), R_8).

- Chapter 22 -

Axiom Systems for First-Order Logic

We now turn to axiom systems for quantification theory (a synonym for first-order logic). In analogy with propositional logic, we call an axiom system for first-order logic *correct* if only valid formulas are provable, and *complete* if all valid formulas are provable. The only acceptable axioms systems for first-order logic are those that are both correct and complete. In analogy with propositional logic, to show that an axiom system for first-order logic is correct, it suffices to show that all axioms of the system are valid and that the inference rules preserve validity (if the premises of the of the rule are valid, so is the conclusion).

There are several axiom systems for first-order logic in the literature. Some of them extend axiom systems for propositional logic to axiom systems for first-order logic by adding axioms and inference rules for the quantifiers. Other axiom systems (e.g., Quine [13]), instead of taking axioms for the propositional part, simply take all tautologies as axioms. This is the course we will take. (Of course, all results of this chapter remain valid if, instead of taking all tautologies as axioms, we take any of the previously considered axiom systems for propositional logic.)

The System $\mathfrak{S}_1$

We now consider the following system $\mathfrak{S}_1$:

Axioms. All tautologies.

Inference Rules

R_1. $\dfrac{(\gamma \Rightarrow \gamma(a)) \Rightarrow X}{X}$.

R_2. $\dfrac{(\delta \Rightarrow \delta(a)) \Rightarrow X}{X}$, provided a does not occur in δ or in X.

That this system is *correct* follows easily from a result of Chapter 19.

Problem 22.1. Prove that.

More interestingly, we have

Theorem 22.1. *The system* $\mathfrak{S}_1$ *is* complete—*every valid sentence is provable in* $\mathfrak{S}_1$.

The proof of Theorem 22.1 1 follows fairly easily from the fact, proved in Chapter 19, that every valid sentence is truth-functionally implied by a regular set (the Regularity Theorem).

Problem 22.2. Prove Theorem 22.1.

The System $\mathfrak{S}_2$

The system $\mathfrak{S}_1$ is something the present author cooked up, mainly as an intermediary between the Regularity Theorem and the system $\mathfrak{S}_2$ to which we now turn, which is much closer to the usual systems in the literature.

Before presenting this system, we wish to point out that any axiom system in which all tautologies are provable and that is closed under Modus Ponens (i.e., such that if X and $X \Rightarrow Y$ are provable, so is Y) must also be closed under *truth-functional implication*—i.e., if $X_1, \ldots, X_n$ are provable and X is truth-functionally implied by the set $\{X_1, \ldots, X_n\}$, then X is also provable in the system, because if X is truth-functionally implied by $\{X_1, \ldots, X_n\}$, then $X_1 \Rightarrow (X_2 \Rightarrow \cdots \Rightarrow (X_n \Rightarrow X) \cdots)$ is a tautology. Since $X_1, \ldots, X_n$ are all provable, then, by repeated use of Modus Ponens, one can successively prove $X_2 \Rightarrow \ldots \Rightarrow (X_n \Rightarrow X) \ldots)$, then $\ldots (X_n \Rightarrow X)$, and then X.

Now for the system $\mathfrak{S}_2$. In the axiom schemes and the inference rules displayed below, it is to be understood that X is any sentence (closed formula), $\phi(x)$ contains no free variables other than x, and $\phi(a)$ is the result of substituting the parameter a for all free occurrences of x in $\phi(x)$.

Axioms.

Group 1. All tautologies.

Group 2. (a) All sentences $\forall x\phi(x) \Rightarrow \phi(a)$.
(b) All sentences $\phi(a) \Rightarrow \exists x\phi(x)$.

Inference Rules.

I. Modus Ponens $\dfrac{X \quad X \Rightarrow Y}{Y}$.

II. (a) $\dfrac{\phi(a) \Rightarrow X}{\exists x\phi(x) \Rightarrow X}$, (b) $\dfrac{X \Rightarrow \phi(a)}{X \Rightarrow \forall x\phi(x)}$,

provided that a does not occur in $\phi(x)$ or in X.

Theorem 22.2 (After Gödel's Completeness Theorem). *The system $\mathfrak{S}_2$ is complete.*

Remark. Gödel [6] proved the completeness of a system not identical with, but closely related to, $\mathfrak{S}_2$. Gödel was the first to prove the completeness of *any* axiom system for quantification theory.

We will prove Theorem 22.2 by showing that everything provable in $\mathfrak{S}_1$ (which we already know to be complete) is also provable in $\mathfrak{S}_2$.

In proving Theorem 22.2, it will save labor to establish the following pretty obvious lemma:

Lemma 22.1.

(a) *For any γ, the sentence $\gamma \Rightarrow \gamma(a)$ is provable in $\mathfrak{S}_2$.*

(b) *For any δ, if $\delta(a) \Rightarrow X$ is provable in $\mathfrak{S}_2$ and if a does not occur in δ or in X, then $\delta \Rightarrow X$ is provable in $\mathfrak{S}_2$.*

Problem 22.3. Prove the above lemma.

Problem 22.4. Now prove the completeness of $\mathfrak{S}_2$ by showing that all provable formulas of $\mathfrak{S}_1$ are provable in $\mathfrak{S}_2$.

Now what about the *correctness* of the system $\mathfrak{S}_2$; how do we know that everything provable in $\mathfrak{S}_2$ is really valid? Well, the axioms are obviously valid, and the rule of Modus Ponens is obviously correct (in that if the premises X and $X \Rightarrow Y$ are valid, so is the conclusion Y). But how do we know that the rules II(a) and II(b) preserve validity?

Problem 22.5. Prove that the rules II(a) and II(b) are correct—i.e., that if a does not occur in either X or $\phi(x)$, then

(a) If $X \Rightarrow \phi(a)$ is valid, so is $X \Rightarrow \forall x \phi(x)$;

(b) If $\phi(a) \Rightarrow X$ is valid, so is $\exists x \phi(x) \Rightarrow X$.

Now that we know that the rules II(a) and II(b) preserve validity, it follows that all the rules of $\mathfrak{S}_2$ do, and since the axioms are all valid, it follows by induction that all provable formulas of $\mathfrak{S}_2$ are valid. Thus the system $\mathfrak{S}_2$ is correct.

The System $\mathfrak{S}_3$

Here is another closely related axiom system, $\mathfrak{S}_3$, which, too, is closely related to some of the standard systems in literature.

Axioms.

Group 1. All tautologies.

Group 2. (Same as those of $\mathfrak{S}_2$.)
(a) All sentences $\forall x \phi(x) \Rightarrow \phi(a)$.
(b) All sentences $\phi(a) \Rightarrow \exists x \phi(x)$.

Group 3. (a) $\forall x(X \Rightarrow \phi(x)) \Rightarrow (X \Rightarrow \forall x \phi(x)$,
(b) $\forall x(\phi(x) \Rightarrow X) \Rightarrow (\exists x \phi(x)) \Rightarrow X)$,
provided x does not occur in X.

Inference Rules.

I. Modus Ponens.

II. $\dfrac{\phi(a)}{\forall x \phi(x)}$ (Generalization Rule).

Problem 22.6. Prove that $\mathfrak{S}_3$ is correct and complete. (Show that everything provable in $\mathfrak{S}_2$ is provable in $\mathfrak{S}_3$.)

Solutions

22.1. The axioms of $\mathfrak{S}_1$, being tautologies, are of course valid. That the inference rule preserves validity is but a re-statement of Part (d) of Theorem 19.1. Therefore the system is correct.

22.2. Suppose X is valid. Then, by the Regularity Theorem of Chapter 19, X is truth-functionally implied by a regular set R such that no critical parameter of R occurs in X. If R is empty, then X is already a tautology, hence is immediately provable in $\mathfrak{S}_1$. If R is not empty, then arrange the elements of R in an *inverse* regular sequence—i.e., a sequence $(r_1, \ldots, r_n)$ such that the sequence $(r_n, \ldots, r_1)$ is a regular sequence. Thus, for each $i \leq n$, r_i is of the form $Q_i \Rightarrow Q_i(a_i)$, and, if Q_i is of type D, then the critical parameter a_i does not appear in any later term, nor in X. Now, since X is truth-functionally implied by R, the formula $(r_1 \wedge \cdots \wedge r_n) \Rightarrow X$ is a tautology, hence so is the truth-functionally equivalent formula $r_1 \Rightarrow (r_2 \Rightarrow \cdots \Rightarrow (r_n \Rightarrow X) \cdots)$. This latter formula is thus provable in $\mathfrak{S}_1$ (it is an axiom), hence by one of the inference rules R_1 or R_2, the formula $r_2 \Rightarrow \cdots \Rightarrow (r_n \Rightarrow X) \cdots$ is provable. If $n > 2$, then anther application of an inference rule gives a proof of $r_3 \Rightarrow \cdots \Rightarrow (r_n \Rightarrow X) \cdots$. In this way we successively eliminate $r_1, \ldots, r_n$ and obtain a proof of X. Thus the following sequence of lines is a proof of X:

$$
\begin{array}{l}
r_1 \Rightarrow (r_2 \Rightarrow \cdots \Rightarrow (r_n \Rightarrow X) \cdots) \\
r_2 \Rightarrow (\cdots \Rightarrow (r_n \Rightarrow X) \cdots) \\
\quad \vdots \\
r_n \Rightarrow X \\
X.
\end{array}
$$

22.3. (a) If γ is the formula $\forall x \phi(x)$, then $\gamma \Rightarrow \gamma(a)$ is the formula $\forall x \phi(x) \Rightarrow \phi(a)$, which is an axiom of $\mathfrak{S}_2$, and hence provable in $\mathfrak{S}_2$. If γ is of the form $\sim\exists x \phi(x)$, then $\gamma \Rightarrow \gamma(a)$ is the formula $\sim\exists x \phi(x) \Rightarrow \sim\phi(a)$, which is truth-functionally implied by (in fact, truth-functionally equivalent to) the axiom $\phi(a) \Rightarrow \exists x \phi(x)$, hence is provable in $\mathfrak{S}_2$ (since $\mathfrak{S}_2$ is closed under truth-functional implication).

(b) Suppose $\delta(a) \Rightarrow X$ is provable in $\mathfrak{S}_2$ and a does not occur in δ or on X. If δ is of the form $\exists x \phi(x)$, then $\delta(a) \Rightarrow X$ is the sentence $\phi(a) \Rightarrow X$, hence $\exists x \phi(x) \Rightarrow X$ is provable in $\mathfrak{S}_2$ by Rule II(a), and this is the sentence $\delta \Rightarrow X$. If δ is of the form $\sim\forall x \phi(x)$, then $\delta(a) \Rightarrow X$ is the sentence $\sim\phi(a) \Rightarrow X$. Since it is provable in $\mathfrak{S}_2$, so is the sentence $\sim X \Rightarrow \phi(a)$, which is truth-functionally equivalent to it. Since a does not occur in X, it also does not occur in $\sim X$, so $\sim X \Rightarrow \forall x \phi(x)$ is provable in $\mathfrak{S}_2$ by Rule II(b). Hence its truth-functional equivalent $\sim\forall x \phi(x) \Rightarrow X$ is provable in $\mathfrak{S}_2$, and this is the sentence $\delta \Rightarrow X$.

22.4. The axioms of $\mathfrak{S}_1$ are also axioms of $\mathfrak{S}_2$, hence immediately provable in $\mathfrak{S}_2$. We must now show that, in each of the inference rules R_1 and R_2, if the premises are provable in $\mathfrak{S}_2$, so is the conclusion.

Re R_1: Suppose that $(\gamma \Rightarrow \gamma(a)) \Rightarrow X$ is provable in $\mathfrak{S}_2$. The sentence $\gamma \Rightarrow \gamma(a)$ is also provable in $\mathfrak{S}_2$, by Lemma 1. These two sentences truth-functionally imply X, hence X is provable in $\mathfrak{S}_2$.

Re R_2: Suppose that $(\delta \Rightarrow \delta(a)) \Rightarrow X$ is provable in $\mathfrak{S}_2$ and that a occurs in neither δ nor X. Then $\sim\delta \Rightarrow X$ and $\delta(a) \Rightarrow X$ are both provable in $\mathfrak{S}_2$, since both are truth-functionally implied by $(\delta \Rightarrow \delta(a)) \Rightarrow X$ (as the reader can easily verify). Since $\delta(a) \Rightarrow X$ is provable and the proviso is met, $\delta \Rightarrow X$ is provable (by Lemma 1). Thus $\sim\delta \Rightarrow X$ and $\delta \Rightarrow X$ are both provable in $\mathfrak{S}_2$, hence so is X.

22.5. (a) Suppose $X \Rightarrow \phi(a)$ is valid, and a does not occur in X or in $\phi(x)$. Then, given any interpretation I of all the predicates and parameters of X and $\phi(x)$, there is at least one value of a such that $\phi(a) \Rightarrow \forall x \phi(x)$ is true under I (since the formula $\exists y(\phi(y) \Rightarrow \forall x \phi(x))$ is valid). Also, for *all* values a, the sentence $X \Rightarrow \phi(a)$ is true under I (since the sentence $X \Rightarrow \phi(a)$ is given to be valid). Thus there is a value of a such that $X \Rightarrow \phi(a)$ and $\phi(a) \Rightarrow \forall x \phi(x)$ are both true under I, hence $X \Rightarrow \forall x \phi(x)$ is true under I. Thus $X \Rightarrow \forall x \phi(x)$ is true under every interpretation I, and hence is valid.

(b) This can be proved in a manner similar to (a). Alternatively, it can be proved as a consequence of (a), as follows: Suppose $\phi(a) \Rightarrow X$ is valid, and a does not occur in X nor in $\phi(x)$. Then $\sim X \Rightarrow \sim\phi(a)$ is valid, hence so is $\sim X \Rightarrow \forall x \sim\phi(x)$ (by (a)). Hence $\sim\forall x \sim\phi(x) \Rightarrow X$ (which is equivalent to $\sim X \Rightarrow \forall x \sim\phi(x)$) is valid, hence so is $\exists x \phi(x) \Rightarrow X$ (since $\sim\forall x \sim\phi(x)$ is equivalent to $\exists x \phi(x)$).

22.6. CORRECTNESS. All axioms of $\mathfrak{S}_3$ are valid. Also, if $\phi(a)$ is valid, then so is $\forall x \phi(x)$, because, if $\phi(a)$ is valid, then, under any interpretation, $\phi(a)$ is true for *all* possible values of the parameters, hence in particular for all choices of a, and hence $\forall x \phi(x)$ is valid. Thus inference rule II of $\mathfrak{S}_3$ preserves validity, and hence the system $\mathfrak{S}_3$ is correct. (*Note:* Although it is true that, if $\phi(a)$ is valid, so is $\forall x \phi(x)$, this does not mean that $\phi(a) \Rightarrow \forall x \phi(x)$ is a valid formula! In general, it isn't, so it should not be taken as an axiom, as has unfortunately been done by at least one irresponsible author.)

COMPLETENESS. All axioms of $\mathfrak{S}_2$ are also axioms of $\mathfrak{S}_3$, hence provable in $\mathfrak{S}_3$. Now for the inference rules of $\mathfrak{S}_2$: Rule I of $\mathfrak{S}_2$ (Modus Ponens) is also an inference rule of $\mathfrak{S}_3$. As for the inference rule II(a), suppose $\phi(a) \Rightarrow X$ is provable in $\mathfrak{S}_3$. Let $\Psi(x)$

be the formula $\phi(x) \Rightarrow X$. Then $\Psi(a)$ is $\phi(a) \Rightarrow X_x(a)$; but $X_x(a)$ is X, since X is closed, and thus $\Psi(a)$ is $\phi(a) \Rightarrow X$. Since $\Psi(a)$ is provable, so is $\forall x \Psi(x)$ (by the generalization rule of $\mathfrak{S}_3$), which is $\forall x(\phi(x) \Rightarrow X)$. Thus $\forall x(\phi(x) \Rightarrow X)$ is provable in $\mathfrak{S}_3$, but so is $\forall x(\phi(x) \Rightarrow X) \Rightarrow (\exists x \phi(x) \Rightarrow X)$ (which is an axiom of $\mathfrak{S}_3$); hence by truth-functional implication, so is $\exists x \phi(x) \Rightarrow X$. Thus Rule II(a) of $\mathfrak{S}_2$ holds in $\mathfrak{S}_3$. As to II(b), suppose $X \Rightarrow \phi(a)$ is provable in $\mathfrak{S}_3$. Then so is $\forall x(X \Rightarrow \phi(x))$ (by the generalization rule), but also, so is $\forall x(X \Rightarrow \phi(x)) \Rightarrow (X \Rightarrow \forall x \phi(x))$; hence $X \Rightarrow \forall x \phi(x)$ is provable in $\mathfrak{S}_3$ (by truth-functional implication). Thus Rule II(b) of $\mathfrak{S}_2$ holds in $\mathfrak{S}_3$.

- Part VI -

More on First-Order Logic

- Chapter 23 -

Craig's Interpolation Lemma

In 1957, William Craig [3] stated and proved a celebrated result known as Craig's Interpolation Lemma, which we state and prove in this chapter. This lemma has many highly significant applications, some of which we will give in the next two chapters.

A sentence Z is called an *interpolant* for a sentence $X \Rightarrow Y$ iff the following two conditions are met:

(1) $X \Rightarrow Z$ and $Z \Rightarrow Y$ are both valid.

(2) Every predicate and parameter of Z occurs in both X and Y.

Craig's Interpolation Lemma states that, if $X \Rightarrow Y$ is valid, then there is an interpolant for it, provided that Y alone is not valid and X alone is not unsatisfiable.

Actually, if we allow t's and f's to be part of our formal language (as we indicated at the end of Chapter 7), and define "formula" accordingly, then the above proviso is unnecessary, because, if $X \Rightarrow Y$ is valid, then if Y itself is valid, then t is an interpolant for $X \Rightarrow Y$ (since $X \Rightarrow t$ and $t \Rightarrow Y$ are then both valid, and there are no predicates or parameters in t), and alternatively if X is unsatisfiable then $X \Rightarrow f$ and $f \Rightarrow Y$ are both valid, and f is then an interpolant for $X \Rightarrow Y$. This is the course we shall take.

Formulas that do not involve t or f will be called *standard* formulas; others, *non-standard* formulas. As shown at the end of Chapter 7, any non-standard formula is reducible to (i.e., is equivalent to) either a standard formula or to t or f. Our use of non-standard formulas is only to facilitate the proof of Craig's Lemma for *standard* sentences of the form $X \Rightarrow Y$.

There is a corresponding interpolation lemma for propositional logic: If $X \Rightarrow Y$ is a tautology of propositional logic, then there is a sentence Z (again called an interpolant for $X \Rightarrow Y$) such that $X \Rightarrow Z$ and $Z \Rightarrow Y$ are valid, and all *propositional variables* of Z occur in both X and Y. For example, q is an interpolant for $(p \wedge q) \Rightarrow (q \vee r)$.

Many proofs of this celebrated lemma have been given. Craig's original proof was a very complicated one involving a special system called "Linear Reasoning" (cf. [3]), but several simplifications have subsequently been given. The proof we give here is due to Melvin Fitting [4] and is in turn a variant of one given in [17].

Biased Tableaux

Preparatory to the proof of Craig's Lemma, we need to consider a slight variant of the tableau method.

In proving a sentence $X \Rightarrow Y$ by a tableau for *unsigned* formulas, we start with $\sim(X \Rightarrow Y)$, then apply the α-rule, adding X and $\sim Y$. After this, any sentence added to the tableau is a descendant of either X or $\sim Y$, but, looking at the finished tableau, we cannot tell which of the added sentences came from X and which from $\sim Y$. What we need now is a bookkeeping device to let us know this. The device we will use is due to Fitting [4].

We think of X as "left" and $\sim Y$ as "right," corresponding to the respective positions of X and Y in $X \Rightarrow Y$. So we introduce the symbols "L" and "R" (suggesting *left* and *right*, respectively) and we define a *biased* sentence as an expression LZ or RZ. Thus in proving $X \Rightarrow Y$ by the biased tableau method, we start the tableau by adding LX and $R{\sim}Y$, and extend the tableau rules, in a straightforward way, to biased sentences. For example, for biased sentences the α-rule is replaced by the two rules

$$\frac{L\alpha}{\begin{array}{c}L\alpha_1\\ L\alpha_2\end{array}} \qquad \frac{R\alpha}{\begin{array}{c}R\alpha_1\\ R\alpha_2\end{array}}$$

Let us use the symbol "π" to stand for either "L" or "R"; then we can succinctly state the rules thus.

$$\frac{\pi\alpha}{\begin{array}{c}\pi\alpha_1\\ \pi\alpha_2\end{array}}$$

The other tableau rules are treated similarly.

$$\begin{array}{c}\pi\beta \\ \swarrow\searrow \\ \pi\beta_1 \quad \pi\beta_2\end{array} \qquad \frac{\pi\gamma}{\pi\gamma(a)} \qquad \frac{\pi\delta}{\pi\delta(a)}$$ provided a is new to the branch.

A tableau thus constructed is called a *biased* tableau. A branch of a biased tableau is called *closed* if it contains some Z and $\sim Z$, ignoring the L and R symbols—in other words, if it contains either some LZ and $L{\sim}Z$, or some RZ and $R{\sim}Z$, or some LZ and $R{\sim}Z$, or $L{\sim}Z$ and RZ. Now, from a closed biased tableau for LX, $R{\sim}Y$, we will see how to obtain an interpolant for $X \Rightarrow Y$.

To begin with, for any set $\{LX_1, \ldots, LX_n, RY_1, \ldots, RY_k\}$ of biased formulas, by an *interpolant for the set* we shall mean an interpolant for the formula $(X_1 \wedge \cdots \wedge X_n) \Rightarrow ({\sim}Y_1 \vee \cdots \vee {\sim}Y_k)$. We note that an interpolant for the set $\{LX, RY\}$ is thus nothing more nor less than an interpolant for $X \Rightarrow {\sim}Y$, and an interpolant for the set $\{LX, R{\sim}Y\}$ is an interpolant for the sentence $X \Rightarrow {\sim}{\sim}Y$, and hence also for the sentence $X \Rightarrow Y$ (why?). Thus, given a valid standard sentence of the form $X \Rightarrow Y$, we are to find an interpolant for the set $\{LX, R{\sim}Y\}$. It is not from the sentence itself that we will find this interpolant; we will find it instead from a biased tableau *proof* of the sentence.

Let us call a set S of biased sentences *covered* if there is an interpolant for it, and let us call a branch of a biased tableau *covered* if the set of biased sentences on the branch is covered; finally, let us call a biased tableau *covered* if all of its branches are covered.

PROOF OF CRAIG'S LEMMA

Continuing from the preparatory work above, we are now to show that, for any valid sentence $X \Rightarrow Y$, if $\mathfrak{T}$ is any closed biased tableau starting with LX, $R{\sim}Y$, then, at each stage of the construction of $\mathfrak{T}$, the tableau at that stage is covered (and hence, so is the initial stage LX, $R{\sim}Y$). We do this in *reverse* order, working our way *up* the tree—that is, we first show that the tableau at the *final* stage (in which all branches are closed) is covered and, next, that, at any stage other than the initial one, if the tableau at that stage is covered, so was the tableau at the stage right before.

Now, at any stage—other than the last—of the construction of a tableau, the next stage results from taking just one branch Θ and either extending it to a single larger branch Θ_1 (by an α, γ or δ-rule), or else splitting it to two branches Θ_1 and Θ_2 (by a β-rule). What is to be shown is that, in the first case, if Θ_1 is covered, so is Θ, and, in the second case, if Θ_1 and Θ_2 are both covered, so is Θ. (In this way we are making our way *up* the tree.) Also, of course, we must show that any closed branch is covered. So we must show that for any set S of biased sentences, the following facts hold:

(0) If S is closed, then it is covered.

(A) (1) If $S \cup \{L\alpha, L\alpha_1, L\alpha_2\}$ is covered, so is $S \cup \{L\alpha\}$.

(2) If $S \cup \{R\alpha, R\alpha_1, R\alpha_2\}$ is covered, so is $S \cup \{R\alpha\}$.

(B) (1) If $S \cup \{L\beta, L\beta_1\}$ and $S \cup \{L\beta, L\beta_2\}$ are both covered, so is $S \cup \{L\beta\}$.

(2) Similarly with "R" in place of "L."

(C) (1) If $S \cup \{L\gamma, L\gamma(a)\}$ is covered, so is $S \cup \{L\gamma\}$.

(2) Similarly with "R" in place of "L."

(D) (1) If $S \cup \{L\delta, L\delta(a)\}$ is covered, where a is a parameter new to $S \cup \{\delta\}$, so is $S \cup \{L\delta\}$.

(2) Similarly with "R" in place of "L."

To reduce clutter, let us use "π" to stand for either "L" or "R." We can then rewrite the above conditions as follows:

(0) Every closed set is covered.

(A) If $S \cup \{\pi\alpha, \pi\alpha_1, \pi\alpha_2\}$ is covered, so is $S \cup \{\pi\alpha\}$.

(B) If $S \cup \{\pi\beta, \pi\beta_1\}$ and $S \cup \{\pi\beta, \pi\beta_2\}$ are both covered, so is $S \cup \{\pi\beta\}$.

(C) If $S \cup \{\pi\gamma, \pi\gamma(a)\}$ is covered, so is $S \cup \{\pi\gamma\}$.

(D) If $S \cup \{\pi\delta, \pi\delta(a)\}$ is covered, where a new to $S \cup \{\delta\}$, so is $S \cup \{\pi\delta\}$.

Actually, it is a bit simpler to prove the following more general conditions:

(0) Same.

(A′) If $S \cup \{\pi\alpha_1, \pi\alpha_2\}$ is covered, so is $S \cup \{\pi\alpha\}$.

(B′) If $S \cup \{\pi\beta_1\}$ and $S \cup \{\pi\beta_2\}$ are both covered, so is $S \cup \{\pi\beta\}$.

(C′) If $S \cup \{\pi\gamma(a)\}$ is covered, then so is $S \cup \{\pi\gamma\}$.

(D′) If $S \cup \{\pi\delta(a)\}$ is covered, where a is new to $S \cup \{\delta\}$, then $S \cup \{\pi\delta\}$ is also covered.

Conditions A′–D′ do indeed imply conditions A–D, respectively. For example, to show that A′ implies A, suppose A′ holds. Now suppose that $S \cup \{\pi\alpha, \pi\alpha_1, \pi\alpha_2\}$ is covered. Let $S_1 = S \cup \{\pi\alpha\}$. Then $S_1 \cup \{\pi\alpha_1, \pi\alpha_2\}$ is the same set as $S \cup \{\pi\alpha, \pi\alpha_1, \pi\alpha_2\}$, hence $S_1 \cup \{\pi\alpha_1, \pi\alpha_2\}$ is covered. Therefore, by A′, $S' \cup \{\pi\alpha\}$ is covered—but $S' \cup \{\pi\alpha\}$ is the same set as $S \cup \{\pi\alpha\}$. Thus if A′ holds, then it follows that if $S \cup \{\pi\alpha, \pi\alpha_1, \pi\alpha_2\}$ is covered, so is $S \cup \{\pi\alpha\}$—in other words, that A holds. Thus A′ does imply A. Similarly, it can be seen that B′, C′, and D′ imply B, C, and D, respectively. And so we will verify 0, A′, B′, C′ and D′.

First ,for the propositional cases 0, A′, and B′, the following facts hold, whose verification we leave to the reader:

Fact 0. An interpolant for $S \cup \{LX, L{\sim}X\}$ is f; an interpolant for $S \cup \{RX, R{\sim}X\}$ is t; an interpolant for $S \cup \{LX, R{\sim}X\}$ is X; an interpolant for $S \cup \{L{\sim}X, RX\}$ is ${\sim}X$.

Fact A′. An interpolant for $S \cup \{\pi\alpha_1, \pi\alpha_2\}$ is also an interpolant for $S \cup \{\pi\alpha\}$.

Fact B′. This is the more interesting propositional case, since we must treat the case $\pi = L$ differently from $\pi = R$.

(a) If X is an interpolant for $S \cup \{L\beta_1\}$ and Y is an interpolant for $S \cup \{L\beta_2\}$, then an interpolant for $S \cup \{L\beta\}$ is $X \vee Y$.

(b) If X is an interpolant for $S \cup \{R\beta_1\}$ and Y is an interpolant for $S \cup \{R\beta_2\}$, then an interpolant for $S \cup \{R\beta\}$ is $X \wedge Y$!

Exercise 23.1. Verify the above facts.

We might note that, at this point, we have proved Craig's Lemma for propositional logic.

Now for the quantifiers: Curiously enough, the case for γ is more complicated than that for δ! In both cases, S is some set $\{LX_1, \ldots, LX_n, RY_1, \ldots, RY_k\}$. To reduce clutter, we let s_1 be the sentence $X_1 \wedge \cdots \wedge X_n$ and s_2 be the sentence ${\sim}Y_1 \vee \cdots \vee {\sim}Y_k$. Then an interpolant for the set $S \cup \{L\gamma(a)\}$ is simply an interpolant for the sentence $(s_1 \wedge \gamma(a)) \Rightarrow s_2$, and an interpolant for the set $S \cup \{L\gamma\}$ is simply an interpolant for the sentence $(s_1 \wedge \gamma) \Rightarrow s_2$. So, given an interpolant X for the sentence $(s_1 \wedge \gamma(a)) \Rightarrow s_2$, we are to find an interpolant for $(s_1 \wedge \gamma) \Rightarrow s_2$. Well, since X is an interpolant for $(s_1 \wedge \gamma(a)) \Rightarrow s_2$, the sentences $(s_1 \wedge \gamma(a)) \Rightarrow X$ and $X \Rightarrow s_2$ are both valid, and all predicates and parameters of X occur both in $s_1 \wedge \gamma(a)$ and in s_2. Now, the sentences $(s_1 \wedge \gamma) \Rightarrow X$ and $X \Rightarrow s_2$ are certainly both valid (since $(s_1 \wedge \gamma)$ logically implies $s_1 \wedge \gamma(a)$), but X may nevertheless fail to be an interpolant for $(s_1 \wedge \gamma) \Rightarrow s_2$, because the parameter a may occur in

X but not in $S_1 \wedge \gamma$! If a does not occur in X, then X is indeed an interpolant for $(S_1 \wedge \gamma) \Rightarrow s_2$, or if a does occur in either s_1 or γ, then again X is an interpolant for $(s_1 \wedge \gamma) \Rightarrow s_2$. But what if a *does* occur in X, but not in $(s_1 \wedge \gamma)$? This is the critical case, and we must now resort to a clever stratagem.

For this case, we take some variable x that does not occur in X, and we let $\phi(x)$ be the result of replacing every occurrence of a in X by x. We assert that the sentence $\forall x \phi(x)$ is an interpolant for $(s_1 \wedge \gamma) \Rightarrow s_2$.

To see this, we first note that the parameter a does not occur in $\forall x \phi(x)$, so all predicates and parameters that occur in $\forall x \phi(x)$ occur both in $s_1 \wedge \gamma$ and in s_2. Next, we note that $\phi(a)$ is X itself—hence, of course, $\forall x \phi(x) \Rightarrow X$ is valid. It remains to be shown that $(s_1 \wedge \gamma) \Rightarrow \forall x \phi(x)$ is valid. Well, since $(s_1 \wedge \gamma) \Rightarrow \phi(a)$ is valid and a does not occur in $s_1 \wedge \gamma$ or in $\phi(x)$, it follows that $(s_1 \wedge \gamma) \Rightarrow \forall x \phi(x)$ is valid (as shown in Problem 22.5). This completes the proof that if $(s_1 \wedge \gamma(a)) \Rightarrow s_2$ has an interpolant, so does $(s_1 \wedge \gamma) \Rightarrow s_2$. Now, our proof did not in any way depend on the nature of the sentences s_1 and s_2; the result would hold for *any* sentences k_1 and k_2 in place of s_1 and s_2. What we have really proved is the following, which will be used again and which we record as

PROPOSITION 23.1. *For any sentences k_1, k_2 and any γ and any parameter a: If $(k_1 \wedge \gamma(a)) \Rightarrow k_2$ has an interpolant, so does $(k_1 \wedge \gamma) \Rightarrow k_2$. More specifically, if X is an interpolant for $(k_1 \wedge \gamma(a)) \Rightarrow k_2$, then either X is an interpolant for $(k_1 \wedge \gamma) \Rightarrow k_2$ or $\forall x \phi(x)$ is such an interpolant, where $\phi(x)$ is the result of substituting a new variable x for a in X.*

Next, we must show that if $S \cup \{R\gamma(a)\}$ has an interpolant, so does $S \cup \{R\gamma\}$. This is tantamount to showing that if $s_1 \Rightarrow (s_2 \vee \sim\gamma(a))$ has an interpolant, then so does $s_1 \Rightarrow (s_2 \vee \sim\gamma)$. More specifically, we will show

PROPOSITION 23.2. *If X is an interpolant for $s_1 \Rightarrow (s_2 \vee \sim\gamma(a))$, then either X is an interpolant for $s_1 \Rightarrow (s_2 \vee \sim\gamma)$, or $\exists x \phi(x)$ is such an interpolant, where $\phi(x)$ is the result of substituting a new variable x for all occurrences of the parameter a in X.*

We could prove Proposition 23.2 from scratch (in much the same manner as Proposition 23.1), but we prefer to establish it as a *corollary* of Proposition 23.1. To do this, let us first note two things: First, that if Z is an interpolant for $X \Rightarrow Y$, then $\sim Z$ is an interpolant for $\sim Y \Rightarrow \sim X$ (as is easily verified). Secondly, if Z is an interpolant for $X \Rightarrow Y$, then, for any sentences X_1 and Y_1, if X_1 is logically equivalent to X and has the same predicates and parameters as X, and if Y_1 is logically equivalent to Y and has the same predicates and parameters as Y, then Z is also an interpolant for $X_1 \Rightarrow Y_1$ (which is really quite obvious!).

O.K., now suppose X is an interpolant for $s_1 \Rightarrow (s_2 \vee \sim\gamma(a))$. Then X is an interpolant for $\sim(s_2 \vee \sim\gamma(a)) \Rightarrow \sim s_1$ (by the first fact noted above). Hence $\sim X$ is an interpolant for $(\sim s_2 \wedge \sim\sim\gamma(a)) \Rightarrow \sim s_1$ (by the second fact noted above), hence is also an interpolant for $(\sim s_2 \wedge \gamma(a)) \Rightarrow \sim s_1$ (again, by the second fact noted above). Now, either $\sim X$ is an interpolant for $(\sim s_2 \wedge \gamma) \Rightarrow \sim s_1$, or it isn't. If it isn't, then let $\phi(x)$ be the result of replacing all occurrences of the parameter a in X by a new variable x, in which case $\sim\phi(x)$ is the result of replacing all occurrences of a in $\sim X$ by x, and so, by Proposition 23.1, $\forall x \sim\phi(x)$ is then such an interpolant! Thus either $\sim X$ or $\forall x \sim\phi(x)$ is an interpolant for $(\sim s_2 \wedge \gamma) \Rightarrow \sim s_1$. Now, if Z is any such interpolant, then $\sim Z$ is an interpolant for $\sim\sim s_1 \Rightarrow \sim(\sim s_2 \wedge \gamma)$ (why?), hence also for $s_1 \Rightarrow (s_2 \vee \sim\gamma)$ (why?). Thus either $\sim\sim X$ or $\sim\forall x \sim\phi(x)$ is an interpolant for $s_1 \Rightarrow (s_2 \vee \sim\gamma)$. If the former, then X is such an interpolant (verify!). If the latter, then $\exists x \phi(x)$ is such an interpolant (verify!).

We have now taken care of the α, β and γ cases. The δ case is relatively simple.

Suppose X is an interpolant for $S \cup \{L\delta(a)\}$ and a is new to S and to δ. Then X is an interpolant for the sentence $(s_1 \wedge \delta(a)) \Rightarrow s_2$, and a does not occur in s_1, s_2 or δ. Since every parameter of X occurs in s_2 and a does not occur in s_2, a cannot occur in X, hence X is already an interpolant for $(s_1 \wedge \delta) \Rightarrow s_2$ (verify!). (The fact is that, since $(s_1 \Rightarrow \delta(a)) \Rightarrow X$ is valid, so is $\delta(a) \Rightarrow (s_1 \Rightarrow X)$, and since a does not occur in δ nor in $(s_1 \Rightarrow X)$, it follows that $\delta \Rightarrow (s_1 \Rightarrow X)$ is valid (Problem 23.5, Chapter 22), hence so is $(s_1 \wedge \delta) \Rightarrow X$.)

The proof that if $S \cup \{R\delta(a)\}$ has an interpolant and a is new to S and δ, then $S \cup \{R\delta\}$ also has one (in fact the same one) is pretty similar, and is left to the reader.

This concludes the proof of Craig's Lemma.

- CHAPTER 24 -

ROBINSON'S THEOREM

One important application of Craig's Lemma is a result of Abraham Robinson's [14], known as Robinson's Consistency Theorem, which we will shortly state and prove. First for some preliminaries.

In what follows, we will be considering sentences (closed formulas) with no parameters. We use letters X, Y, Z to stand for such sentences, and we use the letter S, with or without subscripts, to stand for sets of such sentences.

We recall that an interpretation I is said to *satisfy* a set S iff all elements of S are true under I, and we call S *satisfiable* iff S is satisfied by at least one interpretation. We continue to say that X is a *logical consequence* of S—more briefly, that X is a *consequence* of S, or that X is *implied* by S—iff X is true under all interpretations that satisfy S.

PROBLEM 24.1. Suppose that $S_1 \subseteq S_2$ (S_1 is a subset of S_2). Which, if either, of the following two statements are true?

(1) If X is implied by S_2 then it is implied by S_1.

(2) If X is implied by S_1 then it is implied by S_2.

PROBLEM 24.2 (IMPORTANT!). Show that, for any set S of sentences and for any sentences X and Y, if Y is a logical consequence of $S \cup \{X\}$, then $X \Rightarrow Y$ is a logical consequence of S.

For any set S of sentences, by the *language* of S—symbolized $\mathfrak{L}(S)$—is meant the set of all sentences X such that each predicate of X occurs in at least one element of S. A set S is called *logically closed*—more briefly, *closed*—iff it contains every sentence in $\mathfrak{L}(S)$ that is a logical consequence of S. For any set S, we shall let S^* be the set of all sentences in $\mathfrak{L}(S)$ that are logical consequences of S. Obviously $S \subseteq S^*$ (S is a subset of S^*) and, if S is logically closed, then $S^* \subseteq S$, and hence $S^* = S$. Indeed, S is logically closed iff $S^* = S$. We leave it to the reader to verify that S^* itself is logically closed and hence that $(S^*)^* = S^*$.

PROBLEM 24.3. Which of the following two conditions implies the other?

(1) For any elements $X_1, \ldots, X_n$ of S and any sentence Y in the language of S, if $(X_1 \wedge \cdots \wedge X_n) \Rightarrow Y$ is valid, then Y is in S.

(2) S is logically closed.

In the literature, a closed non-empty set of sentences is called a *theory*, and its elements are called the *theorems* or *provable* elements of the theory.

DISCUSSION. Why are such sets called theories? Well, such sets usually arise by starting with a set A of sentences and adding its elements as additional axioms to some standard axiom system of first-order logic (such as the system $\mathfrak{S}_2$ of Chapter 22). The set of sentences provable in that enlarged system is closed (as is easily verified)—indeed, it is the closure A^* of A.

PROBLEM 24.4. Consider the following statement: For any theory $\mathfrak{T}$, if every element of $\mathfrak{T}$ is satisfiable, then $\mathfrak{T}$ is satisfiable. Is that statement true or false?

PROBLEM 24.5. Prove that, for any theory $\mathfrak{T}$, the following two conditions are equivalent:

(1) $\mathfrak{T}$ is unsatisfiable.

(2) $\mathfrak{T}$ contains *all* sentences in the language of $\mathfrak{T}$.

PROBLEM 24.6. Is it true that if S is closed then S contains all valid sentences that are in the language of S?

A theory $\mathfrak{T}$ is called *complete* iff for every sentence X in the language of $\mathfrak{T}$, either $X \in \mathfrak{T}$ or $(\sim X) \in \mathfrak{T}$.

Now, here is Robinson's Theorem.

THEOREM R (ROBINSON'S CONSISTENCY THEOREM). *For any* complete *theory $\mathfrak{T}$, if theories $\mathfrak{T}_1$ and $\mathfrak{T}_2$ are satisfiable extensions of $\mathfrak{T}$ and the language of $\mathfrak{T}_1 \cap \mathfrak{T}_2$ is the same as the language of $\mathfrak{T}$, then $\mathfrak{T}_1 \cup \mathfrak{T}_2$ is satisfiable.*

This theorem is closely related to Craig's Lemma, and we will derive it as a consequence of Craig's Lemma. We first state and prove some other consequences of Craig's Lemma.

Two theories $\mathfrak{T}_1$ and $\mathfrak{T}_2$ are called *compatible* if there is at least one interpretation that satisfies both $\mathfrak{T}_1$ and $\mathfrak{T}_2$—in other words, iff $\mathfrak{T}_1 \cup \mathfrak{T}_2$ is satisfiable. If one of $\mathfrak{T}_1$, $\mathfrak{T}_2$ contains a sentence whose negation is in the other, then obviously $\mathfrak{T}_1 \cup \mathfrak{T}_2$ is unsatisfiable. Now, a key consequence of Craig's Lemma is

THEOREM 24.1. *If $\mathfrak{T}_1 \cup \mathfrak{T}_2$ is unsatisfiable and if the predicates of $\mathfrak{T}_1$ and the predicates of $\mathfrak{T}_2$ have at least one common member, then one of $\mathfrak{T}_1$, $\mathfrak{T}_2$ must contain a sentence whose negation is in the other.*

PROOF: Assume the hypotheses. If either $\mathfrak{T}_1$ alone or $\mathfrak{T}_2$ alone is unsatisfiable, then it is pretty obvious that one of them must contain a sentence whose negation is in the other (Problem 24.7 below).

PROBLEM 24.7. Why is this?

To continue the proof: The interesting case is that in which $\mathfrak{T}_1$ and $\mathfrak{T}_2$ are both satisfiable. So let us assume this.

Since $\mathfrak{T}_1 \cup \mathfrak{T}_2$ is unsatisfiable, it follows by the Compactness Theorem that some finite subset S of $\mathfrak{T}_1 \cup \mathfrak{T}_2$ is unsatisfiable. This subset must contain at least one element of $\mathfrak{T}_1$ and at least one element of $\mathfrak{T}_2$—otherwise, one of $\mathfrak{T}_1$, $\mathfrak{T}_2$ would be unsatisfiable—so there are elements $X_1, \ldots, X_n$ of $\mathfrak{T}_1$ and elements $Y_1, \ldots, Y_k$ of $\mathfrak{T}_2$ such that the set $\{X_1, \ldots, X_n, Y_1, \ldots, Y_k\}$ is unsatisfiable. We let $X = X_1 \wedge \cdots \wedge X_n$ and $Y = Y_1 \wedge \cdots \wedge Y_k$, so the sentence $X \Rightarrow \sim Y$ is valid. Then, by Craig's Lemma, there is a sentence Z *in the language of both $\mathfrak{T}_1$ and $\mathfrak{T}_2$* such that $X \Rightarrow Z$ and $Z \Rightarrow \sim Y$ are both valid. Since Z is a logical consequence of X and X is a sentence of $\mathfrak{T}_1$ and $\mathfrak{T}_1$ is closed under logical implication, Z must also be a member of $\mathfrak{T}_1$. Since $Z \Rightarrow \sim Y$ is logically valid, so is $Y \Rightarrow \sim Z$, and since Y is in $\mathfrak{T}_2$, so is $\sim Z$. Thus Z is a member of $\mathfrak{T}_1$ whose negation is in $\mathfrak{T}_2$. This concludes the proof.

A theory $\mathfrak{T}'$ is called a *conservative extension* of a theory $\mathfrak{T}$ if, first of all, $\mathfrak{T}'$ is an extension if $\mathfrak{T}$ (i.e., $\mathfrak{T}$ is a subset of $\mathfrak{T}'$) and secondly every sentence of $\mathfrak{T}'$ that is in the language of $\mathfrak{T}$ is in $\mathfrak{T}$ (or, as it is sometimes stated, for any sentence X in the language of $\mathfrak{T}$, if X is provable in $\mathfrak{T}'$ then it is also provable in $\mathfrak{T}$; or, more succinctly, $\mathfrak{T}'$ cannot prove any more sentences in $\mathfrak{L}(\mathfrak{T})$ than $\mathfrak{T}$ can).

As a clever consequence of Theorem 22.1, we get the following result, which will be easily seen to yield Robinson's Consistency Theorem:

THEOREM 24.2. *If theories $\mathfrak{T}_1$ and $\mathfrak{T}_2$ are conservative extensions of a theory $\mathfrak{T}$, and if every sentence that is in the language of both $\mathfrak{T}_1$ and $\mathfrak{T}_2$ is in the language of $\mathfrak{T}$, then the closure of $\mathfrak{T}_1 \cup \mathfrak{T}_2$ is also a conservative extension of $\mathfrak{T}$.*

PROOF: Assume the hypotheses. Now suppose that X is in the closure of $\mathfrak{T}_1 \cup \mathfrak{T}_2$ and is in the language of $\mathfrak{T}$. We are to show that X is in $\mathfrak{T}$. Now, X, being in $(\mathfrak{T}_1 \cup \mathfrak{T}_2)^*$, is a logical consequence of $\mathfrak{T}_1 \cup \mathfrak{T}_2$ and therefore the set $(\mathfrak{T}_1 \cup \mathfrak{T}_2) \cup \{\sim X\}$ is not satisfiable (because $\sim X$ is obviously a consequence of this set, and no sentence and its negation can both be consequences of a satisfiable set), and therefore the set $\mathfrak{T}_1 \cup (\mathfrak{T}_2 \cup \{\sim X\})^*$ is unsatisfiable. (Obviously , $\mathfrak{T}_2 \cup \{\sim X\}$ is a subset of its closure $(\mathfrak{T}_2 \cup \{\sim X\})^*$; hence $\mathfrak{T}_1 \cup (\mathfrak{T}_2 \cup \{\sim X\})$ is a subset of $\mathfrak{T}_1 \cup (\mathfrak{T}_2 \cup \{\sim X\})^*$; hence the set $\mathfrak{T}_1 \cup (\mathfrak{T}_2 \cup \{\sim X\})^*$ is unsatisfiable.) Therefore, by Theorem 22.1, there is a sentence Y in $\mathfrak{T}_1$ whose negation is in $(\mathfrak{T}_2 \cup \{\sim X\})^*$. The sentence Y is in the language of $\mathfrak{T}_1$, hence so is $\sim Y$. The sentence $\sim Y$ must also be in the language of $\mathfrak{T}_2$ (because X, being in the language of $\mathfrak{T}$, is obviously in the language of its extension $\mathfrak{T}_2$, hence the languages of $\mathfrak{T}_2$ and $\mathfrak{T}_2 \cup \{\sim X\}$ are the same, and $\sim Y$ is certainly in the language of $\mathfrak{T}_2 \cup \{\sim X\}$, being in $(\mathfrak{T}_2 \cup \{\sim X\})^*$). Thus $\sim Y$ is in the language of both $\mathfrak{T}_1$ and $\mathfrak{T}_2$, hence is in the language of $\mathfrak{T}$ (by hypothesis). Since X is also in the language of $\mathfrak{T}$, so is $\sim X$, and therefore $(\sim X \Rightarrow \sim Y)$ is in the language of $\mathfrak{T}$.

Now, $\sim Y$ is a logical consequence of $\mathfrak{T}_2 \cup \{\sim X\}$ (being a member of $(\mathfrak{T}_2 \cup \{\sim X\})^*$); hence $(\sim X \Rightarrow \sim Y)$ is a logical consequence of $\mathfrak{T}_2$ (by Problem 24.2!), and hence is a member of $\mathfrak{T}_2$ (since $\mathfrak{T}_2$ is logically closed). Thus $(\sim X \Rightarrow \sim Y)$ is in $\mathfrak{T}_2$ and also is in the language of $\mathfrak{T}$, hence is a member of $\mathfrak{T}$ (since $\mathfrak{T}_2$ is a conservative extension of $\mathfrak{T}$). Thus Y and $(\sim X \Rightarrow \sim Y)$ are both members of $\mathfrak{T}$, hence so is X (since X is a consequence of the set $\{Y, (\sim X \Rightarrow \sim Y)\}$, and hence of the set $\mathfrak{T}$, and $\mathfrak{T}$ is logically closed).

PROBLEM 24.8. Prove the following two facts:

(1) Any conservative extension of a satisfiable theory is satisfiable.

(2) Any satisfiable extension of a *complete* theory is conservative.

PROBLEM 24.9. Now complete the proof of Robinson's Consistency Theorem.

SOLUTIONS

24.1. It is the second statement that is true: Suppose X is implied by S_1. Thus X is true under every interpretation that satisfies S_1. Now let I be any interpretation that satisfies S_2. Then I obviously also satisfies S_1, hence X is true under I. Thus X is true under every interpretation that satisfies S_2, hence is implied by S_2.

24.2. Suppose that Y is a consequence of $S \cup \{X\}$. Now let I be any interpretation that satisfies S. If X is false under I, then $X \Rightarrow Y$ is true under I. On the other hand, if X is true under I, then I satisfies $S \cup \{X\}$, hence Y is true under I (being a consequence of $S \cup \{X\}$, and so again $X \Rightarrow Y$ is true under I. Thus, in either case $X \Rightarrow Y$ is true under I, so $X \Rightarrow Y$ is true under all interpretations that satisfy S.

24.3. The two statements are equivalent: It is easy to see that (2) implies (1), for suppose (2) holds—i.e., that S is logically closed. Now suppose that $X_1, \ldots, X_n$ are elements of S and that Y is in the language of S and that $(X_1 \wedge \cdots \wedge X_n) \Rightarrow Y$ is valid. Then Y is implied by the set $\{X_1, \ldots, X_n\}$ (why?), hence by the superset S (by virtue of Problem 24.1), hence Y is in S (since S is closed). Thus (2) implies (1).

Next, suppose that (1) holds. We are to show that (2) holds, that S is logically closed. Well, suppose that Y is implied by S. By the compactness theorem, Y is then implied by some finite subset $\{X_1, \ldots, X_n\}$ of S. Thus $(X_1 \wedge \cdots \wedge X_n) \Rightarrow Y$ is valid. Hence $Y \in S$ (by (1)). Thus (1) implies (2), and therefore (1) and (2) are equivalent.

24.4. Some readers may well say that the statement is obviously false, but in fact it is true! The key point is that $\mathfrak{T}$ is logically closed. For an *arbitrary* set S of sentences it certainly is not always true that if every element of S is satisfiable, then S is satisfiable (for example, if X is a formula that is neither valid nor unsatisfiable, then X and $\sim X$ are both satisfiable, but the set $\{X, \sim X\}$ is obviously not satisfiable). If S is logically closed, however, it is a different story. Suppose a logically closed set S is unsatisfiable. Then, by the Compactness Theorem, some finite subset $\{X_1, \ldots, X_n\}$ of S is unsatisfiable. Let X be the conjunction $X_1 \wedge \cdots \wedge X_n$ (the order doesn't really matter). Then X is unsatisfiable. The sentence X is obviously a logical consequence of the larger set S (why?), and, since S is closed, X is therefore in S. Thus, for any closed set S, if S is unsatisfiable, then at least one element of S is unsatisfiable. Therefore, for any closed set S, if all elements of S are satisfiable (i.e., no element of S can be unsatisfiable), then S cannot be unsatisfiable—it must be satisfiable.

24.5. If $\mathfrak{T}$ contains all sentences in the language of $\mathfrak{T}$, then, of course, $\mathfrak{T}$ is unsatisfiable. Conversely, suppose that $\mathfrak{T}$ is unsatisfiable. Since there are then no interpretations that satisfy $\mathfrak{T}$, it is vacuously true that every sentence X is true under *all* interpretations that satisfy

$\mathfrak{T}$, so every sentence is implied by $\mathfrak{T}$, hence every sentence in the language of $\mathfrak{T}$ is in $\mathfrak{T}$.

24.6. Yes, it is true. Suppose S is closed and X is a valid sentence in the language of S. Then X is true under *all* interpretations, hence under all interpretations that satisfy S. Thus X is implied by S, hence is in S, since S is closed.

24.7. Suppose $\mathfrak{T}_1$ is unsatisfiable. Let P be a predicate that occurs in at least one element of $\mathfrak{T}_1$ and in at least one element of $\mathfrak{T}_2$. Let X be the valid sentence $\forall x(Px \vee \sim Px)$. X is in the language of $\mathfrak{T}_1$ and in the language of $\mathfrak{T}_2$. Since X is valid and is in the language of $\mathfrak{T}_2$, it is a member of $\mathfrak{T}_2$ (by Problem 24.6). Also, since $\sim X$ is in the language of $\mathfrak{T}_1$ and $\mathfrak{T}_1$ is unsatisfiable, $(\sim X)$ is in $\mathfrak{T}_1$ (all sentences in the language of $\mathfrak{T}_1$ are, by Problem 24.5).

24.8. Let $\mathfrak{T}_2$ be a conservative extension of $\mathfrak{T}_1$.

(1) Suppose that $\mathfrak{T}_2$ is unsatisfiable. Then every sentence in the language of $\mathfrak{T}_2$ is in $\mathfrak{T}_2$ (Problem 24.5), hence every sentence in the language of $\mathfrak{T}_1$ (which is also the language of $\mathfrak{T}_2$) is in $\mathfrak{T}_2$ and hence also in $\mathfrak{T}_1$ (because $\mathfrak{T}_2$ is a conservative extension of $\mathfrak{T}_1$), so $\mathfrak{T}_1$ is unsatisfiable. Thus if $\mathfrak{T}_2$ is unsatisfiable, so is $\mathfrak{T}_1$, and therefore if $\mathfrak{T}_1$ is satisfiable, so is $\mathfrak{T}_2$.

(2) Now suppose $\mathfrak{T}_1$ is complete and $\mathfrak{T}_2$ is satisfiable (which, of course, implies that $\mathfrak{T}_1$ also is). Now let X be any sentence in the language of $\mathfrak{T}_1$ that is an element of $\mathfrak{T}_2$. We must show that $X \in \mathfrak{T}_1$. Since $\mathfrak{T}_2$ is satisfiable and $X \in \mathfrak{T}_2$, $(\sim X)$ cannot be in $\mathfrak{T}_2$, hence it cannot be in $\mathfrak{T}_1$, hence X must be in $\mathfrak{T}_1$ (since $\mathfrak{T}_1$ is complete).

24.9. By virtue of Problem 24.8, if $\mathfrak{T}$ is a complete satisfiable theory, then being a satisfiable extension of $\mathfrak{T}$ is the same thing as being a conservative extension of $\mathfrak{T}$. So the hypothesis of Robinson's Theorem implies the hypothesis of Theorem 2, and hence, by Theorem 24.2, the closure of $\mathfrak{T}_1 \cup \mathfrak{T}_2$ is a conservative extension of $\mathfrak{T}$, hence a satisfiable extension of $\mathfrak{T}$ (again by Problem 24.8); hence its subset $\mathfrak{T}_1 \cup \mathfrak{T}_2$ is satisfiable.

- Chapter 25 -

Beth's Definability Theorem

Another important application of Craig's Interpolation Lemma is that it provides a particularly neat proof of an important result by the Dutch logician Evert Beth [1], to which we now turn.

We consider a finite set S of sentences without parameters and involving only unary predicates $P, P_1, \ldots, P_n$. For any sentence X in the language of S, we write $S \vdash X$ to mean that X is a logical consequence of S (which we recall means that X is true in every interpretation that satisfies S). We say that P is *explicitly definable* from the predicates $P_1, \ldots, P_n$ with respect to S iff there is a formula $\phi(x)$ whose predicates are all among the predicates $P_1, \ldots, P_n$ (and hence P is not a predicate that occurs in $\phi(x)$) such that $S \vdash \forall x(Px \equiv \phi(x))$. Such a formula $\phi(x)$ is called an *explicit definition* of P from $P_1, \ldots, P_n$ with respect to S.

There is another notion of definability, called *implicit* definability. We say P is *implicitly* definable from $P_1, \ldots, P_n$ iff the following condition holds: We take a new one-place predicate P' distinct from any of $P_1, \ldots, P_n$ that does not occur in any element of S, and we let S' be the result of substituting P' for P in every element of S. Then P is called *implicitly definable* from $P_1, \ldots, P_n$ with respect to S iff $S \cup S' \vdash \forall x(Px \equiv P'x)$.

Remarks. Actually, this condition is equivalent to the condition that, for any two interpretations of the predicates $P, P_1, \ldots, P_n$ that both satisfy S, if they agree on each of the predicates $P_1, \ldots, P_n$, then they also agree on P. (Two interpretations I_1 and I_2 are said to *agree* on a predicate Q iff Q is assigned the same value under I_1 and under I_2.) This equivalence is not necessary for the proof of Beth's Theorem, but is of independent interest, and I will say more about it later.

It is relatively easy to prove that if P is explicitly definable from $P_1, \ldots, P_n$ relative to S, then it is implicitly so definable.

PROBLEM 25.1. How is it proved?

Beth's Theorem is the converse.

THEOREM B (BETH'S DEFINABILITY THEOREM). *If P is implicitly definable from $P_1, \ldots, P_n$ with respect to S, then P is explicitly definable from $P_1, \ldots, P_n$ with respect to S.*

This theorem is far from obvious! But Craig's Interpolation Lemma yields a very elegant proof of it, which we now give.

Suppose P is implicitly definable from $P_1, \ldots, P_n$ with respect to S. Let P' and S' be as previously indicated, so we have

$$S \cup S' \vdash \forall x(Px \equiv P'x).$$

Let X be the conjunction of the elements of S (the order doesn't matter), and X' the conjunction of the elements of S', and we have

$$X \wedge X' \vdash \forall x(Px \equiv P'x).$$

Therefore, the sentence $(X \wedge X') \Rightarrow \forall x(Px \equiv P'x)$ is valid, hence, for any parameter a, the sentence $(X \wedge X') \Rightarrow (Pa \equiv P'a)$ is valid, hence so is the sentence

$$(X \wedge X') \Rightarrow (Pa \Rightarrow P'a).$$

But, by propositional logic, this latter sentence is equivalent to the sentence

$$(X \wedge Pa) \Rightarrow (X' \Rightarrow P'a)$$

(verify!), hence this sentence is valid. Now, in this sentence, P does not occur in $X' \Rightarrow P'a$, nor does P' occur in $X \wedge Pa$. Therefore, by Craig's Lemma, there is an interpolant Z for the sentence $(X \wedge Pa) \Rightarrow (X' \Rightarrow P'a)$. All predicates and parameters of Z occur in both $(X \wedge Pa)$ and $(X' \Rightarrow P'a)$, hence neither P nor P' can occur in Z—indeed, all predicates of Z are in the set $P_1, \ldots, P_n$. Also, Z contains no parameters except (possibly) a. Let x be a variable that does not occur in Z, and let $\phi(x)$ be the result of substituting x for all occurrences of a in Z, so $\phi(a) = Z$. Then $\phi(a)$ is an interpolant for $(X \wedge Pa) \Rightarrow (X' \Rightarrow P'a)$. Thus the following two sentences are valid:

(1) $(X \wedge Pa) \Rightarrow \phi(a)$.

(2) $\phi(a) \Rightarrow (X' \Rightarrow P'a)$.

From (1) we get the valid sentence

(1′) $X \Rightarrow (Pa \Rightarrow \phi(a))$.

From (2) we get the valid sentence

(2′) $X' \Rightarrow (\phi(a) \Rightarrow P'a)$.

Since (2′) is valid, so is its notational variant

(2″) $X \Rightarrow (\phi(a) \Rightarrow Pa)$.

If a sentence is valid, it remains valid if we replace any predicate by any a new predicate.

By (1′) and (2″) we get the valid sentence

$$X \Rightarrow (Pa \equiv \phi(a)).$$

Since a does not occur in X, the sentence $X \Rightarrow \forall x(Px \equiv \phi(x))$ is valid. Since X is the conjunction of the elements of S, it then follows that $S \vdash \forall x(Px \equiv \phi(x))$, so the formula $\phi(x)$ explicitly defines P from $P_1, \ldots, P_n$ with respect to S.

Remark. The fact that we assumed S to be finite was not really essential to the proof; a compactness argument can be used to modify the proof for an infinite set.

Let me come back to the earlier remark that P is implicitly definable from $P_1, \ldots, P_n$ with respect to S iff it is the case that any two interpretations of $P, P_1, \ldots, P_n$ that satisfy S and agree on $P_1, \ldots, P_n$ must also agree on P.

Problem 25.2. Prove this.

Solutions

25.1. For this problem and the next, it is to be understood that P' is a predicate distinct from each of $P, P_1, \ldots, P_n$ and that P' occurs in no element of S, and that S' is the result of substituting P' for P in every element of S.

Now suppose that P is explicitly definable from $P_1, \ldots, P_n$ with respect to S; let $\phi(x)$ be a formula such that $S \vdash \forall x(Px \equiv \phi(x))$. Then, of course, also $S' \vdash \forall x(P'x \equiv \phi(x))$. Hence $S \cup S' \vdash \forall x(Px \equiv \phi(x)) \wedge \forall x(P'x \equiv \phi(x)$. Hence $S \cup S' \vdash \forall x(Px \equiv P'x)$ (because $((\forall x(Px \equiv \phi(x)) \wedge \forall x(P'x \equiv \phi(x))) \Rightarrow \forall x(Px \equiv P'x)$ is logically valid, as the reader can verify).

25.2. To reduce clutter, we will use the following abbreviations: We will say that a sequence $(A, A_1, \ldots, A_n)$ of sets *satisfies* S iff all elements of S are true under that interpretation that assigns A to P, A_1 to $P_1, \ldots A_n$ to P_n. We will say that the same sequence satisfies S' iff every element of S' is true under the interpretation that assigns A to P', A_1 to $P_1, \ldots A_n$ to P_n. Finally, we will say that a sequence $(A, A', A_1, \ldots, A_n)$ satisfies $S \cup S'$ iff all elements of $S \cup S'$ are true under the interpretation that assigns A to P, A' to P', A_1 to $P_1, \ldots A_n$ to P_n. The following two facts should be obvious:

FACT 1. $(A, A_1, \ldots, A_n)$ satisfies S iff it satisfies S'.

FACT 2. $(A, A', A_1, \ldots, A_n)$ satisfies $S \cup S'$ iff both $(A, A_1, \ldots, A_n)$ satisfies S and $(A', A_1, \ldots, A_n)$ satisfies S'.

Next, we observe

LEMMA 25.1. *The following two conditions are equivalent:*

I_1. *For any sets $A, A', A_1, \ldots, A_n$, if $(A, A_1, \ldots, A_n)$ satisfies S and $(A', A_1, \ldots, A_n)$ satisfies S, then $A = A'$.*

I_2. *For any sets $A, A', A_1, \ldots, A_n$, if $(A, A', A_1, \ldots, A_n)$ satisfies $S \cup S'$ then $A = A'$.*

PROOF (OF LEMMA 25.1): (a) To show that I_1 implies I_2, suppose I_1 holds. Suppose that $(A, A', A_1, \ldots, A_n)$ satisfies $S \cup S'$. Then, by Fact 2,

(1) $(A, A_1, \ldots, A_n)$ satisfies S;

(2) $(A', A_1, \ldots, A_n)$ satisfies S'.

By (2) and Fact 1 we have

(3) $(A', A_1, \ldots, A_n)$ satisfies S.

From (1) and (3) and I_1, it follows that $A = A'$.

(b) To show that I_2 implies I_1, suppose I_2. Now suppose that (1) $(A, A_1, \ldots, A_n)$ satisfies S, and (2) $(A', A_1, \ldots, A_n)$ satisfies S. Then also, (3) $(A', A_1, \ldots, A_n)$ satisfies S' (by Fact 1). From (1), (3) and Fact 2, we see that $(A, A', A_1, \ldots, A_n)$ satisfies $S \cup S'$. Then $A = A'$ (by hypothesis I_2). Thus if (1) and (2) hold, then $A = A'$, which means that I_1 holds.

Now we can conclude the solution of Problem 25.2, which is to show that P is implicitly definable from $P_1, \ldots, P_n$ with respect to S

if and only if condition I_1 holds. By virtue of the above lemma, it suffices to show that P is implicitly definable from $P_1, \ldots, P_n$ with respect to S if and only if condition I_2 holds. This is what we shall do.

(a) In one direction, suppose that P is implicitly definable from $P_1, \ldots, P_n$ with respect to S—i.e., $S \cup S' \vdash \forall x(Px \equiv P'x)$. We are to show that I_2 holds. Well, suppose that $(A, A', A_1, \ldots, A_n)$ satisfies $S \cup S'$. We are to show that $A = A'$. Let I be the interpretation that assigns $A, A', A_1, \ldots, A_n$ to $P, P', P_1, \ldots, P_n$ respectively. Thus I assigns A to P and A' to P'. Let U be the domain of the interpretation I; thus A and A' are subsets of U. Since I satisfies $S \cup S'$ and $S \cup S' \vdash \forall x(Px \equiv P'x)$, the sentence $\forall x(Px \equiv P'x)$ is true under I. This means that, for every element $k \in U$, the sentence $Pk \equiv P'k$ is true, which means that Pk is true iff $P'k$ is true. But Pk is true iff $k \in A$ (since A is assigned to P under I), and, similarly, $P'k$ is true iff $k \in A'$. Therefore, $k \in A$ iff $k \in A'$ (for every k in U), ,so $A = A'$.

(b) Conversely, suppose that condition I_2 holds. We are to show that P is implicitly definable from $P_1, \ldots, P_n$ with respect to S—i.e., that $S \cup S' \vdash \forall x(Px \equiv P'x)$. Well, let I be any interpretation, in some universe U, that satisfies $S \cup S'$—we are to show that $\forall x(Px \equiv P'x)$ is true under I. By the given condition I_2, the predicates P and P' receive the same value A. Therefore, for any element $k \in U$, the sentence Pk is true under I iff $k \in A$, and also $P'k$ is true under I iff $k \in A$. Thus Pk is true under I iff $P'k$ is true under I, and hence $Pk \equiv P'k$ is true under I. Since this is so for *every* element k in U, $\forall x(Px \equiv P'x)$ is true under I. This completes the proof.

- Chapter 26 -

A Unification

An alternate title for this chapter could be "Be Wise; Generalize!" (Now, where have I seen that before?), or it could be "Quantification Theory in a Nutshell" (the title of one of my papers [22]), or it could be "Abstract Quantification Theory" (the title of another paper of mine in [21]).

The point now is that, in Part I of this volume, the concluding chapter generalized many results about the logic of lying and truth-telling, and we are now in a position to generalize many of the most important results in first-order logic. This chapter should particularly appeal to those readers who have a liking for the abstract!

Let us review some of the more important results in first-order logic that we have seen to be true.

T_1. The Completeness Theorem for Tableaux. Every valid formula is provable by the tableaux method.

T_2. The Skolem-Löwenheim Theorem. For any set S of formulas, if S is satisfiable, then S is satisfiable in a denumerable domain.

T_3. The Compactness Theorem. For any infinite set S of formulas, if every finite subset of S is satisfiable, then S is satisfiable.

T_4. The Regularity Theorem. Every valid sentence X is truth-functionally implied by a regular set R

T_5. The Completeness of the Axiom System $\mathfrak{S}_2$.

T_6. Craig's Interpolation Lemma and its applications.

In 1966, I published a result [20] entitled "A Unifying Principle in Quantification Theory," which generalizes all six above results; each can be established as a corollary of this one principle. In this chapter, I will be presenting, for the first time, a still more general result called "Theorem AM"—a very abstract result about arbitrary sets, rather than just sets of formulas; this theorem makes no reference at all to logical connectives or quantifiers but has applications to first-order logic (as well as to other possible logics).

As a recreational introduction to this theorem, I will present a very elaborate puzzle about variable liars, whose solution is virtually tantamount to proving the Completeness Theorem for Tableaux. Then I will present my generalizing Theorem AM and show that it generalizes both the solution to the puzzle and the results T_1–T_6 of first-order logic.

THE CLUB PROBLEM

We now turn to yet another very curious universe named Vlam, or just V for short. The first curious thing about this universe is that it may have infinitely many inhabitantsand each inhabitant can have denumerably many children. The child-parent relation is well-founded, and hence obeys the Generalized Induction Principle: For a property to hold for all the inhabitants, it is sufficient that it holds for all the childless inhabitants, and that, for any parent X, if all children of X have the property, so does X. (Thus we can have proofs by generalized induction.)

A set of inhabitants is called *agreeable* iff no two members are enemies. Now, the inhabitants of this universe are variable liars—on each day, an inhabitant either lies the entire day or tells the truth the entire day, but he or she might lie on one day and be truthful on another. Here are the lying and truth-telling habits of this universe:

V_1. A father tells the truth on a given day if all his children do.

V_2. A mother tells the truth on a given day if at least one of her children does.

V_3. For any agreeable set of *childless* inhabitants, there is at least one day on which all the members tell the truth.

The inhabitants of V form various clubs. A person x is said to be *compatible* with a set S iff $S \cup \{x\}$ (the result of adjoining x to S) is a club. We are given that for any club C the following three conditions hold:

C_0. C is agreeable.

C_1. For any father in C, each of his children is compatible with C.

C_2. For any mother in C, at least one of her children is compatible with C.

We can then obtain the following result:

THEOREM CL (THE CLUB THEOREM). *For any club C, there is at least one day on which all members of C tell the truth.*

We will later see that this theorem is essentially a disguised version of the Completeness Theorem for Tableaux! We will obtain it as a consequence, or, rather, a special case, of Theorem U below, which is a theorem about sets in general and which generalizes not only the Club Theorem but also all the previously mentioned results T_1–T_6 of first-order logic.

As an informal introduction to what follows, instead of a universe of *people*, we will now consider a set V of arbitrary elements. Instead of the relation "x is a *child* of y," we consider an arbitrary relation "x is a *component* of y" of elements of V. Elements having no components will be called *atomic* (in the club context, these are the childless people); others will be called *non-atomic* or *compound*. Instead of *fathers* and *mothers*, we respectively have *conjunctive* elements and *disjunctive* elements. Instead of the relation "x is an enemy of y," we have an arbitrary relation "x clashes with y." Instead of *clubs* we have certain sets called *consistent* sets.

Let me now be more formal.

BASIC FRAMEWORKS

By a *basic framework* we shall mean a collection of the following items:

(1) A denumerable set V. (Until further notice, the word *element* shall mean element of V, and *set* shall mean subset of V.)

(2) A well-founded relation between elements, which we call "x is a *component* of y." Elements with no components are called *initial* or *atomic*—others are called *non-atomic* or *compound*.

(3) A set of compound elements called *conjunctive* elements, and a set of compound elements called *disjunctive* elements, such that every compound element is either conjunctive or disjunctive (or possibly both). We shall use the letter c to stand for any conjunctive element, and d for any disjunctive element.

(4) A two-place *symmetric* relation x cl y, which we read: "x *clashes with* y." (It is *symmetric* in the sense that x clashes with y if and only if y clashes with x. In general, a relation xRy is called *symmetric* iff xRy is equivalent to yRx.) We call a set *agreeable* iff no two of its elements clash.

(5) A collection $\mathfrak{T}$ of sets called *truth sets* such that the following three conditions hold:

Tr_1. For any truth set T, if all components of a conjunctive element c are in T, then so is c.

Tr_2. For any truth set T, if at least one component of a disjunctive element d is in T, then so is d.

Tr_3. Every agreeable set of atomic elements is a subset of at least one truth set.

This concludes our definition of a basic framework.

We define a set S to be *downward closed*, or *closed downward*, iff, for every c in S, all components of c are in S, and, for every d in S, at least one component of d is in S. We shall call S a *Hintikka set* iff it is agreeable and closed downward.

We define a set S to be *upward closed*, or *closed upward*, iff, for any conjunctive element c, if all components of c are in S, then so is c, and for any disjunctive element d, if any component of d is in S, then so is d.

Lemma 26.1. *If D is closed downward and U is closed upward and all the atomic elements of D are in U, then $D \subseteq U$ (D is a subset of U).*

Problem 26.1. Prove Lemma 26.1. (Easy; use generalized induction.)

Note. Conditions Tr_1 and Tr_2 of item 5 above jointly say that every truth set is closed upward.

Next, for reasons that will be apparent later, we shall call a collection $\mathfrak{C}$ of subsets of V a *consistency collection* iff for every set C in $\mathfrak{C}$ and every conjunctive element c and every disjunctive element d, the following three conditions hold:

C_0. C is agreeable (no two elements of C clash).

C_1. If $c \in C$, then the set $C \cup \{x\}$ is in $\mathfrak{C}$ for every component x of c.

C_2. If $d \in C$, then the set $C \cup \{x\}$ is in $\mathfrak{C}$ for at least one component x of d.

The elements of a consistency collection $\mathfrak{C}$ are called *consistent sets* (relative to $\mathfrak{C}$). In what follows, whenever we refer to a set as "consistent," the term is to be understood as relative to a collection $\mathfrak{C}$ fixed for the discussion.

Let us say that x is *consistent with* a set S iff $S \cup \{x\}$ is a consistent set. Then we can replace C_1 and C_2 by the following:

C_1. If $c \in C$, then every component of c is consistent with C.

C_2. If $d \in C$, then at least one component of d is consistent with C.

Let us note that conditions C_1 and C_2 jointly say that, for any element $C \in \mathfrak{C}$, the set of elements that are consistent with C is closed downward.

Now comes the main result.

THEOREM AM (ABSTRACT MODEL EXISTENCE THEOREM). *For any consistency collection $\mathfrak{C}$ and truth collection $\mathfrak{T}$, any element of $\mathfrak{C}$ is a subset of some element of $\mathfrak{T}$.*

REMARK. If we call the elements of $\mathfrak{C}$ *consistent sets* and the elements of $\mathfrak{T}$ *truth sets* and if we call a set *satisfiable* iff it is a subset of a truth set, then the above theorem can be rephrased: Every consistent set is satisfiable.

Before proving the above theorem, I would like to show you how it yields the Club Theorem as a corollary. So let us revisit the universe Vlam. For the component relation, we take the relation "x is a child of y." Thus the atomic elements are the childless people. For the *conjunctive* elements we take the fathers, and for the *disjunctive* elements we take the mothers. For the *consistent* sets we take the clubs. Then the conditions C_0, C_1, C_2 we gave for the clubs in Vlam are to the effect that the collection $\mathfrak{C}$ of all clubs is a consistency collection (as defined).

Let us now call a set S of inhabitants a *joyful* set iff there is some day on which S is the set of all inhabitants who tell the truth on that day. It follows from conditions V_1 and V_2 we gave for Vlam that every joyful set is closed upward, and from condition V_3 it follows that every agreeable set of childless inhabitants is a subset of some joyful set. We can therefore conclude that the collection of all joyful sets satisfies the defining conditions Tr_1–Tr_3 of a truth collection. It then follows from Theorem AM that every club C is a subset of some joyful set J, and since there is a day on which all inhabitants in J tell the truth, it follows that all inhabitants in C tell the truth on that day.

PROOF (OF THEOREM AM): We first note that it immediately follows from Lemma 26.1 that every Hintikka set is a subset of some element of $\mathfrak{T}$.

PROBLEM 26.2. Why?

Thus it suffices to show that any consistent set C is a subset of some Hintikka set. We shall do this for the case when C is infinite (if C is finite, the proof is even simpler, and the necessary modification is left to the reader).

To begin with, the set V is denumerable, so let E be some enumeration $v_1, v_2, \ldots, v_n, \ldots$ of all the elements of V, fixed for the discussion. For any non-empty subset S of V, by the *first* element of S we shall mean the first element with respect to the enumeration E—i.e., the element v_n such that v_n is in S, but for no $k<n$ is v_k in S. Also, given any property of elements of V that holds for at least one element, by the *first* element having that property we mean the first element of the set of all elements having the property.

The consistent set C is assumed to be denumerable, and we enumerate it in some order $c_1, c_2, \ldots, c_n, \ldots$ (say, in the order in which the elements appear in the enumeration E, but any other order would do as well).

Now we shall construct an infinite sequence $(x_1, x_2, \ldots, x_n, \ldots)$ such that the set $\{x_1, x_2, \ldots, x_n, \ldots\}$ is a Hintikka set that includes all elements of C. We construct this sequence in stages—at each stage we construct a finite segment $(x_1, \ldots, x_n)$ of our desired sequence, and, at the next stage, we extend it to a larger finite sequence $(x_1, \ldots, x_n, \ldots, x_{n+m})$.

First for some preliminary notation: Let Θ be a finite sequence $(x_1, \ldots, x_n)$. We shall say that Θ is *consistent with* C iff the set $C \cup \{x_1, \ldots, x_n\}$ is consistent. Next, for any element y, by Θ, y we mean the sequence $(x_1, \ldots, x_n, y)$, by Θ, y_1, y_2 we mean $(x_1, \ldots, x_n, y_1, y_2)$, and, more generally, by $\Theta, y_1, \ldots, y_m$ we mean $(x_1, \ldots, x_n, y_1, \ldots, y_m)$.

Now for the construction of our desired infinite sequence $(x_1, x_2, \ldots, x_n, \ldots)$, which we denote by Θ.

We take c_1 as the first term of our desired sequence, and we accordingly let Θ_1 be the unit sequence (c_1). This concludes our first stage. Obviously Θ_1 is consistent with C (since $c_1 \in C$). Now suppose we have completed the nth stage and have on hand a finite sequence $(x_1, \ldots, x_k)$, which we call Θ_n, which is consistent with C, and which is such that k is greater than or equal to n. We then extend Θ_n to a finite sequence Θ_{n+1} in a manner depending on the nature of the term x_n.

(1) If x_n is atomic, we let Θ_{n+1} be Θ_n, c_{n+1}.

(2) If x_n is conjunctive and has only finitely many components $y_1, \ldots y_r$, we let Θ_{n+1} be $\Theta_n, y_1, \ldots, y_r, c_{n+1}$.

(3) If x_n is disjunctive, we take Θ_{n+1} to be Θ_n, y, c_{n+1}, where y is the first component of x_n such that Θ_n, y is consistent with C (there is at least one such component by condition C_2 of the definition of a

consistency property). Since Θ_n, y is consistent with C, so is Θ_{n+1} (viz. Θ_n, y, c_{n+1}).

(4) If x_n is conjunctive and has infinitely many components (this is the delicate case!), then we let y be the first component of x_n that is not already a term of Θ_n, and we take Θ_{n+1} to be $\Theta_n, y, x_n, c_{n+1}$ (it is important that we have repeated x_n, because, in that way, the element x_n appears in the final sequence $(x_1, x_2, \ldots, x_n, \ldots)$ infinitely often, and hence all the components of x_n are terms of the sequence).

We see that, for each n, the consistency of Θ_n with C implies the consistency of Θ_{n+1} with C, and also Θ_1 is consistent with C—hence, by induction, Θ_n is consistent with C for all n.

We let H be the set $\{x_1, x_2, \ldots, x_n, \ldots\}$ of all terms of the sequence Θ. Every element c_n of C was introduced into the sequence at the nth stage, hence C is a subset of H. From the way Θ was constructed, it follows that H is downward closed (verify!). Also, since each Θ_n is consistent with C, Θ_n cannot contain any two clashing elements—the set of terms of Θ is agreeable. Therefore, H is agreeable.

PROBLEM 26.3. Why is H agreeable?

Thus H is a Hintikka set that includes C. This completes the proof.

Applications to First-Order Logic

For applications of our Abstract Model Existence Theorem to first-order logic, we consider the basic framework in which the items 1–5 defining a basic framework are specified as follows:

(1) For V, we take either the set of all sentences (closed formulas), with or without parameters, or the set of all signed sentences (both universes are of interest).

(2) The components of a signed sentence are to be as previously defined—i.e., the components of α are α_1 and α_2, the components of β are β_1 and β_2, the components of γ are all the sentences $\gamma(a)$, where a is any parameter, and the components of δ are all the sentences $\delta(a)$, where a is any parameter. Since every sentence has rank, the relation "x is a component of y" is indeed well-founded.

(3) The conjunctive elements are to be the α's and γ's; the disjunctive elements are to be the β's and δ's.

(4) For the relation "*x clashes* with *y*," we take the relation "*x* is the *conjugate* of *y*." Thus the *agreeable* sets are those that contain no sentence and its conjugate.

(5) For the set $\mathfrak{T}$, we take the collection of all truth sets in the domain of the parameters as previously defined—i.e., the collection of all sets S that are closed both upward and downward and that are such that, for any sentence X, either X or its conjugate $\overline{X}$ is in S, but not both. Equivalently, S is a truth set iff there is an interpretation I in the universe of the parameters such that S is the set of all sentences that are true under I.

Obviously, anything we have proved about basic frameworks in general applies to these two frameworks in particular (the framework in which V is the set of unsigned sentences and the framework in which V is the set of signed sentences).

In the notation of first-order logic, a collection Σ of sets of sentences is a *consistency collection*, as previously defined, iff for every element C of Σ the following conditions hold:

C_0. C contains no element and its conjugate.

C_1. (a) For any α in C, the set $C \cup \{\alpha_1\}$ and the set $C \cup \{\alpha_2\}$ are both members of Σ.

(b) For any γ in C and any parameter a, the set $C \cup \{\gamma(a)\}$ is in Σ.

C_2. (a) For any β in C, either $C \cup \{\beta_1\}$ or $C \cup \{\beta_2\}$ is in Σ.

(b) For any δ in C, there is at least one parameter a such that $C \cup \{\delta(a)\}$ is in Σ.

We note that C_1(a) and C_1(b) jointly say that for any conjunctive element c in C and any component x of c, the set $C \cup \{x\}$ is a member of Σ. Conditions C_2(a) and C_2(b) jointly say that for any disjunctive element d of C there is at least one component x of d such that $C \cup \{x\}$ is a member of Σ.

Now, there is another notion of a consistency collection in literature, and both are important. We shall call a collection satisfying C_0–C_2(b) above a *type 1* consistency collection. By a *type 2* consistency collection we shall mean a collection Σ satisfying conditions C_0 through C_2(a), but instead of C_2(b), Σ is to satisfy the following:

C_2. (b*) For any element C in Σ, if δ is in C, then $C \cup \{\delta(a)\}$ is in Σ for every parameter a that is *new* to C (i.e., does not occur in any element of C).

These two notions are of incomparable strengths; a consistency collection of type 1 is not necessarily of type 2, nor is one of type 2 necessarily of type 1. For example, consider a type 2 consistency collection that contains a member C such that every parameter occurs in some element of C—hence no parameter is new to C. Then, for a given δ in C, there is no guarantee that there is a parameter a such that $C \cup \{\delta(a)\}$ is a member of Σ.

Let us call a set S of sentences a *free* set iff infinitely many parameters are new to S. Obviously any type 2 consistency collection of *free* sets is of type 1—in particular, any type 2 collection of finite sets is of type 1. The following fact is quite useful:

FACT 1. For any consistency collection Σ of type 2, the collection Σ' of all *free* elements of Σ is a consistency collection of type 1.

We leave the proof of Fact 1 to the reader (it is really quite simple).

We recall that a set S is said to be *satisfiable* in a domain D iff there is an interpretation I in D such that all elements of S are true under I. A set S is called *denumerably satisfiable* iff it is satisfiable in some denumerable domain. Now, it should be obvious that S is satisfiable in some denumerable domain iff it is satisfiable in any other denumerable domain, because the nature of the elements of the domain doesn't really matter. Thus, in particular, S is denumerably satisfiable iff it is satisfiable in the domain of the parameters—equivalently, iff it is a subset of some truth-set (some element of $\mathfrak{T}$). So, by the Abstract Model Existence Theorem, we at once have, for first-order logic,

THEOREM M (THE MODEL EXISTENCE THEOREM). *For any consistency collection Σ of type 1, every element of Σ is denumerably satisfiable.*

From the Model Existence Theorem and Fact 1 above, we easily get

THEOREM U (THE UNIFICATION THEOREM). *For any type 2 consistency collection Σ, every* free *member of Σ is denumerably satisfiable—in particular, every pure member of Σ (i.e., every member whose elements contain no parameters) is denumerably satisfiable.*

PROBLEM 26.4. Prove Theorem U.

REMARKS. The notion of a consistency collection of type 2 appeared historically before that of type 1. I introduced this notion in [20] and, I believe, it is the first definition ever given of a consistency collection (or consistency property, as I then called it). Theorem U, which I proved in [20], is known as the "Unifying Principle of Quantification Theory." The Model Existence Theorem came later. Both theorems have their uses.

The Model Existence Theorem is particularly useful in the field known as *Infinitary Logic*; cf. Keisler [8]. For first-order logic, I prefer working with the Unification Theorem.

Applications of the Unification Theorem

Following the terminology of [20] and [17], I shall call a property of sets (of sentences) an *analytic consistency property* iff the collection Σ of all sets having that property is a consistency collection of type 2. Thus Theorem U can be restated: For any analytic consistency property P, any set having property P is denumerably satisfiable. We can now get various results in Quantification Theory as consequences of Theorem U by showing various properties to be analytic consistency properties. For example, the theorems T_1–T_6 listed near the beginning of this chapter follow from the fact that the properties given below are indeed analytic consistency properties (their proofs are left to the reader).

(1) Call a set S *tableau-consistent* iff there exists no closed tableau for S. It is easily seen that tableau-consistency is an analytic consistency property, hence, by the Unification Theorem, every tableau-consistent pure set is satisfiable. Therefore, if a pure set S is unsatisfiable, there exists a closed tableau for S. In particular, if a sentence X is valid, the set $\{\sim X\}$ is unsatisfiable, hence there is a closed tableau for $\{\sim X\}$, which means that X is tableau-provable. Thus every valid X is tableau-provable.

(2) Satisfiability itself is easily seen to be an analytic consistency property. Hence, by Theorem U, every pure satisfiable set is denumerably satisfiable (the Skolem-Löwenheim Theorem).

(3) Call a set *F-consistent* iff all of its finite subsets are satisfiable. It is easily seen that F-consistency is an analytic consistency property, and hence, by Theorem U, a pure set is satisfiable provided all of its finite subsets are (the Compactness Theorem).

(4) Call a finite set *A-consistent* iff it has no associate. It can be seen (but not too easily) that A-consistency is an analytic consistency property, and, therefore (again by Theorem U), that for any finite set S, if S has no associate, then S is satisfiable. Hence every finite unsatisfiable set has an associate, and, therefore, every valid formula is truth-functionally implied by some finite regular set. This is the Regularity Theorem.

(5) The completeness of the axiom system $\mathfrak{S}_2$ can be obtained as a consequence of the Regularity Theorem (as we have seen), which in turn is a consequence of Theorem U. Alternatively, the completeness of $\mathfrak{S}_2$ can be obtained directly from Theorem U by defining a finite set $\{x_1, \ldots, x_n\}$ to be *$\mathfrak{S}_2$-consistent* iff $\sim(x_1 \wedge \ldots \wedge x_n)$ (the order doesn't matter) is not provable in $\mathfrak{S}_2$. It can be shown directly that $\mathfrak{S}_2$-consistency is an analytic consistency property, and hence that every $\mathfrak{S}_2$-consistent set is satisfiable. It then follows that if X is valid, the set $\{\sim X\}$ is not $\mathfrak{S}_2$-consistent, and therefore $\sim\sim X$ is provable in $\mathfrak{S}_2$; hence so is X. Thus $\mathfrak{S}_2$ is complete.

(6) Now for Craig's Interpolation Lemma:To begin with, by a *partition* $S_1|S_2$ of a set S is meant a pair of subsets S_1 and S_2 of S such that every element of S is in either S_1 or S_2, but not both—in other words, $S_1 \cup S_2 = S$ and $S_1 \cap S_2 = \emptyset$.

For any sets S_1 and S_2, by an *interpolant* between S_1 and S_2 we shall mean a sentence Z such that all predicates and parameters of Z occur in both S_1 and S_2, and $S_1 \cup \{\sim Z\}$ and $S_2 \cup \{Z\}$ are both unsatisfiable. We note that Z is an interpolant of a sentence $X \Rightarrow Y$ iff Z is an interpolant between $\{X\}$ and $\{\sim Y\}$. We now define a set S to be *Craig-consistent* iff there is some partition $S_1|S_2$ of S such that there exists no interpolant between S_1 and S_2. Well, Craig-consistency can be seen to be an analytic consistency property, hence every Craig-consistent set is satisfiable (by Theorem U). Therefore, if S is unsatisfiable, it is not Craig-consistent, which means that for *every* partition $S_1|S_2$ of S, there exists an interpolant between S_1 and S_2. Now, suppose $X \Rightarrow Y$ is valid. Then the set $\{X, \sim Y\}$ is unsatisfiable. Take the partition $\{X\}|\{\sim Y\}$ of the set $\{X, \sim Y\}$. Then there is an interpolant Z between $\{X\}$ and $\{\sim Y\}$. Such a Z is an interpolant for $X \Rightarrow Y$.

EXERCISE 26.1 (A GRAND EXERCISE!). Verify that the six properties above, which I claimed to be analytic consistency properties, really *are* analytic consistency properties!

SOLUTIONS

26.1. Let $P(x)$ be the property that if $x \in D$ then $x \in U$. We are to show that every x has property P, and hence that $D \subseteq U$. We show that every x has property P by generalized induction on P.

BASIC STEP. Every atomic element has property P by hypothesis.

INDUCTIVE STEP. For any x, we are to show that if all components of x have property P, so does x. Well, suppose that all elements of

x have property P and suppose that $x \in D$. Now, x is either conjunctive or disjunctive. Suppose x is conjunctive. Then all components of x are in D (since D is closed downward), and consequently all components of x are in U (by the inductive hypothesis any component of x which is in D is also in U). Hence $x \in U$ (since U is closed upward). On the other hand, suppose x is disjunctive. Then x has at least one component y in D (since D is closed downward), and y is therefore also in U (again by the inductive hypothesis). Thus x has a component in U, and hence $x \in U$ (since U is closed upward). This completes the induction.

26.2. Let H be a Hintikka set. Let B be the set of atomic elements of H. Since no two elements of H clash (by the definition of a Hintikka set), obviously no two elements of B clash. Hence B is a subset of some element T of $\mathfrak{T}$ (by Tr_3). Thus every atomic element of H is in T, and H is closed downward and T is closed upward; hence $H \subseteq T$, by Lemma 26.1.

26.3. For each n, let S_n be the set of terms of the sequence Θ_n. Obviously, if $n<m$, then $S_n \subseteq S_m$. Now, any element x of H is in some S_n and any element y of H is in some S_m. If $m<n$ then x and y are both in S_n; if $n<m$, then x and y are both in S_m; and, of course, if $n=m$, then x and y are both in S_n. Thus, for any elements x and y in H, there is some n such that S_n contains both x and y. Therefore, if H contained some clashing x and y, then some S_n would, which we have seen is not the case. Therefore H contains no clashing pair of elements.

26.4. Suppose Σ is a consistency collection of type 2. Let Σ' be the collection of all *free* members of Σ. Then Σ' is a consistency collection of type 1 (by Fact 1). Then any free member S of Σ is a member of Σ', and hence is denumerably satisfiable, by Theorem M.

- CHAPTER 27 -

LOOKING AHEAD

At this point you know a good deal about first-order logic, which many regard as the fundamental logic underlying all of mathematics. Many of the formal mathematical systems in existence are formed by adding new axioms (say, about sets, or numbers, or geometric axioms about things like *points* or *lines*)—adding these extra axioms to those of any standard axiom system of quantification theory, such as the system $\mathfrak{S}_2$, which you already know. As I stated in the Preface, you are now in a position to understand any of my technical writings in the fields discussed in this book, as well as the writings of many other logicians. Now, where shall you go from here? I strongly suggest that you next turn your attention to Gödel's famous Incompleteness Theorem, which I will soon tell you about—but first I would like you to look at a few puzzles, which you will later see are not completely irrelevant.

Suppose I put a penny and a quarter down on the table and ask you to make a statement. If the statement is true, then I promise to give you one of the coins, not saying which. But if the statement is false, then I won't give you either coin.

PROBLEM 27.1. There is a statement you can make such that I would have no choice but to give you the quarter (assuming I keep my word). What statement could that be?

PROBLEM 27.2. There is another statement you could make, which would force me to give you *both* coins. What statement could do this?

PROBLEM 27.3. There is yet another statement you could make which would make it impossible for me to keep my word. What statement would work?

Problem 27.4. More drastically, there is a statement you could make such that the only way I could keep my word would be by paying you a billion dollars. What statement would work?

The Incompleteness Phenomenon

In the year 1931, Gödel amazed the entire mathematical world with a paper [7] that began with the following startling words:

> The development of mathematics in the direction of greater precision has led to large areas of it being formalized, so that proofs can be carried out according to a few mechanical rules. The most comprehensive formal systems to date are, on the one hand, the *Principia Mathematica* of Whitehead and Russell and, on the other hand, the Zermelo-Fraenkel system of axiomatic set theory. Both systems are so extensive that all methods of proof used in mathematics today can be formalized in them—i.e., can be reduced to a few axioms and rules of inference. It would seem reasonable, therefore, to surmise that these axioms and rules of inference are sufficient to decide *all* mathematical questions that can be formulated in the system concerned. In what follows it will be shown that this is not the case, but rather that, in both of the cited systems, there exist relatively simple problems of the theory of ordinary whole numbers that cannot be decided on the basis of the axioms.

Gödel then goes on to explain that the situation does not depend on the special nature of the two systems under consideration but holds for an extensive class of mathematical systems.

Roughly speaking, Gödel showed that, for each such system, there had to be a sentence that asserted its own non-provability in the system, and hence was true if and only if it was not provable in the system. This means that the sentence was either true but not provable (in the system) or else false but provable. Under the reasonable assumption that all provable sentences are true, the sentence in question was true but not provable in the system.

I like to illustrate this with the following analogy: One day, a perfectly accurate logician visited the island of knights and knaves. We recall that on this island knights make only true statements and knaves make only false ones, and each inhabitant is either a knight or a knave. The logician was perfectly accurate in that anything he gave a proof of was really true;

he never proved anything false. The logician met a native named Jal who made a statement from which it follows that Jal must be a knight, but the logician could never prove that he is!

PROBLEM 27.5. What statement would work?

The statement that Jal made (as given in the solution to the above problem) is the analogue of Gödel's famous sentence that asserts its own non-provability. A more humorous analogue of Gödel's sentence is this: Consider the sentence

This sentence can never be proved.

We assume that only true sentences can be proved. Now, suppose the boxed sentence is false. Then what it says is *not* the case, which means that the sentence *can* be proved, contrary to our assumption that no false sentences are provable. Therefore the sentence is true. I have now proved that the sentence is true. Since it is true, what it says *is* the case, which means that the sentence cannot be proved. So how come I just proved it?

The fallacy of the above seeming paradox is that the notion of *proof* is not well-defined. Yes, in a given mathematical system, the notion of *proof within that system* is indeed well-defined, but the notion of *proof*, without specifying a formal system, is quite vague and not viable. Now, suppose we have a precise mathematical system—call it $\mathfrak{S}$—which proves various English sentences and proves only true ones. Consider now the following sentence:

This sentence is not provable in system $\mathfrak{S}$.

The paradox disappears! Instead, we now have the interesting fact that the boxed sentence must be true, but not provable in system $\mathfrak{S}$ (for if it were false, it would be provable in system $\mathfrak{S}$, contrary to the given condition that no false sentence is provable in the system).

Now, how did Gödel manage to construct a "self-referential" sentence—a sentence that asserted its own non-provability? I believe the following puzzle embodies the key idea.

Imagine a machine that prints out expressions built from the following three symbols:

$$P, R, \sim.$$

By an *expression* is meant any combination of these symbols (for example, $PP{\sim}RP{\sim}{\sim}$ is an expression, and so is $\sim$, P, or R standing alone). By a *sentence* is meant any expression of one of the following four forms, where X is any expression:

(1) PX;

(2) $\sim PX$;

(3) RPX;

(4) $\sim RPX$.

These sentences are interpreted as follows:

(1) PX means that X is printable (i.e., the machine can print X), and is accordingly called *true* iff X is printable.

(2) $\sim PX$ of course means that X is not printable and is accordingly called *true* iff X is not printable.

(3) RPX means that the expression XX (called the *repeat* of X—hence the letter R) is printable, and is accordingly called *true* iff XX is printable.

(4) $\sim RPX$ is the opposite of RPX, and is thus *true* iff XX is not printable.

We have here an interesting loop. The machine is printing out various sentences that assert what the machine can and cannot print. We are given that the machine is totally *accurate* in that it prints only true sentences. Thus

(1) If PX is printable, so is X.

(2) If $\sim PX$ is printable, then X is not printable.

(3) If RPX is printable, so is XX.

(4) If $\sim RPX$ is printable, then XX is not printable.

We see that if PX is printable, so is X. What about the converse: If X is printable, does it follow that PX is necessarily printable? No, it does not: If X is printable, then PX must be *true*, but it is not given that all true sentences are printable—only that all printable sentences are true. As a matter of fact, there is a true sentence that the machine cannot print!

Problem 27.6. Exhibit a true sentence that the machine cannot print. *Hint:* Construct a sentence that asserts its own non-printability—i.e., that is true if and only if it is non-printable.

Discussion. (To be read after the solution to the above problem.) Obviously, no machine could be accurate that prints a sentence that asserts its own non-printability! This is reminiscent of the scene in "Romeo and Juliet" in which the nurse comes running to Juliet and says "I have no breath!" Juliet replies: "How can you say that you have no breath when you have breath left to say 'I have no breath'?"

To get side-tracked for a moment, let me tell you that, to my surprise, I discovered that this machine has the curious property that there exist two sentences X and Y such that one of the pair must be true but not printable, but there is no way of telling which one it is! In fact, there are two such pairs.

Problem 27.7. Find such sentences X and Y. *Hint:* Construct sentences X and Y such that X asserts that Y is printable and Y asserts that X is not printable.

To remain side-tracked for another moment, I am reminded of another puzzle I once invented that consisted of a proof that either Tweedledee exists or Tweedledum exists, but there is no way of telling which! The proof goes as follows:

(1) Tweedledee does not exist.
(2) Tweedledum does not exist.
(3) At least one sentence in this box is false.

It really follows from the three sentences in the above box that either Tweedledee exists or Tweedledum exists, but there is no way to tell which. I leave the proof of this as an exercise to the reader.

Remark. It was puzzles like these that induced my colleague Professor Melvin Fitting (formerly a student of mine) to introduce me at a math lecture I once gave by saying: "I now introduce Professor Smullyan, who will prove to you that either he doesn't exist or you don't exist, but you won't know which."

Now let me get back on track.

Gödelian Systems

In what follows, the word *number* shall mean a *natural number*—i.e., a whole number that is either zero or positive.

As previously remarked, Gödel stated that his argument holds for a wide class of mathematical systems. What characterizes this wide class? Well, in each such system there is a well-defined set of expressions, some of which are called *sentences* and others called *predicates* (more completely, *numerical predicates*). (Informally, each predicate H is thought of as the name of some set of natural numbers.) With each predicate H and each number n is associated a sentence denoted $H(n)$ (which, informally speaking, expresses the proposition that n belongs to the set named by H). There is a well-defined set of sentences called *true* sentences. With each sentence S is associated a sentence $\overline{S}$ (usually written $\sim S$) called the *negation* of S, and with each predicate H is associated a predicate $\overline{H}$ (usually written $\sim H$) such that, for each number n, $\overline{H}(n)$ is the negation of $H(n)$ and is thus *true* iff $H(n)$ is not true. There is a well-defined procedure for *proving* various sentences. I shall call the system *correct* iff every provable sentence is true.

Gödel assigned to each expression X a (natural) number n, subsequently called the *Gödel number* of X, such that distinct expressions have distinct Gödel numbers. I will call a number n a *sentence-number* iff it is the Gödel number of a sentence, and if so, by S_n I shall mean the sentence whose Gödel number is n. Similarly, I will call n a *predicate-number* if it is the Gödel number of a predicate, in which case I shall let H_n be the predicate whose Gödel number is n.

Now for two key definitions: For any predicate-number n, by n^* I shall mean the Gödel number of the sentence $H_n(n)$. Next, I shall say that a predicate K *diagonalizes* predicate H, or that K is a *diagonalizer* of H, iff for every predicate-number n, the sentence $K(n)$ is true iff $H(n^*)$ is true.

Next, I shall define such a system to be *Gödelian* iff it satisfies the following two conditions:

G_1. Every predicate H has a diagonalizer K.

G_2. There is a predicate P (called a *provability predicate*) such that for every sentence-number n, the sentence $P(n)$ is true iff S_n is provable.

All the systems of Gödel's "wide class" do indeed satisfy conditions G_1 and G_2. And now we have the following "abstract" incompleteness theorem:

Theorem GT (After Gödel, with shades of Tarski). *For any correct Gödelian system, there must be a sentence of the system that is true but not provable in the system.*

Problem 27.8. Prove Theorem GT. (The proof is ingenious, but not very difficult.)

DISCUSSION. In application to various Gödelian systems, the bulk of the work consists in showing that they really are Gödelian. Condition G_1 is usually easy to establish, whereas the establishment of condition G_2 is an enormous undertaking usually involving several dozen preliminary definitions. All this is done in detail in my books *Gödel's Incompleteness Theorems* [18] and *Diagonalization and Self-Reference* [19], in which I give the complete proof for the system named *Peano Arithmetic,* which deals with the arithmetic of natural numbers based on addition and multiplication.

Actually, Theorem GT above departs somewhat from Gödel's original proof, in that the latter did not involve the notion of *truth,* which was made precise only later by Alfred Tarski [27]. Now, I wish to show you an abstract form of an important result due to Tarski [27].

A predicate H is said to *express* the set of all numbers n such that $H(n)$ is true. Thus, for any set A of numbers, H expresses A iff for every number n, the sentence $H(n)$ is true iff $n \in A$. And a number-set A is said to be *expressible* in the system if there is some predicate H that expresses it. The following is an abstract form of Tarski's theorem:

THEOREM T (AFTER ALFRED TARSKI). *For any accurate system satisfying condition G_1 (but not necessarily G_2), the set of Gödel numbers of the* true *sentences is not expressible in the system.*

This theorem is sometimes paraphrased "For accurate systems satisfying condition G_1, truth is not definable in the system."

An alternate formulation of Theorem T is this: A predicate T is called a *truth predicate* (for the system) iff for every sentence-number n, the sentence $T(n)$ is true iff S_n is true. Then Theorem T can be equivalently stated: "If the system is accurate and satisfies condition G_1, then no predicate of the system is a truth predicate."

PROBLEM 27.9. Prove Theorem T.

PROBLEM 27.10. Show how Theorem GT is almost an immediate consequence of Theorem T.

FIXED POINTS

Underlying the proofs of Theorem GT and Theorem T is a principle that is very important in its own right. First, a definition. A sentence S is called a *fixed point* of a predicate H iff the following condition holds: S is true if and only if $H(n)$ is true, where n is the Gödel number of S. Thus a fixed point of H is a sentence S_n such that $H(n)$ is true iff S_n is true.

Theorem F_1 (Fixed Point Theorem). *In any system satisfying condition G_1, each predicate of the system has a fixed point.*

Problem 27.11. Prove the Fixed Point Theorem.

Problem 27.12. Next, show how the Fixed Point Theorem greatly facilitates the proofs of Theorems T and GT.

Fixed points, diagonalization and self-reference all play key roles not only in the works of Gödel and Tarski, but also in the field of recursion theory (a fascinating subject!) and in the field known as *combinatory logic* (an equally fascinating field). I devoted an entire book (*Diagonalization and Self-Reference* [19]) to an in-depth study of these topics and their interrelationships. I strongly suggest that this be the next thing to which you turn your attention. After that, I suggest you study axiomatic set theory and the Continuum Hypothesis. Gödel's work in this area—his proof of the consistency of the Continuum Hypothesis—is, in my opinion, the most remarkable thing he has ever done. A good account of this, as well as an account of Paul Cohen's proof of the consistency of the negation of the Continuum Hypothesis, can be found in the book *Set Theory and the Continuum Problem* by Mel Fitting and myself [26], as well as in several other sources. I can assure you that a study of this subject is most rewarding.

That's all for now. Good reading! I am a bit sad that this book must end. Perhaps one day I might write a sequel. Who knows?

Solutions

27.1. One such statement that works is: "You will not give me the penny." If the statement were false, then what it says would not be the case, which means that I would give you the penny. But my rule was that I cannot give you either coin for a false statement, so the statement can't be false; it must be true. Therefore, it is true that I will not give you the penny, yet I must give you one of the coins for having made a true statement. Hence I have no choice other than to give you the quarter.

27.2. A statement that works is: "You will give me either both coins or neither one." The only way the statement could be false is for me to give you one of the coins but not the other. I cannot, however, give you either coin for a false statement, and therefore your statement cannot be false; it must be true. This mean that I give you either both coins or neither one; but I can't give you neither coin, since your statement is true; hence I must give you both.

27.3. A statement that makes it impossible for me to keep my word is: "You will give me neither coin." If I give you either coin, that makes the statement false, and I have broken my word by giving you a coin for a false statement. On the other hand, if I give you neither coin, that makes your statement true, and I have then failed to give you a coin for a true statement, hence I have again broken my word.

27.4. A statement that would work is: "You will give me neither the penny, nor the quarter, nor a billion dollars." If the statement were true, then I would have to give you one of the coins for a true statement, but doing so would falsify the statement (which says that I will give you neither one of the coins nor a billion dollars). Hence the statement cannot be true; it must be false. Since it is false that I will give you *neither* of the three, I must give you *one* of the three—I must give you either the penny or the quarter or a billion dollars. But I can't give you either the penny or the quarter for a false statement, hence I must give you a billion dollars!

27.5. There are many statements that would work, but the version most relevant to our purposes is that Jal said: "You can never prove that I am a knight." Now, suppose that Jal were a knave. Then his statement would be false (since knaves make only false statements), which would mean that the logician *could* prove that Jal is a knight, and hence would prove something false, contrary to our assumption that the logician never proves false statements. Thus Jal cannot be a knave; he must be a knight. Hence his statement is true, which means that the logician can never prove that Jal is a knight. Thus, Jal really is a knight, but the logician can never prove that he is.

Jal's statement "You can never prove that I am a knight" is the analogue of Gödel's sentence that asserts its own non-provability.

27.6. For any expression X, the sentence $\sim RPX$ is true iff the repeat of X is not printable. In particular, taking $\sim RP$ for X, we see that $\sim RP\sim RP$ is true iff the repeat of $\sim RP$ is not printable; but the repeat of $\sim RP$ is the very sentence $\sim RP\sim RP$! Thus the sentence $\sim RP\sim RP$ is true iff it is not printable. Thus it is either true but not printable or false but printable. The latter alternative is ruled out by the given condition that only true sentences can be printed by this machine. Thus the sentence $\sim RP\sim RP$ is true, but the machine cannot print it.

27.7. We want sentences X and Y such that X is true if and only if Y is printable, and Y is true if and only if X is not printable.

First, suppose we had such sentences. X is either true or not true. Suppose X is true. Then Y really is printable, as X asserts. Since Y is printable, it must be true (since only true sentences are printable), which means that X is not printable, as Y asserts. Thus, if X is true, then it is not printable. Now suppose that X is not true. Then X *falsely* asserts that Y is printable, hence, in reality, Y is not printable. But also, since X is not true, it is not printable; hence Y, which asserts just this fact, must be true! Thus, if X is false, then Y is true but not printable.

In summary, if X is true, then X is true but not printable; and if X is not true, then Y is true but not printable. There is no way to tell whether X is true or not, hence there is no way to tell whether it is X or Y that must be true but not printable by the machine.

Now to exhibit such sentences X and Y. One solution is to take $X{=}P{\sim}RPP{\sim}RP$ and $Y{=}{\sim}RPP{\sim}RP$. X is the sentence PY, hence X asserts that Y is printable. Y asserts that the repeat of $P{\sim}RP$ is not printable, but the repeat of $P{\sim}RP$ is X.

Another solution is to take $X{=}RP{\sim}PRP$ and $Y{=}{\sim}PRP{\sim}PRP$. Clearly, Y asserts that $RP{\sim}PRP$, which is X, is not printable, and X asserts that the repeat of ${\sim}PRP$, which is Y, is printable.

27.8. The predicate P obeys condition G_2. We now consider the predicate $\overline{P}$. By G_1, there is a predicate H that diagonalizes $\overline{P}$. Thus for every predicate-number n, the sentence $H(n)$ is true iff $\overline{P}(n^*)$ is true, i.e., iff $P(n^*)$ is not true, i.e., iff S_{n^*} is not provable. Let k be the Gödel number of the predicate H, so that $H{=}H_k$. Thus for every predicate-number n, the sentence $H_k(n)$ is true iff S_{n^*} is not provable. In particular, taking for n the number k, we see that $H_k(k)$ is true iff S_{k^*} is not provable. But k^* is the Gödel number of $H_k(k)$, so S_{k^*} is the sentence $H_k(k)$. (Surprise!) Thus S_{k^*} is true iff it is not provable (in the system). Under the assumption that only true sentences are provable, the sentence S_{k^*} must be true but not provable in the system.

27.9. We are to show (assuming condition G_1) that no predicate H of the system can be a truth-predicate. To do this, it suffices to show that for any predicate H, there is at least one sentence-number n for which it is *not* the case that $H(n)$ is true iff S_n is true. Such a number n will be said to be a *witness* that H is not a truth predicate.

Well, consider any predicate H. By G_1, there is a predicate K that diagonalizes the predicate $\overline{H}$. Thus for any predicate-number n, the sentence $K(n)$ is true iff $\overline{H}(n^*)$ is true. We let k be the Gödel number of K, and we see that $K(k)$ is true iff $\overline{H}(k^*)$ is true. But k^* is the Gödel number of $K(k)$, and $K(k)$ is the sentence S_{k^*}. Thus $\overline{H}(k^*)$ is true iff S_{k^*} is true, but also $\overline{H}(k^*)$ is true iff $H(k^*)$ is not true, and therefore $H(k^*)$ is not true iff S_{k^*} is true, or, what is the same thing, $H(k^*)$ is true iff S_{k^*} is not true. Thus k^* is a witness that H is *not* a truth predicate.

27.10. If we had proved Tarski's Theorem first, the proof of Theorem GT would have been much simpler: By Tarski's Theorem, there is a witness n that the provability predicate P is not a truth predicate. Thus $P(n)$ is true iff S_n is not true. But also, since P is a provability predicate, $P(n)$ is true iff S_n is provable. Thus S_n is provable iff S_n is not true, or, what is the same thing, S_n is true iff S_n is not provable.

27.11. Assume condition G_1. For any predicate H, let K be a predicate that diagonalizes H. Thus, for any sentence-number n, the sentence $K(n)$ is true iff $H(n^*)$ is true; in particular, $K(k)$ is true iff $H(k^*)$ is true, where k is the Gödel number of the sentence K. Thus k^* is the Gödel number of the sentence $K(k)$, so $K(k)$ is a fixed point of H.

27.12. If we had proved the Fixed Point Theorem first, then the proofs of Theorems T and GT would have been greatly simplified: For any predicate H, the Gödel number of any fixed point of the predicate $\overline{H}$ is obviously a witness that H is not a truth predicate (verify!). This gives us Theorem T. Also, for any provability predicate P, any fixed point of the predicate $\overline{P}$ must be a sentence which is true iff it is not provable (verify!). This gives us Theorem GT.

REMARK. I previously hinted that my penny and quarter puzzle (Problem 27.1) was somehow related to Gödel's Theorem. What I had in mind was this: I thought of the penny as standing for *provability* and the quarter as standing for *truth*. Thus the statement "You will not give me the penny" corresponds to Gödel's sentence, which in effect says: "I am not provable."

REFERENCES

[1] E. W. Beth. "On Padoa's method in the theory of definition." *Indag. Math.* 15 (1953), 330–339.

[2] A. Church. *Introduction to Mathematical Logic I*. Princeton, NJ: Princeton University Press, 1956.

[3] W. Craig. "Linear Reasoning. A New Form of the Herbrand-Gentzen Theorem." *J. Symb. Log.* 22:3 (1957), 250–268.

[4] M. Fitting. *First-Order Logic and Automated Theorem Proving*, second edition. New York: Springer-Verlag, 1996.

[5] G. Frege. *Begriffsschrift, eine der arithmetischen nachgebildete Formelsprache des reinen Denkens*. Halle: Nebert, 1879.

[6] K. Gödel. "Die Vollständigkeit der Axiomen des logischen Funktionalkalküls." *Monatsh. Math. Physik* 37 (1930), 349–360.

[7] K. Gödel. "Über formal unentscheidbare Sätze der Principia Mathematica und verwandter Systeme, I." *Monatsh. Math. Physik* 38 (1931), 173–198.

[8] H. J. Keisler. *Model Theory for Infinitary Logic*. Amsterdam: North-Holland Publishing Company, 1971.

[9] S. Kleene. *Introduction to Metamathematics*. Princeton, NJ: D. Van Nostrand Company, Inc., 1952.

[10] L. Löwenheim. "Uber Möglichkeiten im Relativkalkül.." *Math. Ann.* 76 (1915), 447–470.

[11] J. Łukasiewicz. "Démonstration de la computabilité des axioms de la théorie de la déduction." *Annales de la Société Polonaise de Mathématique* 3 (1925), 149.

[12] J. Nicod. "A Reduction in the Number of Primitive Propositions of Logic." *Proc. Camb. Phil. Soc.* 19 (1917), 32–41.

[13] W. V. Quine. *Mathematical Logic*. New York: W. W. Norton & Company, Inc., 1940.

[14] A. Robinson. "A Result on Consistency and Its Application to the Theory of Definitions." *Indag. Math.* 18 (1956), 47–58.

[15] J. B. Rosser. *Logic for Mathematicians*. New York: McGraw-Hill, 1953.

[16] T. Skolem. "Sur la portée du théorème de Löwenheim-Skolem." *Les Entretiens de Zurich sur les fondements et la méthode des sciences mathématiques* (December 6–9, 1938), pp. 25–52 (1941)

[17] R. M. Smullyan. *First-Order Logic*. New York: Springer-Verlag, 1968. Reprinted by Dover Publications in 1995.

[18] R. M. Smullyan. *Gödel's Incompleteness Theorems*. Oxford, UK: Oxford University Press, 1993.

[19] R. M. Smullyan. *Diagonalization and Self-Reference*. Oxford, UK: Oxford University Press, 1994.

[20] R. M. Smullyan. "A Unifying Principle in Quantification Theory." *Proceedings of the National Academy of Sciences* 45:6 (1963), 828–832.

[21] R. M. Smullyan. "Abstract Quantification Theory." In *Intuitionism and Proof Theory: Proceedings of the Summer Conference at Buffalo, N.Y., 1968*, edited by J. Myhill, R. E. Vesley, and A. Kino, pp. 79–91. Amsterdam: North-Holland, 1971.

[22] R. M. Smullyan. "Quantification Theory in a Nutshell." In *Paul Halmos Celebrating 50 Years of Mathematics*, edited by J. H. Ewing andF. W. Gehring,pp. 297–304. New York: Springer Verlag, 1992.

[23] R. M. Smullyan. *The Lady or the Tiger?* New York: Alfred A. Knopf, 1982.

[24] R. M. Smullyan. *What Is the Name of This Book?* Englewood Cliffs, NJ: Prentice Hall, 1978.

[25] R. M. Smullyan. *To Mock a Mockingbird and Other Logic Puzzles*. New York: Knopf, 1985.

[26] R. M. Smullyan and M. Fitting. *Set Theory and the Continuum Problem*. Oxford, UK: Oxford University Press, 1996.

[27] A. Tarski. "The Semantic Conception of Truth and the Foundations of Semantics." *Philosophy and Phenomenlological Research* 4 (1944), 13–47.

[28] A. N. Whitehead B. and Russell. *Principia Mathematica*, Vols. I–III, secon edition. Cambridge, UK: Cambridge University Press, 1925–1927.

Index

Printed in the United States
by Baker & Taylor Publisher Services